The Essentials
of Political Analysis

The Essentials
of Political Analysis

Third Edition

Philip H. Pollock III
University of Central Florida

CQ PRESS

A Division of SAGE
Washington, D.C.

CQ Press
2300 N Street, NW, Suite 800
Washington, DC 20037

Phone: 202-729-1900; toll-free, 1-866-4CQ-PRESS (1-866-427-7737)

Web: www.cqpress.com

Cover design: Auburn Associates Inc., Baltimore, Maryland
Composition: BMWW, Baltimore, Maryland

☺ The paper used in this publication exceeds the requirements of the
American National Standard for Information Sciences—Permanence of
Paper for Printed Library Materials, ANSI Z39.48-1992.

Printed and bound in the United States of America
12   11   10   5

Library of Congress Cataloging-in-Publication Data

Pollock, Philip H.
   The essentials of political analysis / Philip H. Pollock III. — 3rd ed.
      p. cm.
   Includes bibliographical references and index.
   ISBN 978-0-87289-606-2 (pbk. : alk. paper)
1. Political science—Research—Methodology—Textbooks.   I. Title.

   JA86.P65 2009
   320.072—dc22

                                                      2008038530

To my parents,
Philip H. Pollock Jr.
and Rhoda A. Pollock

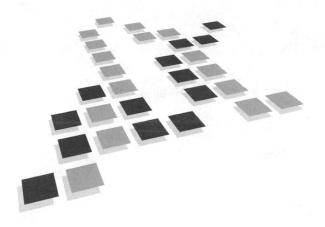

Contents

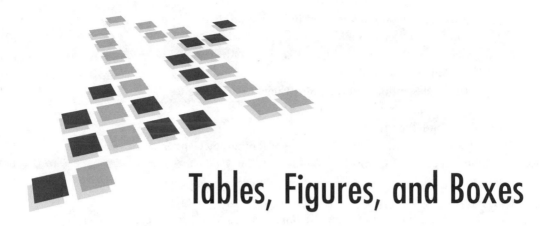

Tables, Figures, and Boxes

Tables

Figures

Boxes

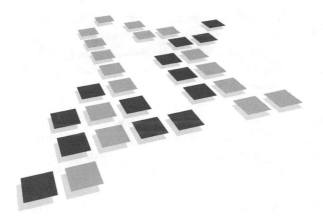

Preface

Students are of two minds about methods. Many students can examine graphic or tabular data and offer a reasonably meaningful description. Provided with a set of procedural guidelines, students display considerable facility at setting up cross-tabulations, comparing percentages or means, sketching bar charts, and writing a paragraph describing the data. At the same time, however, students balk at the idea that inferential statistics can serve as an interpretive tool. They tend to view statistical evidence as an odd element, an additional complication quite separate from their substantive findings. "I'm not really a statistics person" is a familiar refrain. Students can be intrepid interpreters of data yet reluctant practitioners of statistics.

This book cultivates students' nascent analytical abilities and develops their statistical reasoning. Chapters 1 through 5 build descriptive and analytic skills in a (mostly) nonstatistical context. With these essentials in place, students are able to appreciate the pivotal role of inferential statistics—introduced and applied, with increasing sophistication, in Chapters 6 through 9. Because the practical application of methodological concepts enhances students' comprehension, *The Essentials of Political Analysis* contains numerous hypothetical and actual examples. And because students become more adept at describing variables and interpreting relationships between them if they learn elemental graphing techniques, the chapters instruct in the interpretation of graphic displays of political variables. In addition to drawing on phenomena from U.S. politics, examples from comparative politics and international relations are also included. The narrative encourages students to stop and think about the examples, and the exercises at the end of each chapter permit students to apply their newly acquired skills. This volume contains forty-four end-of-chapter exercises.

ORGANIZATION OF THE BOOK AND CHANGES IN THE THIRD EDITION

In structure and content, this edition features significant revisions and enhancements, many of which were suggested by adopters of earlier editions. Chapter 1, which covers concepts and measurement, provides expanded treatment of concept clarification, multidimensional concepts, and reliability and validity. A discussion of levels of measurement (previously in

Chapter 1) now appears in Chapter 2, as does coverage of central tendency, dispersion, and graphic depictions of variables—all topics that appeared in "old" Chapter 3. Chapter 2 also includes a section on additive indexes and an illustration of Likert scaling. Chapter 3 combines the core of old Chapter 2, constructing explanations and framing hypotheses, with the core of old Chapter 3: bivariate cross-tabulation and mean comparison analysis, graphing relationships, and identifying nonlinear patterns.

Chapter 4, on research design, contains some of the book's newest, most heavily revised material. It includes a new discussion of experimental approaches and a revised treatment of the logic of controlled comparison, material that had previously been uncomfortably split between parts of Chapters 2 and 4. An all-new Chapter 5 is wholly devoted to empirical examples of controlled comparisons. Chapter 5 also introduces a heuristic that will help students decide which pattern—spuriousness, additive relationships, or interaction—best characterizes the relationships they analyze and interpret.

Clarifying and defining concepts, understanding measurement error, measuring and describing variables, framing hypotheses and evaluating relationships using cross-tabulation and mean comparison analysis, designing research, setting up and interpreting controlled comparisons—these are among the topics covered and skills honed in the first five chapters. Statistical resources take center stage in Chapters 6 through 9. Chapter 6 covers the foundations of inferential statistics: random sampling and the standard error of the mean. The illustrative examples are more realistic than in earlier editions, and several new figures add clarity to the discussion of the central limit theorem and the normal distribution. Chapter 6 also discusses the Student's t-distribution and demonstrates how to find the standard error of a sample proportion. Students learn to test hypotheses using statistical inference in Chapter 7, which also covers mean differences, differences between proportions, chi-square, and measures of association for nominal and ordinal variables.

This edition focuses more closely on two asymmetrical PRE measures, lambda and Somers' d_{yx}. Chapter 8, "Correlation and Linear Regression," provides an expanded discussion of Pearson's r and features a description of adjusted R-square, widely favored over plain R-square as a measure of explanatory completeness. Chapter 8 also discusses dummy variable regression and interaction effects in multiple regression analysis. Chapter 9, an introduction to binary logistic regression, is the only unchanged carryover from the second edition.

ORGANIZATION OF CHAPTERS AND SPECIAL FEATURES

The Essentials of Political Analysis is organized around a time-honored pedagogical principle: Foreshadow the topic, present the material, and then review the main points. Each chapter opens with a bulleted list of learning objectives, followed by an illustrative example or a schematic road map of the chapter's contents. Key terms appear in bold type throughout the text, and each chapter closes with a summary and a list of the key terms, which are referenced with page numbers. For example, as students begin Chapter 1, "The Definition and Measurement of Concepts," they will be made aware of its six objectives: clarifying the meaning of concepts, identifying multidimensional concepts, writing a conceptual definition, understanding systematic measurement error, understanding random measurement error, and recognizing problems of reliability and validity. The chapter then reminds students of the ubiquity of conceptual questions in political science—for example, "Are women more liberal than men?"—and asks them to consider how political researchers might address such questions.

Following the discussion of the six objectives, the text summarizes the chapter and references each key term.

Abundant tables and figures—about eighty in all—illustrate methodological concepts and procedures. I used hypothetical data in some instances, but most are based on analyses of the American National Election Studies, the General Social Surveys, a dataset containing variables on a large number of countries, and data on the fifty states. Many of the end-of-chapter exercises ask students to analyze actual data as well. A solutions manual is available on the Web to instructors.

COMPANION TEXTS

The Essentials of Political Analysis can be used as a stand-alone text in a political science methods course. Alternatively, it can be supplemented with a workbook, *An SPSS Companion to Political Analysis* or *A Stata Companion to Political Analysis.* The workbooks show students how to use SPSS or Stata to perform the techniques covered in the text: obtaining descriptive statistics, conducting bivariate and multivariate cross-tabulation and mean comparison analyses, running correlation and regression, and performing binary logistic regression. The workbooks also include chapters on statistical significance and measures of association, as well as data transformation procedures. The final chapters of both workbooks provide examples of research projects and help students as they collect and code data, perform original analysis, and write up their findings. Both volumes contain many end-of-chapter exercises—the SPSS book has more than fifty and the Stata book has more than forty. Instructor's solutions manuals provide answers for all exercises. Syntax files for all examples and exercises also are available to adopters.

An SPSS Companion to Political Analysis offers four SPSS data files: selected variables from the 2006 General Social Survey and the 2004 National Election Study, as well as datasets on the 50 states and 191 countries of the world. A text file containing information on the U.S. Senate is used to demonstrate how to code and read data into SPSS. Students work through each chapter's guided examples, using computer screenshots for graphic support. *An SPSS Companion to Political Analysis* accommodates the full version of SPSS as well as the student version and is compatible with SPSS 12.0 or later.

A Stata Companion to Political Analysis contains four Stata datasets: selected variables from the 2004 National Election Study and the 2002 General Social Survey, and datasets on the 50 states and 114 countries of the world. Like the SPSS workbook, a demonstration text file is used to illustrate Stata conventions for data coding and entry. It is compatible with Intercooled Stata or Stata SE, release 8 or later. Although the book was written using Stata 9, a Stata 10 supplement, which covers the Graph Editor, is available at no cost to adopters.

ACKNOWLEDGMENTS

The past insights of a growing number of friendly critics remain imprinted on this edition: Johanna Kristin Birnir of the State University of New York at Buffalo, Pete Furia of Wake Forest University, James Hanley of the University of Illinois at Springfield, Joel Lefkowitz of the State University of New York at New Paltz, and Jay DeSart of Utah Valley State College. I owe special thanks to William Claggett of Florida State University for providing many helpful suggestions on the third edition, and to University of Central Florida colleagues Bruce Wilson and Kerstin Hamann for pointing me toward instructive examples and exercises dealing with comparative politics. I am grateful to Barbara Kinsey for her unerring critique of the newest material in this book, particularly the examples discussed in Chapter 5. I also acknowledge the assistance of the book's reviewers: Brian Fogarty, University of Missouri-St. Louis; Thad Kousser, University of California, San Diego; and Brian Vargus, Indiana University-Purdue University, Indianapolis. Any errors that remain are mine alone.

Everyone at CQ Press has been extraordinarily patient and encouraging. I am especially grateful to Charisse Kiino, acquisitions editor, who helped me get this project off the ground and guided me through its completion. I thank Kerry V. Kern, copyeditor and production editor; Steve Pazdan, managing editor; and Allie McKay, editorial assistant.

Others contributed in different ways to this book. I remain indebted to Theodore J. Eismeier of Hamilton College, who taught me, through many years of productive collaboration, how to meld creative thinking with empirical analysis. Finally, I thank the two most important people in my life, Erin Suzanne Greene and Lauren DeCara Pollock, for their forbearance, inspiration, and love.

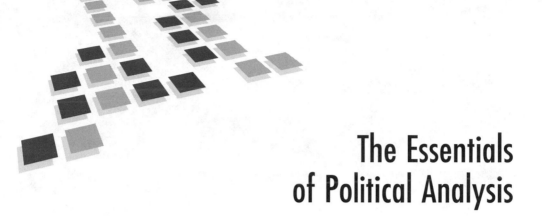

The Essentials
of Political Analysis

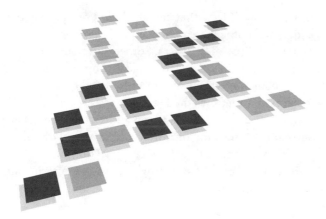

Introduction

The electoral college is perhaps the most unusual electoral institution in the world. With two exceptions[1] every state is a winner-take-all prize: The presidential candidate who wins a simple plurality of popular votes ends up with all the state's electoral votes. This arrangement raises the possibility—realized in the 2000 election—that the winner of the national popular vote can lose the election. Indeed, replacing the electoral college with direct popular election has been a familiar item on the reform agenda for many years.[2] If the United States changed to a new system, what would be the likely consequences? Would the promise of one person-one vote spark citizen involvement, foster intense national campaigns, and mobilize new voters? Or would it weaken most states' political importance and render all but the largest urban areas and richest media markets irrelevant to the election outcome?

Issues of institutional reform are not the only topics that come to mind when elections are being discussed. For example, over the past twenty-five years or so women have become much more likely than men to support the candidate of the Democratic Party. What accounts for this shift? Does the Democratic policy agenda appeal more strongly to women than to men? If so, which policies? We also know that people who earn lower incomes are more likely than higher-income people to vote Democratic. If women, on average, earn lower incomes than men, then maybe the "gender gap" is really an "income gap." If one were to compare women and men with similar incomes, would the gap still show up?

Of course, challenging and important issues are not confined to U.S. politics. The collapse of the Soviet Union ended nearly half a century of predictability in patterns of power and internal governance. Emerging independent states have made new claims of national legitimacy. Yet, as undemocratic governments dissolve, one might ask, what are the prospects that democratic forms will replace them? Is democratic development pushed along by economic relationships, such as free markets and open competition? What roles do political institutions, cultural beliefs, or ethnic antagonisms play? How would one define a democratic political system anyway?

These are the sorts of questions political scientists ask all the time. Researchers observe the sometimes chaotic political scene and create explanations for what they see. They offer hypotheses about political relationships and collect facts that can shed light on the way the political world works. They exchange ideas with other researchers and discuss the merits of various explanations, while refining some and discarding others. Sometimes political scientists describe "What if?" scenarios, using established facts or workable assumptions to make predictions about future facts. (If the presidential electoral system were based on direct election, what would be the likely consequences?) Sometimes the facts that researchers seek are already there, waiting to be described and measured. (What is the income difference between women and men?) Scholars may disagree on the meaning of important ideas and discuss the measurement of complex concepts. (How would one define democracy?) Through it all, political scientists learn to be dispassionate yet skeptical—debating hypotheses, offering alternative explanations or measurements, questioning analyses and results, and illuminating political relationships.

WHAT THIS BOOK IS ABOUT

In this book you will learn essential empirical methods for doing your own political analysis and for critically evaluating the work of others. The first five chapters deal with the logic behind political research. In Chapter 1 we consider how to think clearly about political concepts, as we weigh the challenges involved in measuring concepts in the real world. In Chapter 2 you will learn how to measure variables, the irreducible elements of description and analysis. In Chapter 3 we discuss the features of acceptable explanations in political science, and you will learn to frame hypotheses and make comparisons, the core methodology of political analysis. In Chapters 4 and 5 we cover research design—an overall set of procedures for testing explanations—and describe the logic and practice of controlled comparison, the main method for taking rival explanations into account. In these chapters the emphasis is on the logic of how one goes about adducing facts and evaluating relationships. You will find that the great enterprise of political research has much to do with thinking about concepts, looking at relationships between variables, creating explanations, figuring out patterns, and controlling for competing processes.

You will also find that basic statistical knowledge is a key resource for the researcher—an indispensable skill for interpreting relationships. Suppose, for example, that you were interested in describing the size of the gender gap among voting-age adults. Although you would not enjoy the uncommon luxury of observing the entire population of women and men you wanted to study, you would have access to a sample, a smaller group of women and men drawn at random from the larger population. Two questions would arise. First, how closely does the gender gap in the sample reflect the true gender gap in the unseen population? Second, how strong is the relationship between gender and partisanship? The answer to the first question lies in the domain of inferential statistics, the essentials of which are covered in Chapter 6 and part of Chapter 7. The answer to the second question requires a working knowledge of the most commonly used measures of association, also discussed in Chapter 7. In Chapter 8 we consider linear regression analysis, one of the more sophisticated and powerful methods having wide application in political research. And in Chapter 9 you will learn to use and interpret logistic regression, a specialized but increasingly popular analysis technique. This book uses a lot of examples, many of which are based on mass-level surveys of U.S. public opinion. Of course, your own substantive interests may lie elsewhere: comparative politics, international relations, public policy, judicial politics, state government, or any number of other areas of political research. Rest assured, the essential principles apply.

FACTS AND VALUES IN PERSPECTIVE

Political scientists long have argued among themselves about the great divide between two sorts of questions: questions of fact, *what is*, and questions of value, *what ought to be*. Often this distinction is plain and elementary. To ask whether wealth is equally distributed in the United States is to raise a question of fact, a question that can be addressed through definition and measurement. To ask whether wealth ought to be more equally distributed is to raise a question of value, a question that cannot be answered by empirical analysis. Sometimes, however, the is–ought distinction is not so clear. I might say, for example, that gun ownership is more widespread in the United States than in other countries, and I might assert further that the incidence of gun ownership is connected to gun-related crime. I might therefore offer the opinion that gun ownership ought to be as thoroughly controlled as judicial precedent allows. Fact or value? A bit of both. My opinion about gun regulations is based on assertions about the real world, and these assertions are clearly open to empirical examination. What is the evidence for the connection between guns and crime? Are there plausible alternative explanations? You can see that, to the extent that a value judgment is based on empirical evidence, political analysis can affect opinions by shaping the reasons for holding them. Put another way: Regardless of your personal opinions about political issues, it is important to remain open to new facts and competing perspectives.

Separating one's personal opinion on an issue from objective and open-minded analysis is often easier said than done—and it requires discipline and practice. After all, politics is serious business. And it is compelling *because* it involves differing opinions and the clash of competing values. Consider the discussions and arguments about the tradeoffs between domestic security and civil liberties that you have engaged in or listened to over the past several years. These arguments focus on whether (and in what ways) life in the United States *ought to* change. Many students advocated an emphasis on security—restricting immigration, permitting government authorities more latitude in detaining and arresting suspected terrorists, and relaxing legal protections against electronic surveillance. Other students were skeptical of such measures. They argued that the basic civil liberties of all citizens would be endangered, that the government would interpret such powers too broadly and begin to restrict any speech or activity it deemed a security risk.

How can political analysis help resolve this very serious issue? To be sure, the logic and methods you learn in this book will not show you how to "prove" which competing value—a belief in the desire for security or a belief in civil liberties—is "correct." Yet even in this debate, the protocol of political research can guide your search for the empirical bases of opinions and value judgments. What is the distribution of public opinion on security versus civil liberties? What existing laws need stricter enforcement? What new laws may be required? How has the U.S. government behaved toward its citizens during past national crises? Might not this historical data inform our current predictions about what the government will do? These questions, and countless others, are not easily answered. But they are questions of fact, and, at least in principle, they are answerable. This book is designed to help you frame and address such questions.

THE SCIENTIFIC APPROACH

There is one other way that learning about political research can nurture your ability to analyze political relationships and events—and even to elevate the level of your own political arguments about values. This has to do with an unspoken norm that all scientists follow: *Remain open, but remain skeptical.* All science, political science included, seeks to expand our understanding of the world. To ensure that the pathway to knowledge is not blocked, we must allow entrance to all ideas and theories. Suppose, for example, that I claim that the incidence of property crime is tied to the phases of the moon. According to my "moon theory," crime increases and recedes in a predictable pattern, increasing during the new moon and decreasing during the full moon. Laughable? Maybe. But the "remain open" tenet of scientific inquiry does not permit me to be silenced. So the moon theory gains entrance. Once on the pathway, however, any idea or theory must follow some "be skeptical" rules of the road. There are two sorts of rules. Some rules deal with evaluating questions of fact. These are sometimes called "What?" questions. Other rules deal with evaluating questions of theory. These are sometimes called "Why?" questions.

On questions of fact, scientific knowledge is not based on common sense, mysticism, or intuition. It is based on empirical observation and measurement. These observations and measurements, furthermore, must be described and performed in such a way that any other scientist could repeat them and obtain the same results. Scientific facts are empirical and reproducible. Thus, if I were to claim that the moon theory occurred to me in a dream, my results would be neither empirical nor reproducible. I would fail the fundamental rules for evaluating "What?" questions. If, by contrast, I were to describe an exhaustive examination of crime rate figures, and I could show a strong relationship between these patterns and phases of the moon, then I am still on the scientific path. Another researcher, following in my procedural footsteps, would get the same results.

On questions of theory, scientific knowledge must be explanatory and testable. An idea is explanatory if it describes a causal process that connects one set of facts with another set of facts. In science, explanation involves causation. If I were to propose that moon phases and crime rates go together because criminals are reverse werewolves, only coming out when the moon is new, I would be on shaky ground. I would be relying on a fact that is neither empirical nor reproducible, plus my "explanation" would lack any sense of process or causation. But suppose I said that criminals, like all individuals, seek to minimize the risks associated with their chosen activity. A full-moon situation would represent greater risk, a greater probability of being seen and arrested. A new-moon situation would represent lower risk, a lower probability of being detected. This idea is explanatory. Using plausible assumptions about human behavior, it describes why the two sets of facts go together. One level of the causal process (greater risk) produces one outcome (lower crime rates), whereas a different level of the causal process (lower risk) produces another outcome (higher crime rates).

An idea is testable if the researcher describes a set of conditions under which the idea would be rejected. A researcher with a testable idea is saying, "If I am correct, I will find such and such to be true. If I am incorrect, I will not find such and such to be true." Suppose a skeptical observer (skeptics abound in the scientific world!), upon reading my moon theory, should say: "Your explanation is very interesting. But not all full-moon situations involve higher risk as you have defined it. Sometimes the sky is heavily overcast, creating just as much cover for criminal activity as a new-moon situation. What would the crime rate be in full moon–overcast situations?" This observer is proposing a test, a test I must be willing to accept. If my idea is correct, I should find that full moon–overcast conditions produce crime rates similar to new-moon conditions. If my idea is incorrect, I would not find this similarity. Suppose my idea fails this test. Is that the end of the road for the moon theory? Not necessarily, but I would have to take my failure into account as I rethink the causal process that I proposed originally. Suppose my idea passes this test. Would that confirm the correctness of my theory? No, again. There would be legions of skeptics on the pathway to knowledge, offering alternative theories and proposing new tests.

CONCLUSION

As you can see, political research is an ongoing enterprise. Political analysis requires clarity, questioning, intellectual exchange, and discipline. Yet it also involves openness, creativity, and imagination. Compared with politics itself, which is enormously dynamic and frequently controversial, political analysis may seem rather stodgy. The basic logic and methods—measuring and describing variables, coming up with theories, testing hypotheses, understanding statistical inference, and gauging the strength of relationships—have not changed in many years. (For example, one of the techniques you will read about, chi-square, has been in use for more than a century.) This is a comforting thought. The skills you learn here will be durable. They will serve you now and in the future as you read and evaluate political science. You will bring a new critical edge to the many other topics and media you encounter—election or opinion polls, journalistic accounts about the effects of medical treatments, or policy studies released by organizations with an ax to grind. And you will learn to be self-critical, clarifying the concepts you use and supporting your opinions with empirical evidence.

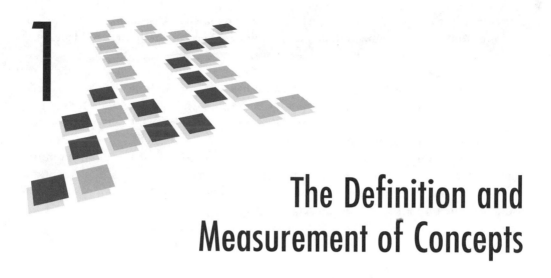

The Definition and Measurement of Concepts

LEARNING OBJECTIVES

In this chapter you will learn:
- How to clarify the meaning of concepts
- How to identify multidimensional concepts
- How to write a definition for a concept
- How systematic error affects the measurement of a concept
- How random error affects the measurement of a concept
- How to recognize problems of reliability and validity

Think for a moment about all the political variety in the world. People vary in their party affiliation: Some are Democrats, some Republicans, and many (self-described Independents) profess no affiliation at all. Some nations are democracies, whereas others are not. Even among democracies there is variety: parliamentary systems, presidential systems, or a combination. Would-be presidential nominees run the ideological gamut from conservatism to liberalism. Each of the terms just mentioned—*party affiliation, democracy, conservatism, liberalism*—refers to ideas that help us discuss and describe the world. It is virtually impossible to converse about politics without using ideas such as these. Ideas, of course, are not concrete. You cannot see, taste, hear, touch, or smell "partisanship," "democracy," or "liberalism." Each of these is a **concept**, an idea or mental construct that represents phenomena in the real world. Some concepts are quite complicated: "globalization," "power," "democratization." Others, such as "political participation" or "social status," are somewhat simpler.

Simple or complicated, concepts are everywhere in political debate, in journalistic analysis, in ordinary discussion, and, of course, in political research. How are concepts used? In partisan or ideological debate—debates about values—concepts can evoke powerful symbols with which people easily identify. A political candidate, for example, might claim that his or her agenda will ensure "freedom," create "equality," or foster "self-determination" around the globe. These are evocative ideas, and they are meant to be. In political research, concepts are not used to stir up value-laden symbols. Quite the opposite. In empirical political science, concepts refer to facts, not values. So when political researchers discuss ideas like "freedom," "equality," or "self-determination," they are using these ideas to summarize and label observable phenomena, characteristics in the real world.

The primary goals of political research are to describe concepts and to analyze the relationships between them. A researcher may want to know, for example, if social trust is declining or increasing in the United States, whether political elites are more tolerant of dissent than are ordinary citizens, or whether economic development causes democracy. The tasks of describing and analyzing concepts—social trust, political elites, tolerance of dissent, economic development, democracy, and any other concepts that interest us—present formidable obstacles. A **conceptual question**, a question expressed using ideas, is frequently unclear and thus is difficult to answer empirically. A **concrete question**, a question expressed using tangible properties, can be answered empirically. In her path-breaking book, *The Concept of Representation*, Hanna Pitkin describes the challenge of defining concepts such as "representation," "power," or "interest." She writes that instances "of representation (or of power, or of interest) . . . can be observed, but the observation always presupposes at least a rudimentary conception of what representation (or power, or interest) *is*, what *counts as* representation, where it leaves off and some other phenomenon begins."[1] We need to somehow transform concepts into concrete terms, to express vague ideas in such a way that they can be described and analyzed.

The same concept can, and often does, refer to a variety of different concrete terms. "Are women more liberal than men?" What is the answer: yes or no? "It depends," you might say, "on what you mean by *liberal*. Do you mean to ask if women are more likely than men to support abortion rights, gun control, government support of education, spending to assist poor people, environmental protection, affirmative action, gay and lesbian rights, funding for drug rehabilitation, or what? Do you mean all these things, some of these things, none of these things, or completely different things?" "Liberal," for some, may mean support for gun control. For others, the concept might refer to support for environmental protection. Still others may think the real meaning of liberalism is support for government spending to assist the poor.

A **conceptual definition** clearly describes the concept's measurable properties and specifies the units of analysis (people, nations, states, and so on) to which the concept applies. For example, consider the following conceptual definition of liberalism: Liberalism is the extent to which individuals support increased government spending for social programs. This statement clarifies a vague idea, liberalism, by making reference to a measurable attribute—support for government spending. Notice the words, "the extent to which." This phrase suggests that the concept's measurable attribute—support for government spending—varies across people. Someone who supports government spending has "more liberalism" than someone who does not support government spending. It is clear, as well, that this particular definition is meant to apply to individuals.[2] As you can see, in thinking about concepts and defining them, we keep an eye trained on the empirical world: What are the concrete, measurable characteristics of this concept? Conceptual definitions will be covered in depth in the first part of this chapter.

Having clarified and defined a concept, we must then describe an instrument for measuring the concept in the real world. An **operational definition** describes the instrument to be used in measuring the concept and putting a conceptual definition "into operation." How might we go about implementing the conceptual definition of liberalism? Imagine crafting a series of ten or twelve survey questions and administering them to a large number of individuals. Each question would name a specific social program: funding for education, assistance to the poor, spending on medical care, support for childcare subsidies, and so on. For each program, individuals would be asked whether government spending should be decreased, kept the same, or increased. Liberalism could then be operationally defined as the number of times a respondent said "increased." Higher scores would denote more liberal attitudes and lower scores would denote less liberal attitudes. As this example suggests, an operational definition provides a procedural blueprint, a measurement strategy. Yet, in describing a measurement strategy, we keep an eye trained on the conceptual world: Does this operational definition accurately reflect the meaning of the concept? In this chapter we consider problems that can emerge when researchers decide on an operational definition. In Chapter 2 we take a closer look at variables, the concrete measurements of concepts.

CONCEPTUAL DEFINITIONS

The first step in defining a concept is to clarify its empirical meaning. To clarify a concept, we begin by making an inventory of the concept's concrete properties. Three problems often arise during the inventory-building process. First, we might think of empirical attributes that refer to a completely different concept. Second, the inventory may include conceptual terms, with attributes that are not measurable. Third, the empirical properties may represent different dimensions of the concept. After settling on a set of properties that best represent the concept, we write down a definition of the concept. This written definition communicates the subjects to which the concept applies and suggests a measurement strategy. Let's illustrate these steps by working through the example introduced above: liberalism.

Clarifying a Concept

The properties of a concept must have two characteristics. They must be concrete, and they must vary. Return to the question posed earlier: "Are women more liberal than men?" This is a conceptual question because it uses the intangible term *liberal* and, thus, does not readily admit to an empirical answer. But notice two things. First, the conceptual term "liberal" certainly represents measurable characteristics of people. After all, when we say that a person or group of people is "liberal," we must have some attributes or characteristics in mind. Second, the question asks whether liberalism varies between people. That is, it asks whether some people have more or less of these attributes or characteristics than other people. In clarifying a concept, then, we want to describe characteristics that are concrete and that vary. What, exactly, are these characteristics?

A mental exercise can help you to identify characteristics that are concrete and that vary. Think of two subjects that are polar opposites. In this example, we are interested in defining liberalism among individuals, so we would think of two polar-opposite people. At one pole is a person who has a great deal of the concept's characteristics. At the other pole is a person who has the perfect opposite of the characteristics. What images of a perfectly liberal person do you see in your mind's eye? What images of a perfect opposite, an antiliberal or conservative, do you see?[3] In constructing these images, be open and inclusive. Here is an example of what you may come up with:

A liberal:	A conservative:
Has low income	Has high income
Is younger	Is older
Supports social justice	Opposes social justice
Opposes the free market	Supports the free market
Supports government-funded health care	Opposes government-funded health care
Opposes tax cuts	Supports tax cuts
Opposes restrictions on abortion	Supports restrictions on abortion
Supports same-sex marriage	Opposes same-sex marriage

Brainstorming polar opposites is an open-ended process, and it always produces the raw materials from which a conceptual definition can be built. Once the inventory is made, however, we need to become more critical and discerning. Consider the first two characteristics. According to the list, a liberal "has low income" and "is younger," whereas a conservative "has high income" and "is older." Think about this for a moment. Are people's incomes and ages really a part of the concept of liberalism? Put another way: Can we think about what it means to be liberal or conservative without thinking about income and age? You would probably agree that we could. To be sure, liberalism may be related to demographic factors, such as income and age, but the concept is itself distinct from these characteristics. This is the first problem to look for when clarifying a concept. Some traits seem to fit with the portraits of the polar-opposite subjects, but they are not essential parts of the concept. Let's drop the nonessential traits and reconsider our newly abbreviated inventory:

A liberal:	A conservative:
Supports social justice	Opposes social justice
Opposes the free market	Supports the free market
Supports government-funded health care	Opposes government-funded health care
Opposes tax cuts	Supports tax cuts
Opposes restrictions on abortion	Supports restrictions on abortion
Supports same-sex marriage	Opposes same-sex marriage

According to the list, a liberal "supports social justice" and "opposes the free market." A conservative "opposes social justice" and "supports the free market." Neither of these items should be on the list. Why not? Because neither one is measurable. Both terms are themselves concepts, and we cannot use one concept to define another. When constructing an inventory, imagine that a skeptical observer is looking over your shoulder, pressing you to specify concrete, measurable traits. How, exactly, would you determine whether someone supports free markets? How would you define social justice? If your initial response is, "I can't define it, but I know it when I see it"—to paraphrase an infamous remark about pornography—then you need to dig deeper for concrete elements.[4] This is the second problem to look for when clarifying a concept. Some descriptions seem to fit the portraits of the polar-opposite subjects, but these descriptions are themselves vague, conceptual terms. Let's drop the conceptual terms from the inventory.

A liberal:	A conservative:
Supports government-funded health care	Opposes government-funded health care
Opposes tax cuts	Supports tax cuts
Opposes restrictions on abortion	Supports restrictions on abortion
Supports same-sex marriage	Opposes same-sex marriage

One could reasonably argue that all these traits belong on an empirical inventory of liberalism. One can think of observable phenomena that would offer tangible measurements, including checkmarks on a questionnaire gauging opinion on different government policies, the display of bumper stickers or yard signs, monetary contributions to issue groups, or a number of other overt behaviors. But examine the list carefully. Can the attributes be grouped into different types? Are some items similar to each other and, as a group, different from other items? You may have already noticed that supports/opposes government-funded health care and opposes/supports tax cuts refer to traditional differences between those who favor a larger public sector and more social services (liberals) and those who favor a more limited governmental role (conservatives). The other items, opposes/supports abortion restrictions and supports/opposes same-sex marriage, refer to more recent disputes between those who favor personal freedoms (liberals) and those who support proscriptions on these behaviors (conservatives). This example illustrates the third problem to look for when clarifying a concept. All the traits fit with the portraits of the polar-opposite subjects, but they may describe different dimensions of the concept.

A **conceptual dimension** is defined by a set of concrete traits of similar type. Some concepts, such as liberalism, are multidimensional. A **multidimensional concept** has two or more distinct groups of empirical characteristics. In a multidimensional concept, each group contains empirical traits that are similar to each other. Furthermore, each group of traits is qualitatively distinct from other groups of traits. To avoid confusion, the different dimensions need to be identified, labeled, and measured separately. Thus the traditional dimension of liberalism, often labeled *economic liberalism*, subsumes an array of similar attributes: support for government-funded health care, aid to poor people, funding for education, spending for infrastructure, and so on. The moral dimension, often labeled *social liberalism*, includes policies dealing with gay and lesbian rights, abortion, the legalization of marijuana, the teaching of evolution, and prayer in schools. By grouping similar properties together, the two dimensions can be separately labeled—economic liberalism and social liberalism—and separately measured.[5]

Many ideas in political science are multidimensional concepts. For example, in his seminal work, *Polyarchy*, Robert A. Dahl points to two dimensions of democracy: contestation and inclusiveness.[6] Contestation refers to attributes that describe the competitiveness of political systems—for example, the presence or absence of frequent elections or whether a country has legal guarantees of free speech. Inclusiveness refers to characteristics that measure how many people are allowed to participate, such as the presence or absence of restrictions on the right to vote or conditions on eligibility for public office. Dahl's conceptual analysis has proven to be an influential guide for the empirical study of democracy.[7]

Many political concepts have a single dimension. The venerable social science concept of social status or socioeconomic status (SES), for example, has three concrete attributes that vary across people: income, occupation, and education. Yet it seems reasonable to say that all three are empirical manifestations of one dimension of SES.[8] Similarly, if you sought to clarify the concept of cultural fragmentation, you may end up with a polar-opposite list of varied but dimensionally similar characteristics of polities: many/few major religions practiced, one/several languages spoken, one/many racial groups, and so on. For each of these concepts, SES and cultural fragmentation, you can arrive at a single measure by determining whether people or polities have a great deal of the concept's characteristics.

A Template for Writing a Conceptual Definition

A conceptual definition must, at a minimum, communicate three things:
1. The variation within a measurable characteristic or set of characteristics
2. The subjects or groups to which the concept applies
3. How the characteristic is to be measured

Following is a workable template for stating a conceptual definition that meets all three requirements:

> The concept of _____ is defined as the extent to which _____ exhibit the characteristic of _____.

For a conceptual definition of economic liberalism, we would write:

> The concept of <u>economic liberalism</u> is defined as the extent to which <u>individuals</u> exhibit the characteristic of <u>supporting government spending for social programs</u>.

The first term, *economic liberalism*, when combined with the words "the extent to which," restates the concept's label and communicates the polar-opposite variation at the heart of the concept. The second term, *individuals*, states the subjects to whom the concept applies. The third term, *supporting government spending for social programs*, suggests the concept's measurement. Let's consider the template in more detail.

By referring to a subject or group of subjects, a conceptual definition conveys the units of analysis. A **unit of analysis** is the entity (person, city, country, county, university, state, bureaucratic agency, etc.) we want to describe and analyze; it is the entity to which the concept applies. Units of analysis can be either individual-level or aggregate-level. When a concept describes a phenomenon at its lowest possible level, it is using an **individual-level unit of analysis**. Most polling or survey research deals with concepts that apply to individual persons and is perhaps the most common individual-level units of analysis you will encounter. Individual-level units are not always persons, however. If you were conducting research on the political themes contained in the Democratic and Republican Party platforms over the past several elections, the units of analysis would be the individual platforms from each year. Similarly, if you were interested in finding out whether environmental legislation was a high priority in Congress, you might examine each bill that is introduced as an individual unit of analysis.

Much political science research deals with the **aggregate-level unit of analysis**, which is a collection of individual entities. Neighborhoods or census tracts are aggregate-level units, as are congressional districts, states, and countries. A university administrator who wondered if student satisfaction is affected by class size would gather information on each class, an aggregation of individual students. Someone wanting to know whether states with lenient voter registration laws had higher turnout than states with stricter laws could use legal statistics and voting data from fifty aggregate-level units of analysis, the states. Notice that collections of individual entities, and thus overall aggregate levels, can vary in size. For example, both congressional districts and states are aggregate-level units of analysis—both are collections of individuals within politically defined geographic areas—but states usually represent a higher level of aggregation because they are composed of more individual entities.

Notice too that the same concept often can be defined at both the individual and aggregate levels. Dwell on this point for a moment. Just as economic liberalism can be defined for individual persons, economic liberalism can be defined for states by aggregating the numbers of state residents who support or oppose government spending: The concept of economic liberalism is defined as the extent to which states exhibit the characteristic of having residents who support government spending for social programs. This conceptual definition makes perfect sense. One can imagine comparing states that have a large percentage of pro-spending residents with states having a lower percentage of pro-spending residents. For statistical reasons, however, the relationship between two aggregate-level concepts usually cannot be used to make inferences about the relationship at the individual level. Suppose we find that states with larger percentages of college-educated people have higher levels of economic liberalism than states with fewer college graduates. Based on this finding, we could not conclude that college-educated individuals are more likely to be economic liberals than are individuals without a college degree.

A classic problem, known as the **ecological fallacy,** arises when an aggregate-level phenomenon is used to make inferences at the individual level. W. S. Robinson, who coined the term more than fifty years ago, illustrated the ecological fallacy by pointing to a counter-intuitive fact: States with higher percentages of foreign-born residents had higher rates of English-language literacy than states with lower percentages of foreign-born residents. At the individual level, Robinson found the opposite pattern, with foreign-born individuals having lower English literacy than native-born individuals. What accounted for these paradoxical findings? The aggregate-level pattern was produced by the tendency for immigrants to settle in states whose native-born residents had comparatively high levels of language proficiency.[9] The ecological fallacy is not new. Indeed, Emile Durkheim's towering study of religion and suicide, published in 1897, may have suffered from it.[10] The main point here is that a proper conceptual definition needs to specify the units of analysis. Researchers must be careful when drawing conclusions based on the study of aggregate-level units of analysis.

OPERATIONAL DEFINITIONS

In suggesting how the concept is to be measured, a conceptual definition points the way to a clear operational definition.[11] An operational definition describes explicitly how the concept is to be measured empirically. Just how would we determine the extent to which people hold opinions that are consistent with economic liberalism? What procedure would produce the truest measure of social liberalism? Suppose we wanted to quantify Dahl's inclusiveness dimension of democracy. We would need to devise a metric that combines the different concrete attributes of inclusiveness. Exactly what form would this metric take? Would it faithfully reflect the conceptual dimension of inclusiveness, or might our measure be flawed in some way? This phase of the measurement process, the step between conceptual definition and operational definition, is often the most difficult to traverse. To introduce some of these difficulties, we describe an example from public opinion research, the study of the concept of political tolerance.

Political tolerance is important to many students of democracy because, arguably, democratic health can be maintained only if people remain open to different ways of thinking and solving problems. If tolerance is low, then democratic procedures will be weakly supported, and the free exchange of ideas might be threatened. Political tolerance is a rather complex concept, and a large body of research and commentary is devoted to it.[12] For our more limited purpose here, consider the following conceptual definition:

> The concept of political tolerance is defined as the extent to which individuals exhibit the characteristic of expressing a willingness to allow basic political freedoms for unpopular groups.

Awkward syntax aside, this is a serviceable definition, and it has been the starting point for a generation of scholars interested in studying the concept. Beginning in the 1950s, the earliest research "operationalized" political tolerance by asking large numbers of individuals if certain procedural freedoms (for example, giving a speech or publishing a book) should be extended to members of specific groups: atheists, communists, and socialists. This seemed like a reasonable operational definition because, at the time at least, these groups represented ideas outside the conformist mainstream and were generally considered unpopular. And the main finding was somewhat unsettling: Whereas those in positions of political leadership expressed high levels of tolerance, the public-at-large appeared much less willing to allow basic freedoms for these groups.

Later research, however, pointed to important slippage between the conceptual definition, which clarified and defined the important properties of political tolerance, and the operational definition, the procedure used to measure political tolerance. The original investigators had themselves chosen which unpopular groups were outside the mainstream, and these groups tended to have a left-wing or left-leaning ideological bent. The researchers were therefore gauging tolerance only toward leftist groups. Think about this measurement problem. Consider a scenario in which a large number of people are asked to "suppose that an admitted communist wanted to make a speech in your community. Should he be allowed to speak, or not?" For the question's designers, the key words are "wanted to make a speech." Thus people who respond "allowed to speak" are measured as having a larger amount of political tolerance than are those who say "not allowed to speak." But it could be that for some respondents—it is impossible to know how many—the key word is "communist." These respondents might base their answers on how they feel about communists, not on how willing they are to apply the principle of free speech. Ideological liberals, who may regard communists as less threatening than other groups, would be measured as more tolerant than ideological conservatives, who regard communists as more threatening than other groups. In sum, although the operational goal was to gauge tolerance, this measurement strategy also measured respondents' ideological sympathies.

A better measurement strategy, one more faithful to the concept, would allow respondents *themselves* to name the groups they most strongly oppose—that is, the groups most unpopular with or disliked by each person being surveyed. Individuals would then be asked about extending civil liberties to the groups they had identified, not those picked beforehand by the researchers. Think about why this is a superior approach. Consider a scenario in which a large number of people are presented with a list of groups: racists, communists, socialists, homosexuals, white separatists, and so on. Respondents are asked to name the group they "like the least." Now recast the earlier survey instrument: "Suppose that [a member of the least-liked group] wanted to make a speech in your community. Should he be allowed to speak, or not?" Because the respondents themselves have selected the least liked group, the investigators can be confident that those who say "allowed to speak" have a larger amount of tolerance than those who say "not allowed to speak." Interestingly, this superior measurement strategy led to equally unsettling findings: Just about everyone, elites and non-elites alike, expressed rather anemic levels of political tolerance toward the groups they liked the least.[13]

Measurement Error

As the tolerance example suggests, we want to devise an operational instrument that maximizes the congruence or fit between the definition of the concept and the empirical measure of the concept. Let's use the term *intended characteristic* to refer to the conceptual property we want to measure. The term *unintended characteristic* will refer to any other property or attribute that we do not want our instrument to measure. The researcher asks, "Does this operational instrument measure the intended characteristic? If so, does it measure *only* that characteristic? Or might it also be gauging an unintended characteristic?" Students of political tolerance are interested in asking individuals a set of questions that accurately gauge their willingness to extend freedoms to unpopular groups. The first measurement of tolerance did not accurately measure this intended characteristic. Why not? Because it was measuring a characteristic that it was not supposed to measure: individuals' attitudes toward left-wing groups. To be sure, the original measurement procedure was tapping an intended characteristic of tolerance. After all, a thoroughly tolerant person would not be willing to restrict the freedoms of any unpopular group, regardless of the group's ideological leanings, whereas a completely intolerant person would express a willingness to do so. When the conceptual definition was operationalized, however, an unintended characteristic, individuals' feelings toward leftist groups, also was being measured. Thus the measurement strategy created a poor fit, an inaccurate link, between the concept of tolerance and the empirical measurement of the concept.

Two sorts of error can distort the linkage between a concept and its empirical measure. Serious problems arise when **systematic measurement error** is at work. Systematic error introduces consistent, chronic distortion into an empirical measurement. Often called measurement bias, systematic error produces operational readings that consistently mismeasure the characteristic the researcher is after. The original tolerance measure suffered from systematic measurement error, because subjects with liberal ideological leanings were consistently (and incorrectly) measured as more tolerant than were ideologically conservative subjects. Less serious, but still troublesome, problems occur when **random measurement error** is present. Random error introduces haphazard, chaotic distortion into the measurement process, producing inconsistent operational readings of a concept. To appreciate the difference between these two kinds of error, and to see how each affects measurement, consider an example.

Suppose that a math instructor wishes to test the math ability of a group of students. This measurement is operationalized by ten word problems covering basic features of math. First let's ask, "Does this operational instrument measure the intended characteristic, math ability?" It seems clear that *some* part of the operational measure will capture the intended characteristic, students' actual knowledge of math. But let's press the measurement question a bit further: "Does the instructor's operational instrument measure *only* the intended characteristic, math ability? Or might it also be gauging a characteristic that the instructor did not intend for it to measure?" We know that, quite apart from mathematical competence, students vary in their verbal skills. Some students can read and understand the math problems more quickly than others. Thus the exam is picking up an unintended characteristic, an attribute it is not supposed to measure—verbal ability.

You can probably think of other characteristics that would "hitch a ride" on the instructor's test measure. In fact, a large class of unintended characteristics is often at work when human subjects are the units of analysis. This phenomenon, dubbed the **Hawthorne effect**, inadvertently measures a subject's response to the knowledge that he or she is being studied. Test anxiety is a well-known example of the Hawthorne effect. Despite their actual grasp of a subject, some students become overly nervous simply by being tested, and their exam scores will be systematically depressed by the presence of test anxiety.[14]

The unintended characteristics we have been discussing, verbal ability and test anxiety, are sources of systematic measurement error. Systematic measurement error refers to factors that produce consistently inaccurate measures of a concept. Notice two aspects of systematic measurement error. First, unintended characteristics such as verbal ability and test anxiety are durable, not likely to change very much over time. If the tests were administered again the next day or the following week, the test scores of the same students—those with fewer verbal skills or more test anxiety—would yield consistently poor measures of their true math ability. Think of two students, both having the same levels of mathematical competence but one having less verbal ability than the other. The instructor's operational instrument will report a persistent difference in math ability between these students when, in fact, no difference exists. Second, this consistent bias is inherent in the measurement instrument. When the instructor constructed a test using word problems, a measure of the unintended characteristic, verbal ability, was built directly into the operational definition. The source of systematic error resides—often unseen by the researcher—in the measurement strategy itself.

Now consider some temporary or haphazard factors that might come into play during the instructor's math exam. Some students may be ill or tired, while others may be well rested. Students sitting near the door may be distracted by commotion outside the classroom, while those that are sitting farther away may be unaffected. Commuting students may have been delayed by traffic congestion caused by a fender-bender near campus, and so, arriving late, they may be pressed for time. The instructor may make errors in grading the tests, accidentally increasing the scores of some students and decreasing the scores of others.

These sorts of factors—fatigue, commotion, unavoidable distractions—are sources of random measurement error. Random measurement error refers to factors that produce inconsistently inaccurate measures of a concept. Notice two aspects of random measurement error. First, unintended characteristics such as commotion and grading errors are not durable, and they are not consistent across students. They may or may not be present in the same student if the test were administered again the next day or the following week. A student may be ill or delayed by traffic one week, well and on time the next. Second, chance events certainly can affect the operational readings of a concept, but they are not built into the operational definition itself. When the instructor constructed the exam, he did not build traffic accidents into the measure. Rather, these factors intrude from outside the instrument. Chance occurrences introduce haphazard, external "noise" that may temporarily and inconsistently affect the measurement of a concept.

Reliability and Validity

We can effectively use the language of measurement error to evaluate the pros and cons of a particular measurement strategy. For example, we could say that the earliest measure of political tolerance, though perhaps having a small amount of random error, contained a large amount of systematic error. The hypothetical math instructor's measurement sounds like it had a dose of both kinds of error—systematic error introduced by durable differences between students in verbal ability and test anxiety, and random error that intruded via an array of haphazard occurrences. Typically, researchers do not evaluate a measure by making direct reference to the amount of systematic error or random error it may contain. Instead they discuss two criteria of measurement: reliability and validity. However, reliability and validity can be understood in terms of measurement error.

The **reliability** of a measurement is the extent to which it is a consistent measure of a concept. A perfectly reliable measure gives the same reading every time it is taken. Applying the ideas we just discussed, we see that a completely reliable measure is one that contains no random error. A measure need not be free of systematic error to be reliable. It just needs to be consistent. Consider a nonsensical example that nonetheless illustrates the point. Suppose a researcher gauges the degree to which the public approves of government spending by using a laser measuring device to precisely record respondents' heights in centimeters, with higher numbers of centimeters denoting stronger approval for spending. This researcher's measure would be quite reliable because it would contain very little random error and would therefore be consistent. But it would clearly be gauging a concept completely different from opinions about government spending. In a more realistic vein, suppose the math instructor recognized the problems caused by random occurrences and took steps to greatly reduce these sources of random error. Certainly his measurement of math ability would now be more consistent, more reliable. However, it would not reflect the true math ability of students because it would still contain systematic error. More generally, although reliability is a desirable criterion of measurement—any successful effort to purge a measure of random error is a good thing—it is a weaker criterion than validity.

The **validity** of a measurement is the extent to which it records the true value of the intended characteristic and does not measure any unintended characteristics. A valid measure provides a clear, unobstructed link between a concept and the empirical reading of the concept. Framed in terms of measurement error, the defining feature of a valid measure is that it contains no systematic error, no bias that consistently pulls the measurement off the true value. Suppose a researcher gauges opinions toward government spending by asking each respondent to indicate his or her position on a 7-point scale, from "spending should be increased" on the left to "spending should be decreased" on the right. Is this a valid measure? A measure's validity is harder to establish than is its reliability. But it seems reasonable to say that this measurement instrument is free from systematic error and thus would closely reflect respondents' true opinions on the issue. Or suppose the math instructor tries to alleviate the sources of systematic error inherent in his test instrument—switching from word problems to a format based on mathematical symbols, and perhaps easing anxiety by shortening the exam or lengthening the allotted time. These reforms would reduce systematic error, strengthen the connection between true math ability and the measurement of math ability, and thus enhance the validity of the test.

Suppose we have a measurement that contains no systematic error but contains some random error. Would this be a valid measure? Can a measurement be valid but not reliable? Although we find conflicting scholarly answers to this question, let's settle on a qualified yes.[15] Instead of considering a measurement as either not valid or valid, think of validity as a continuum, with "not valid" at one end and "valid" at the other. An operational instrument that has serious measurement bias, lots of systematic error, would reside at the "not valid" pole, regardless of the amount of random error it contains. The early measure of political tolerance is an example. An instrument with no systematic error and no random error would be at the "valid" end. Such a measure would return an accurate reading of the characteristic that the researcher intends to measure, and it would do so with perfect consistency. The math instructor's reformed measurement process—changing the instrument to remove systematic error, taking pains to reduce random error—would be close to this pole. Now consider two measures of the same concept, neither of which contains systematic error, but one of which contains less random error. Because both measures vanquish measurement bias, both would fall on the "valid" side of the continuum. But the more consistent measure would be closer to the "valid" pole.

Evaluating Reliability

Methods for evaluating reliability are designed around this assumption: If a measurement is reliable, it will yield consistent results. In everyday language, "consistent" generally means, "stays the same over time." Accordingly, some approaches to reliability apply this measure-now-measure-again-later intuition. Other methods assess the internal consistency of an instrument and thus do not require readings taken at two time points. First we will describe methods based on over-time consistency. Then we will turn to approaches based on internal consistency.

In the **test-retest method** the investigator applies the measure once and then applies it again to the same units of analysis. If the measurement is reliable, then the two results should be the same or very similar. If a great deal of random measurement error is present, then the two results will be very different. For example, suppose we construct a 10-item instrument to measure individuals' levels of economic liberalism. We create the scale by asking each respondent whether spending should or should not be increased on ten government programs. We then add up the number of programs on which the respondent says "increase spending." We administered the questionnaire and then readministered it at a later time to the same people. If the scale is reliable, then each person's score should change very little over time. The alternative-form method is similar to the test-retest approach. In the **alternative-form method** the investigator administers two different but equivalent versions of the instrument— one form at time point 1 and the equivalent form at time point 2. For our economic liberalism example, we would construct two 10-item scales, each of which elicited respondents' opinions on ten government programs. Why go to the trouble of devising two different scales? The alternative-form method remedies a key weakness of the test-retest method: In the second administration of the same questionnaire, respondents may remember their earlier responses and make sure that they give the same opinions again. Obviously, we want to measure economic liberalism, not memory retention.

Methods based on over-time consistency have two main drawbacks. First, these approaches make it hard to distinguish random error from true change. Suppose that, between the first and second administrations of the survey, a respondent becomes more economically liberal, perhaps scoring a 4 the first time and a 7 the second time. Over-time methods of evaluating reliability assume that the attribute of interest—in this case, economic liberalism—is unchanging. Thus the observed change, from 4 to 7, is assumed to be random error. The longer the time period between questionnaires, the bigger this problem becomes.[16] A second drawback is more practical: Surveys are expensive projects, especially when the researcher wants to administer an instrument to a large number of people. The test-retest and alternative-form approaches require data obtained from panel studies. A **panel study** contains information on the same units of analysis measured at two or more points in time. Respondents a, b, and c are interviewed at time 1; respondents a, b, and c are interviewed again at time 2. Data from cross-sectional studies are more the norm in social research. A **cross-sectional study** contains information on units of analysis measured at one point in time. Respondents a, b, and c are interviewed—that's it. Though far from inexpensive, cross-sectional measurements are more easily obtained than panel measures. As a practical matter, then, most political researchers face the challenge of evaluating the reliability of a measurement that was made using cross-sectional data.[17] Internal consistency methods are designed for these situations.

One internal consistency approach, the **split-half method,** is based on the idea that an operational measurement obtained from half of a scale's items should be the same as the measurement obtained from the other half. In the split-half method the investigator divides the scale items into two groups, calculates separate scores, and then compares the measurements. If the items are reliably measuring the same concept, then the two sets of scores should be the same. Following this technique, we would break our ten government spending questions into two groups of five items each, calculate two scores for each respondent, and then compare the scores. Plainly enough, if we have devised a reliable instrument, then the respondents' scores on one 5-item scale should match closely their scores on the other 5-item scale. A more sophisticated internal consistency approach, **Cronbach's alpha,** is a natural methodological extension of the split-half technique. Instead of evaluating consistency between separate halves of a scale, Cronbach's alpha compares consistency between pairs of individual items and provides an overall reading of a measure's reliability.[18] Imagine a perfectly consistent measure of economic liberalism. Every respondent who says "increase spending" on one item also says "increase spending" on all the other items, and every respondent who says "do not increase spending" on one item also says "do not increase spending" on every other item. In this scenario, Cronbach's alpha would report a value of 1, denoting perfect reliability. If responses to the items betray no consistency at all—opinions about one government program are not related to opinions about other programs—then Cronbach's alpha would be 0, telling us that the scale is completely unreliable. Of course, most measurements' reliability readings fall between these extremes.

It is easy to see how the methods of evaluating reliability help us to develop and improve our measures of concepts. To illustrate, let's say we wish to measure the concept of social liberalism, the extent to which individuals accept new moral values and personal freedoms. After building an empirical inventory, we construct a scale based on support for five policies: same-sex marriage, marijuana legalization, abortion rights, stem cell research, and physician-assisted suicide. Our hope is that by summing respondents' five positions, we can arrive at a reliable operational reading of social liberalism. With all five items included, the scale has a Cronbach's alpha equal to .6. Some tinkering reveals that, by dropping the physician-assisted suicide item, we can increase alpha to .7, an encouraging improvement that puts the reliability of our measure near the threshold of acceptability.[19] The larger point to remember is that the work you do at the operational definition stage often helps you to refine the work you did at the concept clarification stage.

Evaluating Validity

Reliability is an important and sought-after criterion of measurement. Most standardized tests are known for their reliability. The SAT, the Law School Admission Test (LSAT), the Graduate Record Examination (GRE), among others, all return consistent measurements. But the debate about such tests does not center on their reliability. It centers, instead, on their validity: Do these exams measure what they are supposed to measure and only what they are supposed to measure? Critics argue that because many of these tests' questions assume a familiarity with white, middle-class culture, they do not produce valid measurements of aptitudes and skills. Recall again the earliest measurements of political tolerance, which gauged the concept by asking respondents whether basic freedoms should be extended to specific groups: atheists, communists, and socialists. Because several different studies used this operationalization and produced similar findings, the measure was a reliable one. The problem was that a durable unintended characteristic, the respondents' attitudes toward left-wing groups, was "on board" as well, giving a consistent if inaccurate measurement of the concept.

The challenge of assessing validity is to identify durable unintended characteristics that are being gauged by an operational measure, that is, to identify the sources of systematic measurement error. To be sure, some sources of systematic error, such as verbal skills or test anxiety, are widely recognized, and steps can be taken to ameliorate their effects. In most situations, however, less well-known factors might be affecting validity. How can these problems be identified? There are two general ways to evaluate validity.

In the **face validity** approach, the investigator uses informed judgment to determine whether an operational procedure is measuring what it is supposed to measure. "On the face of it," the researcher asks, "are there good reasons to think that this measure is not an accurate gauge of the intended characteristic?" In the **construct validity** approach, the researcher examines the empirical relationships between a measurement and other concepts to which it should be related. Here the researcher asks, "Does this measurement have relationships with other concepts that one would expect it to have?" Let's look at an example of each approach.

Responses to the following agree-disagree question have been used by survey researchers to measure the concept of political efficacy, the extent to which individuals believe that they can have an effect on government: "Voting is the only way that people like me can have any say about how the government runs things." According to the question's operational design, a person with a low level of political efficacy would see few opportunities for influencing government beyond voting and thus would give an "agree" response. A more efficacious person would feel that other avenues exist for "people like me" and so would tend to "disagree." But examine the survey instrument closely. Using informed judgment, address the face validity question: Are there good reasons to think that this instrument would not produce an accurate measurement of the intended characteristic, political efficacy? Think of an individual or group of individuals whose sense of efficacy is so weak that they view any act of political participation, including voting, as an exercise in political futility. At the conceptual level, one would certainly consider such people to have a low amount of the intended characteristic. But how might they respond to the survey question? Quite reasonably, they could say "disagree," a response that would measure them as having a large amount of the intended characteristic. Taken at face value, then, this survey question is not a valid measure.[20] This example underscores a general problem posed by factors that affect validity. We sometimes can identify potential sources of systematic error and suggest how this error is affecting the operational measure. Thus people with low and durable levels of efficacy might be measured, instead, as being politically efficacious. However, it is difficult to know the size of this effect. How many people are being measured inaccurately? A few? Many? It is impossible to know.

On a more hopeful note, survey methodologists have developed effective ways of weakening the chronic distortion of measurement bias, even when the reasons for the bias, or its precise size, remain unknown. For example, consider the systematic error that can be introduced by the order in which respondents answer a pollster's questions. Imagine asking people the following two questions about abortion: (1) "Do you think it should be possible for a pregnant woman to obtain a legal abortion if she is married and does not want any more children?" (2) "Do you think it should be possible for a pregnant woman to obtain a legal abortion if there is a strong chance of serious defect in the baby?" The first item receives a substantially higher percentage of "yes" responses when it is asked first than when it is asked after the second item.[21] A palliative is available for such question-order effects: Randomize the order in which the questions appear in a survey. In this way, systematic measurement error is transformed into random measurement error. Random measurement error may not be cause for celebration among survey designers, but, as we have seen, random error is easier to deal with than systematic error.[22]

The second approach to evaluating validity, construct validity, assesses the association between the measure of a concept and another concept to which it should be related. This is a reasonable approach to the problem. For example, if the GRE is a valid measure of students' readiness for graduate school, then GRE scores should be strongly related to subsequent grade point averages earned by graduate students. If the GRE is an inaccurate measure of readiness, then this relationship will be weak.[23]

Here is an example from political science. For many years, the American National Election Study has provided a measurement of the concept of party identification, the extent to which individuals feel a sense of loyalty or attachment to one of the major political parties. This concept is measured by a 7-point scale. Each person is classified as a Strong Democrat, Weak Democrat, Independent-leaning Democrat, Independent—no partisan leanings, Independent-leaning Republican, Weak Republican, or Strong Republican. Applying the face validity approach, this measure is difficult to fault. Following an initial gauge of direction (Democrat, Independent, Republican), interviewers meticulously lead respondents through a series of probes, recording gradations in the strength of their partisan attachments: strongly partisan, weakly partisan, independent-but-leaning partisan, and purely independent.[24] Durable unintended characteristics are not readily apparent in this measurement strategy. But let's apply the construct validity approach. If the 7-point scale accurately measures strength of party identification, then it should bear predictable relationships to other concepts.

For example, we would expect strongly partisan people, whether Democrats or Republicans, to engage in much campaigning during an election—displaying bumper stickers, wearing buttons, trying to persuade others how to vote, attending rallies, or perhaps donating money to one of the parties. By the same token, we would expect weak partisans to engage in fewer of these activities, independent-leaners fewer still, and independents the fewest of all. That is the logic of construct validity. If the 7-point scale is a valid measure of partisan strength, then it should relate to clearly partisan behaviors (campaign activities) in an expected way. How does the concept of party identification fare in this test of its validity?

Table 1-1 shows the empirical relationship between the 7-point party identification measurement and a measurement of campaigning. The numbers in the table were calculated by figuring out the percentage of Strong Democrats who engaged in at least one campaign activity during the election, the percentage of Weak Democrats who engaged in at least one

Table 1-1 The Relationship between Party Identification and Campaign Activity

Party identification	Percentage engaging in at least one campaign activity[a]
Strong Democrat	53
Weak Democrat	34
Independent-leaning Democrat	43
Independent	28
Independent-leaning Republican	47
Weak Republican	43
Strong Republican	57

Source: Nancy Burns, Donald R. Kinder, Steven J. Rosenstone, Virginia Sapiro, and the American National Election Studies, American National Election Study, 2000: Pre- and Post-Election Survey [Computer file], 2nd version. Ann Arbor, Mich.: University of Michigan, Center for Political Studies [producer], 2001; Inter-university Consortium for Political and Social Research [distributor], 2002.

[a] Campaign activities include: trying to influence the vote choice of others; displaying a button, sticker, or sign; going to meetings or rallies; performing campaign work; contributing money to a candidate; contributing money to a party; contributing money to any other political group that supported or opposed candidates.

activity, and so on for each of the groups along the party identification scale. Notice that, as expected, people at the strongly partisan poles, Strong Democrats and Strong Republicans, were the most likely to report campaign behavior. More than half of each of these groups engaged in at least one campaign act. And, again as expected, pure independents were the least likely to engage in partisan activity. But, beyond these expectations, is anything amiss here? Notice that Weak Democrats, measured as having stronger party ties than independent-leaning Democrats, were less likely to campaign than were independent-Democratic leaners. A similar comparison at the Republican side of the scale—Weak Republicans compared with Independent-Republican leaners—shows the same thing: Weak partisans behaved in a less partisan manner than people measured as Independent with partisan leanings.

Scholars who have examined the relationships between the 7-point scale and other concepts also have found patterns similar to that shown in Table 1-1.[25] In applying the construct validity approach, we can use empirical relationships such as that displayed in Table 1-1 to evaluate an operational measure. What would we conclude from this example about the validity of this measurement of partisanship? Clearly the measure is tapping some aspect of the intended characteristic. After all, the scale "behaves as it should" among strong partisans and pure independents. But how would one account for the unexpected behavior of weak partisans and independent leaners? What durable unintended characteristic might the scale also be measuring? Some have suggested that the scale is tapping two durable characteristics: one's degree of partisanship (the intended characteristic) and one's degree of independence (an unintended characteristic). These scholars believe that the two concepts, partisanship and independence, should be measured separately.[26] There is, to put it mildly, spirited debate on this and other questions about the measurement of party identification.[27]

Rest assured that debates about validity in political science are not academic games of "gotcha," with one researcher proposing an operational measure and another researcher marshaling empirical evidence to shoot it down. Rather, the debate is productive. It is centered on identifying potential sources of systematic error, and it is aimed at improving the quality of widely used operational measures. It bears emphasizing, as well, that although the problem of validity is a concern for the entire enterprise of political analysis, some research is more prone to it than others. A student of state politics could obtain a valid measure of the concept of state-supported education fairly directly, by calculating a state's per-capita spending on education. A congressional scholar would validly measure the concept of party cohesion by figuring out, across a series of votes, the percentage of times a majority of Democrats opposed a majority of Republicans. In these examples, the connection between the concept and its operational definition is direct and easy to recognize. By contrast, researchers interested in individual-level surveys of mass opinion, as the above examples illustrate, often face tougher questions of validity.

SUMMARY

In this chapter we introduced the essential features of concepts and measurement. A concept is an idea, a mental image that cannot be measured or quantified. A main goal of social research is to express concepts in concrete language, to identify the empirical properties of concepts so that they can be analyzed and understood. This chapter described a heuristic that may help you to clarify the concrete properties of a concept: Think of polar-opposite subjects, one of whom has a great deal of the concept's properties and the other has none of the properties. The properties you specify should not themselves be concepts, and they should not describe the characteristics of a different concept. It may be, as well, that the concept you are interested in has more than one dimension.

This chapter described how to write a conceptual definition, a statement that communicates variation within a characteristic, the units of analysis to which the concept applies, and how the concept is to be measured. Important problems can arise when we measure a concept's empirical properties—when we put the conceptual definition into operation. Our measurement strategy may be accompanied by a large amount of random measurement error, error that produces inconsistently incorrect measures of a concept. Random error undermines the reliability of the measurements we make. Our measurement strategy may contain systematic measurement error, which produces consistently incorrect measures of a concept. Systematic error undermines the validity of our measurements. Although measurement problems are a persistent worry for social scientists, all is not lost. Researchers have devised productive approaches to enhancing the reliability and validity of their measures.

KEY TERMS

aggregate-level unit of analysis (p. 12)
alternative-form method (p. 18)
concept (p. 7)
conceptual definition (p. 8)
conceptual dimension (p. 11)
conceptual question (p. 8)
concrete question (p. 8)
construct validity (p. 20)
Cronbach's alpha (p. 19)
cross-sectional study (p. 19)
ecological fallacy (p. 13)
face validity (p. 20)

Hawthorne effect (p. 16)
individual-level unit of analysis (p. 12)
multidimensional concept (p. 11)
operational definition (p. 8)
panel study (p. 19)
random measurement error (p. 15)
reliability (p. 17)
split-half method (p. 19)
systematic measurement error (p. 15)
test-retest method (p. 18)
unit of analysis (p. 12)
validity (p. 17)

EXERCISES

1. Suppose a researcher wanted to study the role of religious belief, or religiosity, in society. This researcher sets up an inventory of properties, contrasting the mental images of a religious person and a nonreligious person. Inventory items a, b, and c are as follows:

A religious person:	A nonreligious person:
a. Opposes moral relativism	Supports moral relativism
b. Regularly attends religious services	Never attends religious services
c. Supports school prayer	Opposes school prayer

   A. Two of the items in the inventory do not belong on the list. (i) Which two items do not belong? (ii) This chapter discussed certain problems that arise when constructing an empirical inventory. Making specific reference to these problems, explain why each of the items you chose in (i) does not belong.
   B. Examine the inventory item that does belong on the list. Think up and write down one additional property that is similar to the item that belongs on the list.
   C. Based on your conceptual work in parts A and B: (i) Identify which of the following definitions—x, y, or z—is the best conceptual definition of religiosity. (ii) Explain your choice.
   Definition x: The concept of religiosity is defined as the extent to which individuals exhibit the characteristic of supporting government policies that promote the public observance of faith.
   Definition y: The concept of religiosity is defined as the extent to which individuals exhibit the characteristic of demonstrating that religion is an important part of their everyday lives.
   Definition z: The concept of religiosity is defined as the extent to which individuals exhibit the characteristic of advocating absolutist doctrines.

2. *Finding 1:* An examination of state-level data on public opinion toward immigration restrictions reveals that, as states' percentages of Hispanic residents increase (and states' percentages of non-Hispanic residents decline), public support for restrictive immigration laws increases. *Conclusion:* Hispanics are more likely to support restrictive immigration laws than are non-Hispanics.
   A. For the purposes of this exercise, assume that Finding 1 is correct—that is, assume that Finding 1 accurately describes the data. Is the conclusion supported? Making specific reference to a problem discussed in this chapter, explain your answer.
   B. Suppose that, using individual-level data, you compared the immigration opinions of Hispanic and non-Hispanic residents in each state. *Finding 2:* Hispanic individuals are less likely to support restrictive immigration laws than are non-Hispanic individuals. Explain how Finding 1 and Finding 2 can both be correct.

3. The Acme Widget Factory wants to determine the effects of different working conditions on worker productivity. One group of workers is placed in a pleasant atmosphere with plenty of natural lighting and comfortable workstations. A second group is isolated in an uncomfortable setting with dim lighting and awkward ergonomics. At the end of the study, widget production in the second group was *higher* than widget production in the first group.
   A. A company spokesman admits that "something doesn't seem quite right about our study, but we're not sure what that 'something' is." Making specific reference to a measurement problem discussed in this chapter, explain why the Acme study obtained the results it obtained.
   B. What sort of measurement error is more important in accounting for the Acme study results—random measurement error or systematic measurement error? Explain your answer.

4. A researcher has devised a measurement for gauging the level of political knowledge of individuals. The operational measure is a 100-point scale, ranging from 0 (low knowledge) to 100 (high knowledge). For the purposes of this exercise, assume that you know—but the researcher does not know—the "true" level of knowledge for four people. These true values appear in the table below. The researcher will administer the instrument at two time points and record each respondent's operational readings next to the question marks in the columns labeled "Time 1" and "Time 2." Construct and label two tables just like the one you see here.

Respondent	"True" political knowledge	Researcher's operational measurements of respondent's political knowledge	
		Time 1	*Time 2*
1	50	?	?
2	60	?	?
3	70	?	?
4	80	?	?

A. Suppose that the researcher's operational instrument had some small amount of random measurement error and some small amount of systematic measurement error. Imagine what the researcher's operational readings would look like under such a scenario. Fabricate hypothetical political knowledge measurements for each respondent that would fit this scenario. Write your fabricated measurements in the first table you have constructed.

B. Suppose that the researcher's operational instrument had some small amount of random measurement error and a large amount of systematic measurement error. Fabricate hypothetical political knowledge measurements for each respondent that would fit this scenario. Write your fabricated measurements in the second table you have constructed.

C. (i) Which of the following statements is correct: a, b, or c? (ii) Explain your answer.
   a. The operational readings recorded in A are more valid measures of political knowledge than are the readings recorded in B.
   b. The operational readings recorded in B are more valid measures of political knowledge than are the readings recorded in A.
   c. The operational readings recorded in A and the readings recorded in B are equally valid measures of political knowledge.

5. Two candidates are running against each other for a seat on the city commission. You would like to obtain a valid measurement of which candidate has more pre-election support among the residents of your neighborhood. Your operational measure: Obtain a precise count of the yard signs supporting each candidate. The candidate with a greater number of yard signs will be measured as having greater pre-election support than the candidate having fewer yard signs.

A. This measurement strategy has low face validity. Describe two reasons why the measurement strategy will not produce a valid measurement of residents' pre-election support for the candidates.

B. Putting your yard-sign counting days behind you, you resolve to develop a more valid measurement of pre-election support for the candidates. (i) Describe a measurement strategy that would produce a valid measure of pre-election support for candidates running for election. (ii) Describe how you would establish the construct validity of your measurement.

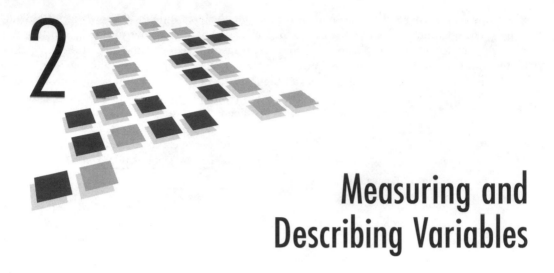

2

Measuring and Describing Variables

LEARNING OBJECTIVES

In this chapter you will learn:
- How to recognize the essential features of a variable
- How to determine a variable's level of measurement
- How to describe the central tendency of a variable
- How to describe the amount of dispersion in a variable

The operational definition of a concept provides a blueprint for its measurement. When we follow through on the plan, when we construct what the blueprint describes, we end up with a variable. Variables provide the raw materials for describing and analyzing the social and political world. A **variable** is an empirical measurement of a characteristic. We might measure marital status among a group of adults by asking each respondent to choose the category that describes them: married, widowed, divorced, separated, or never married. These five categories are the values of the variable, marital status. A person who responds "married" is measured as having a different value on the variable than someone who says "divorced." Similarly, when an application form requests, "Age: _____," it is asking for a value of the variable, age. Someone who writes "20" has a value on the variable that is 9 measurement units (years) younger than someone who writes "29."

All variables share certain features. Every variable has one name and at least two values. Furthermore, if computer analysis is to be performed, then a variable's values must be coded numerically. So, a name, two or more values, and numeric codes—these are the gross anatomy of any variable. However, as the simple examples of marital status and age suggest, important differences exist between variables. Some characteristics, such as age, can be measured with greater precision than others, such as marital status. Accordingly, the values and numeric codes of some variables contain more information than do the values and codes of other variables. After a preliminary dissection of an exemplar variable, we turn to a discussion of levels of measurement, the amount of information conveyed by a variable's values and codes. A variable's level of measurement determines how precisely we can describe it. In this chapter we consider the two cornerstones of description: central tendency and dispersion. Central tendency refers to the typical or "average" value of a variable. Dispersion refers to the amount of variation or "spread" in a variable's values. You will find that central tendency and dispersion are not separate aspects of a variable. They work together to provide a complete description of a variable.

MEASURING VARIABLES

Figure 2-1 displays the key features of a variable and introduces essential terminology. It may also help clear up confusion about variables. Like all variables, marital status has one name, and, like all variables, it has at least two values—in this case, five: married, widowed, divorced, separated, and never married. It is not uncommon to get confused about the distinction between a variable's name and its values. As Figure 2-1 illustrates, the descriptors "married" and "widowed" are not different variables. They are different values of the same variable, marital status. Here is a heuristic that will help you become comfortable with the distinction between a variable's name and values. Think of one unit of analysis and ask this question, "What is this unit's _____?" The word that fills in the blank is never a value. It is always a variable's name. To complete the question this way—"What is this person's divorced?"—makes no sense. But the question, "What is this person's marital status?" makes perfect sense. Ask the question, fill in the blank, and you have the name of a variable—in this case, marital status. Answer the question and you have one of the variable's values: "What is this person's marital status?" "divorced" (or "married," "widowed," "separated," "never married"). The value, "divorced," is one of the values of the variable named marital status.[1]

A variable must have at least two possible values. Beyond this fundamental requirement, however, variables can differ in how precisely they measure a characteristic. Notice that the values of marital status enable us to place people into different categories—and nothing more. Furthermore, the numeric codes that are associated with each value of marital status, code 1 through code 5, merely stand for the different categories—and nothing more. "Married" is different from "never married," and "1" is different from "5." Marital status is an example of a nominal-level variable, a variable whose values and codes only distinguish different categories of a characteristic. Now imagine what the anatomy of a different variable, age, would look like. Just as with marital status, age would have one name. And age would have an array of different values, from "18" to (say) "99." Now move from values to codes. What would be the numeric codes for age? The numeric codes for age would be identical to the values of age. Age is an example of an interval-level variable, which has values that tell us

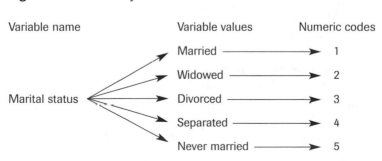

Figure 2-1   Anatomy of a Variable

Variable name	Variable values	Numeric codes
	Married	1
	Widowed	2
Marital status	Divorced	3
	Separated	4
	Never married	5

the exact quantity of a characteristic. The values of interval-level variables convey more information than do the values of nominal-level variables. Another type of variable, ordinal-level variables, conveys more information than nominal variables but less information than interval variables. Let's take a closer look at levels of measurement.

Levels of Measurement

Researchers distinguish three levels of precision with which a variable measures an empirical characteristic. Nominal variables are the least precise. A **nominal-level variable** communicates differences between units of analysis on the characteristic being measured. As discussed above, marital status is a nominal variable. Its values allow us to separate people into different categories. As with most variables, the values of nominal variables are frequently recorded by numeric codes. It is important to remember, however, that these codes do not represent quantities. They merely record differences. Thus, we might measure religious denomination by five values: Protestant, Catholic, Jewish, other religion, and no religious affiliation. For convenience, we could choose numeric codes 1, 2, 3, 4, and 5 to represent those categories. But we could just as easily choose 27, 9, 56, 12, and 77. The numeric codes do not themselves have inherent meaning. They derive their function from the simple fact that they are different. The values and codes of a nominal variable tell us that subjects having one value, such as Protestant, differ from subjects that have another value, such as Catholic. Gender (female/male), region (South/North/Midwest/West), race (white/black), country of origin (born in the United States/not born in the United States), union membership (member/not a member), employment sector (government employee/private sector employee)—all these are examples of characteristics that are measured by nominal variables. In each case, the values, and the numeric codes associated with the values, only represent different categories of the measured characteristic.

Ordinal variables are more precise than nominal-level variables. An **ordinal-level variable** communicates relative differences between units of analysis. Ordinal variables have values that can be ranked. Plus, the ranking is reflected in the variable's numeric codes. Consider an ordinal-level variable named "support for school prayer," which can take on four possible values: strongly oppose, oppose, support, and strongly support. Notice that, just as with nominal variables, the values permit you to classify respondents into different categories. A person who says "strongly oppose" would be measured as being different from a person who says "oppose." But notice that, unlike nominal variables, the values permit you to distinguish the *relative amount* of the characteristic being measured. Someone who says "strongly support" has a higher level of support for school prayer than someone who says "support." The values of ordinal variables have numeric codes that reflect the relative amounts of the measured characteristic. For convenience and simplicity, "strongly oppose" could be coded 1, "oppose" 2, "support" 3, and "strongly support" 4. These codes impart the ranking that underlies the values, with 1 being least supportive and 4 being most supportive. As this example suggests, when it comes to the measurement of attitudes among individuals, ordinal variables are almost always used.

Interval variables give the most precise measurements. An **interval-level variable** communicates exact differences between units of analysis. Age measured in years, for example, is an interval-level variable, since each of its values—18 years, 24 years, 77 years, and so on—measures the exact amount of the characteristic. These quantitative values, furthermore, enable you to measure the precise difference between two units of analysis. How much difference exists between a subject with 24 years and a subject with 18 years? Exactly 6 years. Because the values of an interval variable are the exact numeric quantities of the characteristic being measured, the variable's values do not need to be represented by a separate set of numeric codes. What would be the point? The values themselves tell you all you need to know. If someone were to ask you, "What distance do you drive each day?" your response could be gauged easily by an interval-level value, such as "16 miles." Notice that this value is a not simply a number. It is a number that communicates the exact quantity of the characteristic. The researcher would easily determine that your response is different from someone else's answer (such as "15 miles"), that you drive farther each day (because 16 miles is greater than 15 miles), and that the two responses are separated by exactly one unit (1 mile). Interval-level variables are considered the highest level of measurement because their values do everything that nominal and ordinal values do—they allow the researcher to place units of analysis into different categories, and they permit units to be ranked on the measured characteristic—plus they gauge fine differences between units of analysis.[2]

It is not difficult to think of interval-level variables in everyday life: the liquid volume of a can of soda, the number of weeks in a semester, the score of a baseball game, or the percentage of one's time devoted to studying. When political researchers are using aggregate-level units of analysis, interval variables are common, as well. A student of state politics might measure the percentage of eligible voters who turned out in the gubernatorial election, the number of days before an election that state citizens may register, or the size of the state's education budget. A student of comparative politics might record the number of years that the same regime has been in power in a country or the percentage of the country's budget spent on national defense. A student of interest groups may want to know membership size, number of years since the group's founding, or the cost of joining.

When political researchers are analyzing individual-level units of analysis, nominal and ordinal variables are much more common than interval variables. Ordinal-level variables abound in social research, especially survey research. Questions gauging approval or disapproval of government policies or social behaviors—handgun registration laws, immigration reform, welfare spending, abortion rights, homosexuality, marijuana use, child-rearing practices, and virtually any others that you can think of—are almost always framed by ordinal values.

Additive Indexes

Given the wealth of ordinal variables, particularly in the measurement of attitudes, additive indexes are common sights in social research. An **index** is an additive combination of ordinal variables, each of which is identically coded, and all of which are measures of the same concept. An additive index, also called a *summative scale* or *ordinal scale*, provides a more precise and more reliable measurement of a characteristic. A simple index of economic liberalism was described in Chapter 1: Present respondents with 10 government programs and add up the number of times they say "increase spending." Scores on the index would range from 0 to 10. Of course, a more finely-tuned measure could be achieved by coding responses to each program with a three-value ordinal: "decrease spending" coded 0, "keep the same" coded 1, and "increase spending" coded 2. Summing codes across 10 programs, index scores could range from 0 to 20.

Both of the examples just mentioned—the 0 to 10 index and the 0 to 20 index—loosely fit the Likert approach to index construction. As conceived by its originator, Rensis Likert, a **Likert scale** is an additive index of 5- or 7-value ordinals, each of which captures the strength

Table 2-1   Items in a Likert Scale

1. Our society should do whatever is necessary to make sure that everyone has an equal opportunity to succeed.
2. We have gone too far in pushing equal rights in this country.
3. One of the big problems in this country is that we don't give everyone an equal chance.
4. This country would be better off if we worried less about how equal people are.
5. It is not really that big a problem if some people have more of a chance in life than others.
6. If people were treated more equally in this country we would have many fewer problems.

Sources: Variables V045212, V045213, V045214, V045215, V045216, and V045217 from University of Michigan, Center for Political Studies, American National Election Study, 2004: Pre- and Post-Election Survey [Computer file]. ICPSR04245-v1. Ann Arbor, Mich.: University of Michigan, Center for Political Studies, American National Election Study [producer], 2004. Ann Arbor, Mich.: Inter-university Consortium for Political and Social Research [distributor].

Note: After looking at each statement, respondents are asked: "Do you agree strongly, agree somewhat, neither agree nor disagree, disagree somewhat, or disagree strongly with this statement?"

and direction of agreement (or disagreement) with a declarative statement.[3] Consider Table 2-1, which displays six Likert-type items used to create a measure of egalitarianism, the extent to which individuals believe that government and society should reduce differences between people. For each statement, respondents are asked whether they "agree strongly," "agree somewhat," "neither agree nor disagree," "disagree somewhat," or "disagree strongly." Notice that agreement denotes greater egalitarianism for items 1, 3, and 6 and less egalitarianism for items 2, 4, and 5. Question designers do this to combat the tendency of respondents to agree with survey statements, a nagging but fixable source of systematic measurement error. The quantity and redundancy of the questions—in how many ways can the interviewer ask the same thing?—are signatures of Likert scaling. The similarity of the questions ensures that only one concept is measured. And, generally speaking, the greater the number of questions, the more reliable the scale.[4]

Likert scales are perhaps the most encountered members of a larger class of ordinal-scaling techniques.[5] Obviously, methodologists have made much progress in developing better ordinal-based measurements. But one of the worst-kept secrets in social research is that investigators routinely describe and analyze these scales as if they were interval-level measures. This is one of the longest-debated practices in social science methodology. Two questions arise. Is it appropriate to treat an ordinal scale as an interval variable? Why does it matter? To the first question: If certain conditions are met, the answer is yes.[6] To the second question: It matters because a variable's level of measurement determines how precisely we can describe it. We turn now to a discussion of the essentials of descriptive statistics.

DESCRIBING VARIABLES

The best understood descriptive statistic is a familiar denizen of everyday life: the average. The world seems defined by averages. When your college or university wants to summarize your entire academic career, what one number does it use? What is the average tuition cost of higher education institutions in your state? When people go on vacation, how many days do they typically stay? What is the most popular month for weddings? What make of automobile do most people drive? Political research, too, has a passion for the typical. How much does a congressional candidate commonly spend on a campaign? Do people who describe themselves as Republicans have higher incomes, on average, than Democrats? What opinion do most people hold on the abortion issue? Affirmative action? Immigration reform?

When it comes to describing variables, averages are indispensable. However, political researchers rarely use the term *average* in the same way it is used in ordinary language. They refer to a variable's **central tendency**—that is, its typical or average value. A variable's central tendency may be measured in three ways: the mode, the median, or the mean. The appropriate measure depends on the variable's level of measurement.

The most basic measure of central tendency is the **mode**. The mode of a variable is the most common value of the variable, the value that contains the largest number of cases or units of analysis. The mode may be used to describe the central tendency of any variable. For nominal-level variables, however, it is the only measure that may be used. For describing variables with higher levels of measurement—that is, ordinal or interval—the **median** comes into play. The median is the value of a variable that divides the cases right down the middle—with half of the cases having values below and half having values above the median. The central tendency of an ordinal-level variable may be measured by the mode or median. For interval-level variables, a third measure, the **mean**, also may be used to describe central tendency. The mean comes closest to the everyday use of the term *average*. In fact, a variable's mean *is* its arithmetic average. When we sum all the cases' individual values on a variable and divide by the number of cases, we arrive at the variable's mean value. All these measures of central tendency—the mode, the median, and the mean—are workhorses of description, and they are the main elements in making comparisons and testing hypotheses.

Yet there is more to describing a variable than reporting its measure of central tendency. A political variable also is described by its **dispersion**, the variation or spread of cases across its values. A variable's dispersion is sometimes its most interesting and distinctive feature. When we say that opinions on gun control are "polarized," for example, we are describing their variation, the particular way opinions are distributed across the values of the variable—many people support gun control, many people oppose it, and only a few take a middle position. To say that general "consensus" exists among Americans that capitalism is preferable to communism is to denote little variation among people or widespread agreement on one option over another. When scholars of comparative politics discuss the level of economic equality in a country, they are interested in the variation or dispersion of wealth. Is there little variation, with most economic resources being controlled by a few? Or is the distribution more equitable, with economic resources dispersed across many or most citizens?

Compared with the overworked average—the go-to summary and simplifier—references to a variable's dispersion are uncommon if not rare in everyday life. Variation is underemphasized in social science, too.[7] Here we will discuss the meaning and appropriate uses of the measures of central tendency—mode, median, and mean. We also will explore nonstatistical approaches to describing a variable's dispersion.

Central Tendency and Dispersion

Let's begin by looking at three variables that measure different characteristics of individuals: where they live (region of residence, a nominal variable), how religious they are (frequency of

Table 2-2 Region of Residence (tabular)

Region	Frequency	Percentage
North	711	15.8
Midwest	1,038	23.0
South	1,745	38.7
West	1,016	22.5
Total	4,510	100.0

Source: James A. Davis, Tom W. Smith, and Peter V. Marsden, General Social Surveys, 1972–2006 [Cumulative File] [Computer file]. ICPSR04697-v2. Chicago, Ill.: National Opinion Research Center [producer], 2007; Storrs, Conn.: Roper Center for Public Opinion Research, University of Connecticut/ Ann Arbor, Mich.: Inter-university Consortium for Political and Social Research [distributors], 2007-09-10.

Note: Displayed data are from 2006, hereafter referred to as "2006 General Social Survey."

Table 2-3 Attendance at Religious Services (tabular)

Attendance	Frequency	Percentage	Cumulative percentage
Never or less than once a year	1,322	29.4	29.4
Once a year	571	12.7	42.1
Several times a year	502	11.2	53.3
Once a month	308	6.9	60.2
2–3 times a month	380	8.5	68.7
Nearly every week	240	5.3	74.0
Every week or more	1,168	26.0	100.0
Total	4,491	100.0	

Source: 2006 General Social Survey.

Note: Question: "How often do you attend religious services?" The GSS records nine response categories. Table 2-3 combines "Never" (22.7 percent) with "Less than once a year" (6.7 percent) and "Every week" (18.7 percent) with "Several times a week" (7.3 percent).

attendance at religious services, an ordinal variable), and how much television they watch (hours per day watching TV, an interval variable). In the survey results we are using here, there are four possible values of region (North, Midwest, South, and West), and there are seven possible values of religious attendance (never or less than once a year, once a year, several times a year, once a month, 2–3 times a month, nearly every week, and every week or more). Individuals' daily TV viewing habits can range from 0 to 24 hours. What are the typical values of region, church attendance, and TV watching? Are most people "typical," or are they widely distributed across the values of these variables?

First consider the nominal and ordinal variables, region and religious attendance. Table 2-2 shows how a large sample of individuals responded to the region question in the 2006 General Social Survey. Table 2-3 depicts the religious attendance variable. Each table is a **frequency distribution**, a tabular summary of a variable's values. Frequency distributions are commonly used in data presentations of all kinds—from survey research and journalistic polls to marketing studies and corporate annual reports. The first column of each frequency distribution lists the variable's values. The second column reports the number, or **raw**

Figure 2-2 Region of Residence (graphic)

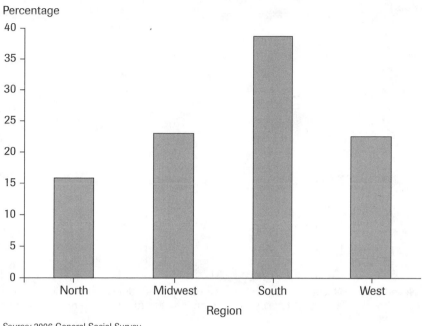

Source: 2006 General Social Survey.

frequency, of individuals giving each response. The raw frequencies are totaled at the bottom of the column. This is the **total frequency**. The third column reports the percentage of cases falling into each value of the variable.[8] The equation to figure the percentage for each value is

Percentage for each value = (raw frequency/total frequency) × 100.

Frequency distributions for ordinal-level variables usually contain an additional column. Unlike nominal variables, which measure differences between units of analysis, ordinal variables tell us the relative amount of the characteristic being measured. This higher level of precision allows us to determine **cumulative percentage**, or the percentage of cases at or below any given value of the variable. So, in Table 2-3, 42.1 percent of respondents attend religious services yearly or never, and 60.2 percent attend monthly or less frequently.

A picture, to use an old cliché, is worth a thousand words. This adage aptly applies to frequency distributions, which are often presented in the form of a **bar chart**, a graphic display of data. For interval-level variables, as we will see, graphic applications are even more important for describing a variable. Figures 2-2 and 2-3 show bar charts for the frequency distributions of, respectively, region of residence and attendance at religious services. Bar charts are visually pleasing and elegant. The variable's values are labeled along the horizontal axis and percentages (or, alternatively, raw frequencies) along the vertical. So, the height of each bar clearly depicts the percentage or number of cases having that value of the variable.[9]

Examine the frequency distributions and bar charts for a few moments. Suppose you had to write a paragraph about each variable. Religious attendance, in particular, has something of an offbeat appearance. How would you describe it?

Figure 2-3 Attendance at Religious Services (graphic)

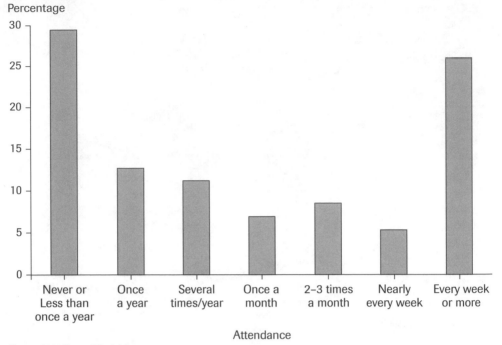

Percentage

Source: 2006 General Social Survey.
Note: Question: "How often do you attend religious services?"

Among the 4,491 respondents, nearly one-third (29.4 percent) rarely or never attend religious services. Although the lowest level of attendance is most common, more than one-fourth of the sample (26.0 percent) were highly observant, attending once a week or more. Most respondents fall into one of the two extremes, and the rest are scattered across the middle values of the variable. There may be a tilt toward lower religiosity: half of the sample (53.3 percent) reported that they attend several times a year or less frequently.

You will find that, in describing the information contained in a frequency distribution or a graphic, you are naturally drawn toward enriching the description by citing a variable's most prominent features. The sentences above noted two such features: the typical response to the religious attendance question and the extent to which responses were dispersed across the variable's values. In describing a variable's typical or average value we are reporting the distribution's central tendency. In describing variety across the variable's values we are reporting the amount of dispersion in the variable.

As noted earlier, the most basic measure of central tendency is the mode, which is defined as the most common value of a variable. The mode of the religiosity variable (Table 2-3) is "never or less than once a year," since the largest percentage of people are in that category. The mode for region (Table 2-1) is "South," the value with the highest percentage. Note that the mode itself is not a percentage or a frequency. It is always a value. A good description takes the following form: "Among the [units of analysis], the mode is [modal value], with [percentage of units] having this value." In the example of region: "Among the 4,510 individuals in the survey, the mode is South, with 38.7 percent having this value."

A frequency distribution having two different values that are heavily populated with cases is called a **bimodal distribution**. Many nominal and ordinal variables have a singular central tendency that is clearly captured by the mode. Region, for example, has a single peak that is easily identified. Religious attendance, by contrast, is a bimodal variable, and a rather interesting one at that. One of its responses ("never or less than once a year") has the highest percentage of cases and, thus, is technically the mode of the distribution. But the flipside response ("every week or more") has nearly as many cases, giving the distribution two very prominent, and very different, response categories. You have to exercise judgment in deciding whether a variable is bimodal. Of course, the percentages of the two values should be similar. Also, in the case of ordinal variables, the two modes should be separated by at least one nonmodal category. Because the two modes of religiosity are separated by five response categories ("once a year," "several times a year," "once a month," "2–3 times a month," and "nearly every week"), the distribution is clearly bimodal. If, instead, the two modes were "never or less than once a year" and "once a year"—that is, if the responses were concentrated in two similar values—we would want to use a single mode to describe the central tendency of the distribution.

The mode may be used in describing any variable, but for nominal-level variables it is the only measure of central tendency that can be used. Again consider the frequency distribution for region (Table 2-2). Notice that, because region is a nominal variable, the rows could be arranged in any order. It would make just as much sense to list South first, then West, Midwest, and North. But for religiosity the situation is different. Since we can order or rank respondents according to their frequency of attendance, it is possible to find a true center of the distribution. For ordinal variables—and for interval variables, too—it is possible to find the value that bisects the respondents into equal percentages, with 50 percent of the cases having higher values of the variable and 50 percent having lower values. The middle-most value of a variable is the median.

The median is a specialized member of a larger family of locational measures referred to as percentiles. Anyone who has taken a standardized college-entrance exam, such as the SAT, is familiar with this family. A **percentile** reports the percentage of cases in a distribution that lie below it. This information serves to locate the position of an individual value relative to all other values. If a prospective college entrant's SAT score puts him in, say, the 85th percentile on the SAT, that person knows that 85 percent of all other test-takers had lower scores on the exam (and 15 percent had higher scores). The median is simply the 50th percentile, the value that divides a distribution in half.

So, what is the median of religious attendance? This is where the cumulative percentage column of Table 2-3 comes into play. We can see that "never or less than once a year" is not the median, since only 29.4 percent of the cases have this value; nor is it "once a year," since 42.2 percent lie at or below this value. Clearly, too, if we jump to "once a month" we have moved too high in the order, since 60.2 percent of the respondents are at or below this value. The median is within "several times a year," since the true point of bisection must occur in this value. How do we know this? Because 53.3 percent of the cases fall in or below this value. Some researchers put a finer point on the computation of the median, but for our purposes the median of an ordinal variable is the value of the variable that contains the median. Thus, we would say that the median is "several times a year."

Now take a step back and reconsider the big picture. Would you say that "several times a year" is the typical response people gave when asked how frequently they attended religious services? Probably not. The example of religious attendance underscores the importance of considering the amount of variation in a variable before deciding if its measure of central tendency is really average or typical.

Here is a general rule: The greatest amount of dispersion for any variable occurs when the cases are equally spread among all values of the variable. Conversely, the lowest amount of dispersion occurs when all the cases are in one value of the variable. If one-seventh (about

14 percent) of the cases fell into each category of the religious attendance variable, this variable would have maximum dispersion (maximum spread). The cases would be exactly evenly spread across all values of the variable. If, by contrast, all respondents reported the same level of religious observance, this variable would have no dispersion at all (minimum spread). By the same token, the region variable would have maximum dispersion if each of its four values (North, Midwest, South, and West) contained 25 percent of the cases; that is, if all the bars in Figure 2-2 were the same height. It would have no dispersion if one value, such as "West," contained all the cases.

Another indication of dispersion, for ordinal variables at least, is provided by comparing the mode and median. If the mode and median are separated by more than one value, then the cases are more spread out than if the mode and median fall in the same value of the variable. By either of these guidelines, religious attendance can be said to have a high degree of dispersion. If dispersion is the most prominent feature of a nominal or ordinal variable, then it would be misleading to use a mode or median to describe it. In such a case, which the religiosity variable exemplifies, we would base our description on the variable's dispersion.

Odd as it may sound, the median's main strength is that it is impervious to the amount of variation in a variable. It simply locates the middle-most score. For bimodal distributions, as we have just seen, this can be a problem. But for single-peaked distributions, those having one mode, the median's resistance to extremely high or low values is a definite asset—an asset that often stands it in good stead with its interval-level sibling, the mean.

Recall that an interval-level variable gives us precise measurements. Unlike nominal variables, whose values merely represent differences, and ordinal variables, whose values can be ranked, the values of interval variables communicate the exact amount of the characteristic being measured. What is more, since interval-level variables are the highest form of measurement, each of the "lower" measures of central tendency—the mode and median—also may be used to describe them. Table 2-4 reports the frequency distribution for the TV-watching variable. What is the mode, the most common number of hours? The most heavily populated value is 2 hours, so 2 is the mode. And what is the median, the middle-most value? Just as with the religiosity variable, we read down the cumulative percentages until we hit the 50th percentile. Since 54.3 percent of the respondents watch television for 2 hours or less, 2 is the median.

What about the mean? As noted at the beginning of the chapter, the mean is the arithmetic center of an interval-level distribution. The mean is obtained by summing the values of all units of analysis and dividing the sum by the number of units. In calculating the mean number of TV-watching hours per day, we add up the hours watched by all respondents (which, according to the survey, is 5,833 hours), and then divide this sum by the number of respondents (1,987). The result: 2.94 hours. Thus, the mode is 2 hours, the median is 2 hours, and the mean is close to 3 hours. Now let's get the big picture and see which of the three measures of central tendency—the mode, the median, or the mean—best conveys the typical number of hours.

Figure 2-4 displays a bar chart of the TV variable. It was constructed just like the earlier bar charts for region and religious attendance. The values of the variable appear along the horizontal axis. The percentages of respondents falling into each value are reported on the vertical axis. Examine Table 2-4 and Figure 2-4. How would you describe this variable? Respondents are grouped fairly densely in the lower range—about 70 percent watch 3 hours or less—and their numbers thin out quickly beyond 5 hours. Only a die-hard 10 percent watch 6 hours or more. But notice that the individuals in the upper values of this variable give the bar chart a longer tail on the right-hand side than on the left-hand side. In fact, the

Table 2-4 Hours Watching TV (tabular)

Hours	Frequency	Percentage	Cumulative Percentage
0	79	4.0	4.0
1	422	21.2	25.2
2	577	29.0	54.3
3	337	17.0	71.2
4	226	11.4	82.6
5	136	6.8	89.4
6	99	5.0	94.4
7	23	1.2	95.6
8	34	1.7	97.3
9	4	.2	97.5
10	23	1.2	98.6
12	14	.7	99.3
13	1	.1	99.4
14	7	.4	99.7
15	2	.1	99.8
18	2	.1	99.9
24	1	.1	100.0
Total	1,987	100.0	

Source: 2006 General Social Survey.
Note: Question: "On the average day, about how many hours do you personally watch television?"

Figure 2-4 Hours Watching TV (graphic)

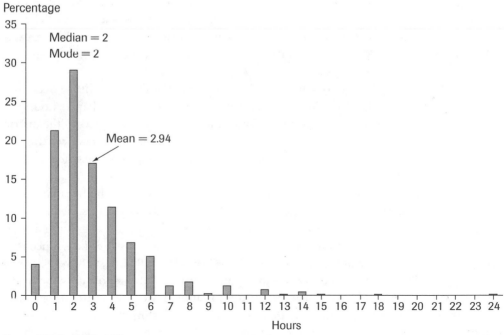

Source: 2006 General Social Survey.
Note: Question: "On the average day, about how many hours do you personally watch television?"

Table 2-5 Number of Programs with "Too Little" Government Spending (tabular)

Number of programs	Percentage	Frequency	Cumulative Percentage
0	17	1.5	1.5
1	41	3.6	5.1
2	56	4.9	10.0
3	99	8.7	18.8
4	129	11.4	30.1
5	163	14.4	44.5
6	171	15.1	59.6
7	155	13.7	73.2
8	126	11.1	84.3
9	119	10.5	94.8
10	59	5.2	100.0
Total	1,135	100.0	

Source: 2006 General Social Survey.

Note: Question: "We are faced with many problems in this country, none of which can be solved easily or inexpensively. I'm going to name some of these problems, and for each one I'd like you to tell me whether you think we're spending too much money on it, too little money, or about the right amount." Displayed data record the number of "too little" responses to: "Improving and protecting the environment," "Improving and protecting the nation's health," "Solving the problems of the big cities," "Halting the rising crime rate," "Dealing with drug addiction," "Improving the nation's education system," "Improving the conditions of Blacks," "Welfare," "Social Security," and "Assistance for childcare."

data in Figure 2-4 are skewed—that is, they exhibit an asymmetrical distribution of cases. Distributions with a longer, or skinnier, right-hand tail have a **positive skew**; those with a skinnier left-hand tail have a **negative skew**. The mean is sensitive to skewness. In this example, respondents in the upper values pull the mean upward, away from the concentration of individuals who watch 2 hours a day.

Extreme values may have an obvious effect on the mean, but they have little effect on the median. In keeping with its reputation for disregarding dispersion, the median (2 hours) reports the value that divides the respondents into equal-size groups, unfazed by the distribution's positive skew. For this reason, the median is called a **resistant measure of central tendency**, and you can see why it sometimes gives a more faithful idea of the true center of an interval-level variable.

Here is a general rule: When the mean of an interval-level variable is higher than its median, the distribution has a positive skew. When the mean is lower than its median, the distribution has a negative skew. Because the mean number of hours is 2.94 and the median number of hours is 2, the distribution depicted in Figure 2-4 has a positive skew. Since the mean and median are rarely the same, an interval-level variable will almost always have some skewness. This being the case, should we simply ignore the mean and report only the median of an interval-level variable? How much skewness is too much? Most computer programs provide a statistical measure of skewness that can help the analyst in this regard.[10] As a practical matter, however, you have to exercise judgment in deciding how much is too much.

Figure 2-5 Number of Programs with "Too Little" Government Spending (graphic)

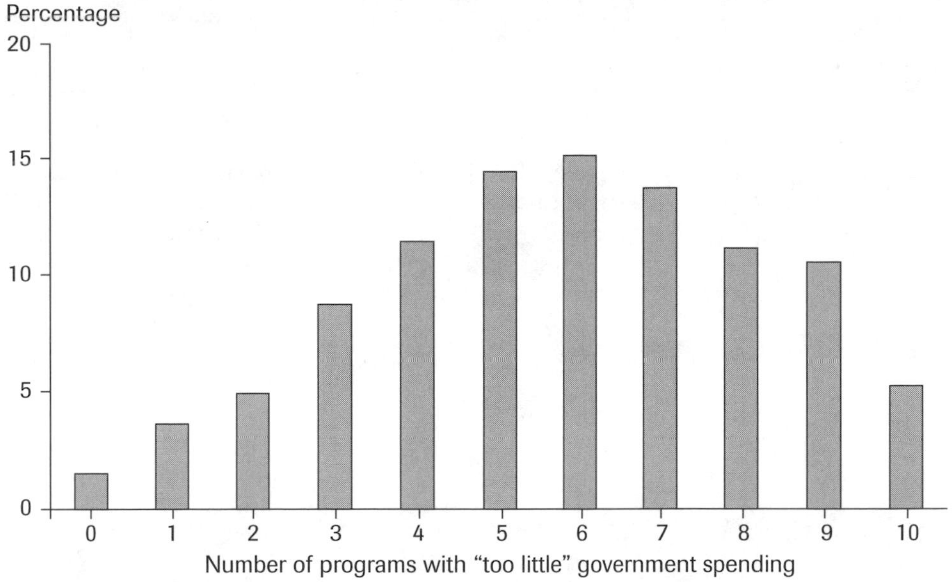

Percentage

Number of programs with "too little" government spending

Source: 2006 General Social Survey.

Consider Table 2-5 and Figure 2-5, which portray the frequency distribution and bar chart of respondents' opinions about government spending on ten programs. The General Social Survey routinely includes a series of questions that ask individuals whether the government is spending "too little," "about the right amount," or "too much" in a number of policy areas. Table 2-5 and Figure 2-5 report the number of programs on which, in the respondents' opinions, the government is spending "too little." This variable, which we could use as a measure of economic liberalism, can take on any value between 0 (the respondent does not think the government is spending too little on any of the programs) and 10 (the respondent thinks the government is spending too little on all ten programs). Higher scores, then, would suggest greater economic liberalism. Examine the frequency distribution and bar chart. Is this distribution skewed? Yes, it is. The mean is 5.78, which is lower than the median and mode, both of which are 6. When the mean is lower than the median, a negative skew is usually the culprit. The skinnier tail toward lower values also suggests negative skewness, which pulls the mean down, slightly off the median of 6. But make a judgment call. Would it be misleading to use the mean value, 5.78, as the central tendency of this distribution? In this case, the mean serves as a good gauge of central tendency.

SUMMARY

Variables are perhaps more variable than you had realized before reading this chapter. Table 2-6 provides a thumbnail summary of key differences in variables, by level of measurement. Let's review these points, beginning with the nominal-ordinal-interval distinctions, a persistent source of confusion. The confusion can usually be cleared up by recalling the difference between a variable's name and a variable's values. A variable's name will tell you the characteristic being measured by the variable. But a variable's values will tell you the variable's

Table 2-6 Measuring and Describing Variables

Level of Measurement		
Nominal (example: region)	*Ordinal* (example: religious attendance)	*Interval* (example: hours watching TV)
Precision: Values allow you to: • separate cases into different categories of the characteristic.	Values allow you to: • separate cases into different categories of the characteristic. • rank cases according to the relative amount of the characteristic.	Values allow you to: • separate cases into different categories of the characteristic. • rank cases according to the relative amount of the characteristic. • determine the exact amount of the characteristic.
Central Tendency: Mode	Mode Median	Mode Median Mean
Dispersion: Low: • one mode prominent. • bar chart single-peaked • noticeably fewer cases in nonmodal categories	Low: • mode and median same or similar • bar chart single-peaked • most cases cluster around median, few cases in extreme values	Low: • median and mean similar and clearly "typical" • bar chart single-peaked • cases cluster around mean, few cases in extreme values
High: • bimodal or multiple modes • bar chart not single-peaked • cases spread out across values	High: • mode and median separated by at least one nonmodal value • bar chart not single-peaked • cases spread out across values	High: • median and mean may be different; mean clearly not "typical" • bar chart not single-peaked • cases spread out across values
		Skewness: Negative skew: • mean lower than median • skinnier left-hand tail • using mean would clearly mislead Positive skew: • mean higher than median • skinnier right-hand tail • using mean would clearly mislead

level of measurement. To figure out a variable's level of measurement, focus on the values and ask yourself this question: Do the values tell me the exact amount of the characteristic being measured? If the answer is "yes," then the variable is measured at the interval level. If the answer is "no," ask another question: Do the values allow me to say that one unit of analysis has more of the measured characteristic than another unit of analysis? If the answer is "yes," then the variable is measured at the ordinal level. If the answer is "no," then the variable is measured at the nominal level. Let's apply these steps to an example.

Survey researchers and demographers are interested in measuring geographic mobility, the extent to which people have moved from place to place during their lives. What values are used to measure this variable? Typically, respondents are asked this question: "Do you currently live in the same city that you lived in when you were 16 years old? Do you live in the same state but a different city? Or do you live in a different state?" So the values are "same city," "same state but different city," and "different state." Look at these values and follow the steps. Do the values tell you the exact amount of geographic mobility, the characteristic being measured? No, the values are not expressed in an interval unit, such as miles. So this is not an interval-level variable. Do the values allow you to say that one person has more of the measured characteristic than another person? Can you say, for example, that someone who still lives in the same city has more or less of the characteristic, geographic mobility, than someone who now lives in the same state but in a different city? Yes, the second person has been more geographically mobile than the first. Because the values permit us to tell the relative difference between the individuals, this variable is measured at the ordinal level.

A variable's level of measurement, as we have seen, determines how completely it can be described. We have also seen that describing a variable requires a combination of quantitative knowledge and informed judgment. Table 2-6 offers some general guidelines for interpreting central tendency and dispersion.

For nominal variables, find the mode. Using a bar chart as a visual guide, ask yourself these questions: Is the distribution single peaked with a prominent mode? Is there more than one mode? Visualize what the bar chart would look like if the cases were spread evenly across all values of the variable. What percentage of cases would fall into each value of the variable if it had maximum variation? Compare this mental image to the actual distribution of cases. Would you say that the variable has a large amount of dispersion? A moderate amount? Or are the cases concentrated in the modal value?

For ordinal variables, find the mode and median. Examining the bar chart, mentally construct a few sentences describing the variable. Just as with nominal variables, imagine a maximum dispersion scenario: Does the actual spread of cases across the variable's values approximate maximum variation? With ordinal variables, you also can compare the modal and median values. Are the mode and median the same, or very close, in value? If so, the central tendency of the variable can be well described by its median. If the mode and median are clearly different values, then it probably would be misleading to make central tendency the focus of description. Instead, describe the variable's dispersion.

For interval variables, find the mode, median, and mean. Because frequency distributions for interval variables tend to be inelegant, a bar chart is essential for getting a clear picture.[11] Consider the three measures of central tendency and examine the shape of the distribution. If the mode, median, and mean fall close to each other on the variable's continuum, and the cases tend to cluster around this center of gravity, then use the mean to describe the average value. Just as with nominal and ordinal variables, a diverse spread of cases denotes greater variation. Interval variables also allow evaluations of symmetry. Is the mean a lot higher or lower than the median? If so, then the distribution may be skewed. Describe the source of skewness. Examine the bar chart and decide whether using the mean would convey a distorted picture of the variable. For badly skewed variables, use the median as the best representation of the distribution's center.

KEY TERMS

bar chart (p. 33)
bimodal distribution (p. 34)
central tendency (p. 31)
cumulative percentage (p. 33)
dispersion (p. 31)
frequency distribution (p. 32)
index (p. 29)
interval-level variable (p. 28)
Likert scale (p. 29)
mean (p. 31)
median (p. 31)

mode (p. 31)
negative skew (p. 38)
nominal-level variable (p. 28)
ordinal-level variable (p. 28)
percentile (p. 35)
positive skew (p. 38)
raw frequency (p. 32)
resistant measure of central tendency (p. 38)
total frequency (p. 33)
variable (p. 26)

EXERCISES

1. A list of terms follows. For each term: (i) State whether the term is a variable name or a variable value. (ii) State the level of measurement. *Example:* support for same-sex marriage. (i) variable name. (ii) ordinal.

   A. lenient
   B. military dictatorship
   C. election year
   D. election percentage
   E. Asian
   F. sophomore
   G. 67 seconds
   H. gender

2. Below are the raw frequencies for two variables: a measure of how much freedom or autonomy individuals have in their jobs (part A), and a measure of opinions about a requirement that would regulate automobile manufacturers (part B). The job autonomy variable, which is based on how much control people have in deciding how to do their jobs, has four values: low autonomy, medium-low autonomy, medium-high autonomy, and high autonomy. The auto manufacturer variable is based on how strongly respondents favor or oppose requiring carmakers to make cars and trucks that use less gasoline. For this survey question, respondents could strongly favor the requirement, favor, neither favor nor oppose, oppose, or strongly oppose such a requirement.[12]

   For each variable: (i) State the variable's level of measurement. (ii) Explain how you know. (iii) Construct a frequency distribution, including frequencies, percentages, and cumulative percentages. (iv) Sketch a bar chart. (v) Identify the mode. (vi) Identify the median. (vii) State whether the variable has high dispersion or low dispersion. (viii) Explain how you know.

   A. Raw frequencies for the job autonomy variable: low autonomy, 450; medium-low, 388; medium-high, 454; high autonomy, 405.
   B. Raw frequencies for car manufacturer requirement: strongly favor requirement, 645; favor, 197; neither favor nor oppose, 52; oppose, 14; strongly oppose requirement, 9.

3. Senator Foghorn described a political rival, Dewey Cheatum, this way: "Dewey Cheatum is a very polarizing person. People either love him or hate him." Suppose a large number of voting-age adults were asked to rate Dewey Cheatum on a 10-point scale. Respondents could give Cheatum a rating ranging between 1 (they strongly disapprove of him) and 10 (they strongly approve of him).

   A. If Senator Foghorn's characterization of Dewey Cheatum is correct, what would a bar chart of approval ratings look like? Sketch a bar chart that would fit Foghorn's description.

B. Still assuming that Senator Foghorn is correct, which of the following sets of values, Set 1 or Set 2, is more plausible? Set 1: mean, 5; median, 5; mode, 5. Set 2: mean, 5; median, 5, mode, 2. Explain your choice. Why is the set of numbers you have chosen more plausible than the other set of numbers?

C. Now assume that the data show Senator Foghorn to be incorrect, and that the following characterization best describes the bar chart of approval ratings: "Dewey Cheatum is a consensus-builder, not a polarizer. He generally elicits positive ratings from most people, and there is little variation in these ratings." Sketch a bar chart that would fit this new description. Invent plausible values for the mean, median, and mode of this distribution.

4. Below is a horizontal axis that could be used to record the values of an interval-level variable having many values, ranging from low values on the left to high values on the right. Draw and label three horizontal axes just like the one shown below.

Low   High

A. Imagine that this variable has a negative skew. What would the distribution of this variable look like if it were negatively skewed? On the first axis you have drawn, sketch a curve depicting a negative skew.

B. Imagine that this variable has a positive skew. On the second axis you have drawn, sketch a curve depicting a positive skew.

C. Imagine that this variable has no skew. On the third axis you have drawn, sketch a curve depicting no skew.

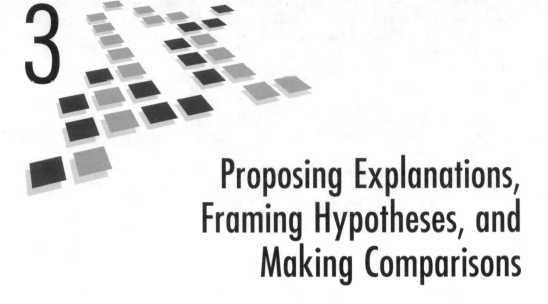

3

Proposing Explanations, Framing Hypotheses, and Making Comparisons

LEARNING OBJECTIVES

In this chapter you will learn:
- How to distinguish an acceptable explanation from an unacceptable explanation
- The difference between an independent variable and a dependent variable
- How to write a hypothesis stating the relationship between an independent variable and a dependent variable
- How to test hypotheses using cross-tabulation analysis and mean comparison analysis
- How to describe linear and nonlinear relationships between variables

The first goal of political research is to define and measure concepts. And you now know the three stages in the measurement process: clearly defining the concept to be measured, determining how to measure the concept accurately, and selecting variables that measure the concept precisely. We begin the measurement process with a vague conceptual term and we end up with an empirical measurement of a concrete characteristic. The measurement process is designed to answer "What?" questions. *What* is political tolerance, and *what* is a valid measurement of this concept?

This chapter concerns the second goal of political research: to propose and test explanations for political phenomena. This goal is not defined by *what*? It is defined by *why*? *Why* are some people more politically tolerant than other people? The empirical landscape is teeming with "Why?" questions. In fact, such questions occur to us quite naturally whenever we observe some difference between people. Why do some people attend religious services weekly whereas others never attend? This is the sort of intellectual activity you engage in all the time—trying to explain the behavior of people and things you observe in the world. Some observed phenomena seem trivial and merely pique our curiosity. Why do some students prefer to sit in the back of the class whereas others prefer the front? Other phenomena occasion more serious thought. News footage of opposing groups of anti-abortion protesters and prochoice demonstrators confronting each other outside an abortion clinic, for example, might raise the question: Why is one group so vociferously opposed to abortion and the other so ardently in favor of keeping it legal? Or consider the fact, routinely noted by political commentators, that only about half of all eligible voters show up at the polls on Election Day. Why do some people vote whereas others do not? And, thinking of elections, why do some voters cast their ballots for the Democrat, some for the Republican, and still others for third-party candidates?

All these questions—and any other "Why?" questions you can think of—have two elements. First, each question makes an explicit observation about a characteristic that varies. Each cites a variable. Explanation in political research begins by observing a variable—a difference that we want to understand. Individuals' abortion beliefs are "anti-abortion" or "prochoice." Students' seating preferences vary between "back" and "front." Voting turnout takes on the values "voted" or "did not vote." And so on, for any "Why?" question. Second, each question implicitly requests a causal explanation for the observed differences. Each can be recast into the form "What causes differences between subjects on this variable?" What causes differences between people in abortion beliefs? What causes differences between students' seating preferences? What causes differences between eligible voters in turnout? Explanation in political research involves causation. An explanation for differences in turnout, for example, might propose that education plays a causal role. As people's level of education increases, they become more aware of politics, and they develop a sense of political efficacy, a belief that they can have an impact on political outcomes. Thus, by this proposed explanation, education causes awareness and efficacy to increase, which cause turnout to increase. This explanation proposes that education is causally linked to turnout.

Proposing explanations defines the creative essence of political research. It is creative because it invites us to think up possible reasons for the observed differences between subjects. It allows us to range rather freely in our search, proposing causal variables that come to mind and describing how those variables might account for what we see. Proposing explanations, however, is not an "anything goes" activity. An explanation must be described in such a way that it can be tested with empirical data. It must suggest a **hypothesis**, a testable statement about the empirical relationship between cause and effect. The education-turnout explanation, for example, suggests a hypothetical relationship between education and voting. If the explanation is correct, then people with less education will vote at lower rates than people with more education. If the explanation is not correct, then level of education and turnout will not be empirically related.[1]

A hypothesis, then, is a conditional statement. It tells us what we should find when we look at the data. When we examine the relationship using empirical data, we are testing the hypothesis. Testing hypotheses defines the methodological essence of political research. It is methodological because it follows a set of procedures for determining whether the hypothesis is incorrect. How would we test the education-turnout hypothesis? Using empirical data, we would compare the turnout rate of people having less education with the turnout rate of people having more education. It is this empirical comparison that tests the hypothesis.

In this chapter we first look at how to construct an explanation. Observing variables in political life, as we have already seen, seems to be a natural and straightforward activity. And essentially it is. In the context of an explanation, however, we view a variable in a particular way—as the effect of some unknown cause. The variable that represents the effect in a causal explanation is called the **dependent variable**. We consider how to propose an acceptable explanation for a dependent variable and how to avoid unacceptable explanations. An acceptable explanation often requires a fair amount of thought—and perhaps some imagination. An acceptable explanation has to be plausible and always requires a long paragraph. Odd as it may sound, however, an acceptable explanation does not have to be correct. In fact, most of the examples of acceptable explanations in this chapter were constructed without any regard for their correctness. The distinguishing feature of an acceptable explanation is not its correctness or incorrectness, but whether it can be *tested* to find out if it is incorrect.

In the second section of this chapter we discuss how to frame a hypothesis. A hypothesis is directly based on the explanation that has been proposed. In framing a hypothesis, the researcher selects a variable that represents the casual factor in the explanation. Like any variable, this variable measures differences between subjects on a characteristic. In the context of an explanation, however, we view this variable in a particular way—as the cause of the dependent variable. The variable that represents a causal factor in an explanation is called the **independent variable**. A hypothesis is a testable statement of the proposed relationship between the independent variable, which measures the cause, and the dependent variable, which measures the effect. A hypothesis states that, as subjects' values on the independent variable change, their values on the dependent variable should be found to change, too, and in just the way suggested by the explanation. Hypotheses are the workhorses of explanation in political science. In this chapter you will learn a foolproof template for writing a testable hypothesis.

In the third section of this chapter you will learn the essential method of testing a hypothesis: comparing values of the dependent variable for subjects having different values on the independent variable. Suppose we theorize that education causes turnout. We think that people with more education will be more likely to vote than will people with less education. To test this hypothesis, we would compare turnout rates (dependent variable) for subjects having different levels of education (independent variable). The method of making comparisons is always the same, but the appropriate hypothesis-testing technique depends on the levels of measurement of the independent and dependent variables. If the independent variable and the dependent variable are measured by categories—that is, if they are nominal-level or ordinal-level variables—then cross-tabulation analysis is performed. If the independent variable is categorical and the dependent variable is interval-level, then mean comparison analysis is performed.

In this chapter you also will learn how to identify linear and nonlinear relationships between independent and dependent variables. In a linear relationship, a change in the independent variable is associated with a consistent change in the dependent variable. In testing the education-turnout idea, for example, suppose we find that 40 percent of low-education people voted, compared with 55 percent of those with medium education and 70 percent among the high-education group. This would be a linear pattern because each change in the independent variable is associated with a 15-point increase in the dependent variable. But suppose we instead find a 40 percent turnout rate among the low-education group, compared with 55 percent among those with medium education and 60 percent in the high-education group. This would be a nonlinear pattern because changes in the independent variable do not have consistent effects on the dependent variable. A change in education from low to medium is associated with a 15-point increase in turnout, but a change from medium to high has only a 5-point effect. In this chapter we will discuss some common patterns of linear and nonlinear relationships.

PROPOSING EXPLANATIONS

Most of us are quite adept at discovering interesting variables. Several examples have already been given at the beginning of this chapter. Political scientists observe and measure variables all the time. Consider this: Many people in the United States think that the government should make it more difficult to buy a gun, yet many think the government should keep things as they are or even make the rules easier.[2] If we were to cast this variable, support for gun restrictions, into a "Why?" question, it becomes a dependent variable, the measurement of a characteristic we wish to explain: "Why do many people favor restrictions and many people oppose them? What causes differences between people on this dependent variable?" Many of us get stymied right away. "That's easy," you might suggest, "People favor tough laws because they support gun control. People oppose such laws because they don't support gun control." This is not an enlightening answer, because it is a tautology—a circular statement that is necessarily true. Good explanations are never circular. Here is a second try. "Gun opinions have something to do with political party affiliations. Democrats and Republicans have different opinions about gun control." This answer is more hopeful. It cites another variable, partisanship, and it says that this new variable has "something to do with" the dependent variable, gun-control opinions. But this statement is still a poor explanation. The connection it makes between two characteristics—party and opinions—is far too vague. Good explanations are never vague. In what way, exactly, do party differences cause or produce different gun opinions? And what is the tendency of this effect? Do Democrats tend to oppose gun control, whereas Republicans tend to be in favor of it? Or do tendencies run the other way, with Democrats more in favor than Republicans? A good explanation connects two variables and provides a detailed description of the causal linkages between them. A good explanation requires some thinking. And it requires some writing:

> When people enter early adulthood, they have only very basic orientations toward politics. Partisanship is one of these orientations. In much the same way that children adopt the religious denomination of their parents, children tend to adopt the party loyalties of their parents. A person who is raised in a Democratic household is likely to identify with the Democrats, whereas a person raised in a Republican household is likely to identify with Republicans. These partisan orientations may be basic, but they become useful later in life when people are deciding their positions on important political issues. Democrats will look at the issue positions of Democratic opinion leaders—Democratic members of Congress, for example—and adopt those issue positions themselves. Republicans will look to Republican opinion leaders. Gun control is a good case in point. Gun control is one issue that divides Democratic and Republican opinion leaders. Democratic opinion leaders, from the presidential and congressional levels to state and local governments, have advocated stricter measures. Republicans have opposed new gun-control measures. The opinions of ordinary citizens have followed this lead. Therefore, Democrats will be more likely than Republicans to favor gun control.

This explanation is far better than the earlier attempts. What makes this a good explanation? First, it describes a connection between the dependent variable, gun-control opinions, and a causal variable, partisanship. In the context of an explanation, the causal variable is called the independent variable: differences in gun-control opinions depend on differences in partisanship.[3] Second, it asserts the direction or tendency of this difference. As the values of the independent variable change from Democrat to Republican, the dependent variable will change in a specific way: Democrats will be more likely than Republicans to favor tougher gun restrictions. Third, it is testable. If we find that Democrats and Republicans do not differ in their gun opinions, then we can seriously question or discard the explanation. So the explanation connects an independent variable with a dependent variable, asserts the tendency of the connection, and suggests a testable empirical relationship.

But notice that the explanation does more than simply propose a testable empirical relationship. It says that the relationship between partisanship and gun-control opinions is one consequence of a general causal process through which opinion leaders shape the attitudes of citizens. Equipped with crude but durable partisan affections developed early in life, people decide their own positions on issues by following the lead of partisan elites. This explanation suggests the existence of other relationships in the empirical data. A scholar of childhood socialization should find evidence of nascent party attachments among children. Furthermore, these early attachments will coincide with those of the children's parents. A student of media should find that people who pay close attention to political news will easily connect their own issue position (for example, supporting gun control) with vote choice (a vote for a Democrat over a Republican). Those who are tuned out to politics, by contrast, will be less likely to know how to use their own opinion to make political choices. A scholar of public opinion should find that changes in citizen attitudes on important issues follow—and do not precede—crystallization of the positions of opinion leaders. According to this general process, for example, the battle lines on immigration reform are first defined by party leaders, and then these positions are echoed by Democrats and Republicans in the mass public. Now, do we know if any of these assertions are correct? No, we do not. But each of these relationships is testable. Each suggests what we should find if the explanation is correct. A good explanation describes a general causal process that suggests several testable relationships. A good explanation arouses curiosity: "Well, *what else* should we find?"

Many very interesting explanations in social science began in just the way illustrated by this example. A researcher observes a phenomenon, develops a causal explanation for it, and then asks, "What else should I find?" In his provocative book *Bowling Alone*, Robert Putnam begins with a seemingly innocuous observation: Although the pastime of bowling has steadily increased in popularity among individuals over the past several decades, the number of bowling leagues has declined precipitously.[4] So the individual enjoyment of the sport has increased, but the collective enjoyment of the sport has declined. This observation, as one reviewer of Putnam's work says, may be "a matter of no small concern to bowling-alley proprietors whose revenues depend heavily on the convivial sharing of beer and pretzels."[5] But is the decline in organized bowling being produced by a general causal process that has other consequences? Putnam argues, in part, that generational change, the replacement of older generations with younger cohorts, is causally linked to the erosion of community and social groups of all kinds. People born in the years preceding World War II, the "long civic generation," are more likely than younger generations to engage in organized social interaction. The effects of generational change can be seen in the lost vitality of all sorts of community groups—parent-teacher associations, civic booster groups, charitable organizations, religious congregations, and, of course, bowling leagues. Older cohorts are more likely than younger generations to favor these social activities. The effects can also be seen in the sorts of activities that are on the rise— memberships in far-flung groups that do not require social interaction, disembodied Internet chat, and, of course, the individual enjoyment of bowling. Older cohorts are less likely than younger people to favor these activities. Thus, Putnam connects a dependent variable, the decline of bowling leagues, with an independent variable, the changing age composition of American society. He describes the tendency of the relationship: Older generations are more likely than younger generations to join bowling leagues. More important, Putnam develops a general explanation that suggests many other relationships between generation and membership in different sorts of organizations. Much of his book is devoted to examining these relationships to find out whether they are correct.[6]

Consider another example that demonstrates well the role of creativity and imagination in thinking up explanations. Malcolm Gladwell's 2000 book *The Tipping Point* also begins with a single, curious observation—about shoes. Hush Puppies, comfortable if unstylish suede footwear, suddenly began appearing on the feet of young and well-appointed males on the streets and in the clubs of Manhattan, where Gladwell lived and worked. Where did this strange trend originate? Gladwell did some checking around. By 1994, Hush Puppies sales had dipped to 30,000 pairs, sold mostly in mom-and-pop retail stores. Wolverine, the shoes' manufacturer, considered the brand all but dead. But later that year Hush Puppies began showing up in an unlikely place, resale shops in Manhattan's Soho and Greenwich Village districts. In 1995, sales rocketed to 430,000 pairs—then to more than 1.5 million pairs in 1996—and Hush Puppies became the accessory of choice among New York's fashion elite. "How does a thirty-dollar pair of shoes," Gladwell wonders, "go from a handful of downtown Manhattan hipsters and designers to every mall in America in the space of two years? . . . Why is it that some ideas or behaviors or products start epidemics and others don't?"[7]

Gladwell proposes that the Hush Puppies fad was the end result of the same sort of causal process that produces epidemics of infectious diseases. Ideas, consumer products, and biological bugs all require the same causal conditions in order to spread. For one thing, a small group of highly infectious people, whom Gladwell calls "connectors," must transmit the infection to many other people beyond a closely knit community. In the case of sexually transmitted diseases, such as the AIDS-causing virus HIV, connectors are highly promiscuous individuals who have weak personal ties to their many partners. In the case of an idea or fad, connectors are people who have an extraordinary number of casual social contacts. These contacts may or may not know each other directly, but they all know the connector.[8] Thus, Gladwell links a dependent variable, the success or failure of contagion, to an independent variable, the presence or absence of connectors. He asserts that his causal explanation helps to account for a host of social results, from the Hush Puppies craze to the precipitous decline in New York's crime rate, the persistence of teen smoking, and even the resounding success of Paul Revere's fabled midnight ride.[9] Interesting and imaginative, to be sure, but is Gladwell's explanation testable?

From the standpoint of proposing a testable relationship, Gladwell's explanation for the success of Paul Revere's famous word-of-mouth campaign is especially instructive. Of course, Paul Revere fits the description of classic connector: "a fisherman and a hunter, a card-player and a theater-lover, a frequenter of pubs and a successful businessman," an organizer and communicator, a recognized and trusted link between the secret and disparate revolutionary societies that emerged after the Boston Tea Party.[10] It is little wonder that he successfully mobilized local militia in the towns along his route. What makes Gladwell's explanation convincing, though, is that it also accounts for the *failure* of Revere's comrade in arms, William Dawes, a fellow revolutionary who set out on the same night, carrying the same alarming message over the same distance and similar countryside and through the same number of towns. The same "disease"—the news that the "Redcoats are coming!"—and the same context—vigilant local militia awakened in the middle of the night—produced a completely different result: a mere handful of men from Dawes's circuit showed up to fight the British the next day. "So why," Gladwell asks, "did Revere succeed where Dawes failed?"[11] William Dawes, like Revere, was a committed and courageous revolutionary. Unlike Revere, however, Dawes was not socially well-connected. He was an ordinary person with a normal circle of friends and acquaintances, known and trusted by only a few people, all of whom were within his own social group. This historical event thus provides a test of Gladwell's explanation. One value of the independent variable, the presence of a connector, produced one value of the dependent variable, the successful contagion of an idea. A different value of the independent variable, the absence of a connector, produced a different value of the dependent variable, the failure of contagion.

FRAMING HYPOTHESES

All these examples illustrate the creative aspect of proposing explanations. Within the generous boundaries of plausibility, researchers can describe causal explanations that link cause with effect, specify tendency, and suggest a variety of other consequences and relationships. Yet we have also seen that any relationship suggested by an explanation must be testable. A testable relationship is one that tells us what we should find when we examine the data. A testable relationship proposes a hypothetical comparison. The partisanship–gun opinions explanation tells us what we should find when we examine the gun-control opinions of Democrats and Republicans. It suggests a hypothetical comparison. If we separate subjects on the values of the independent variable, party, and compare values on the dependent variable, gun-control opinions, we should find a difference: Democrats will be more likely than Republicans to fall into the "more restrictive" category of the dependent variable.

The deceptively simple necessity of proposing a testable comparison is central to the methodology of political research. This comparison is formalized by a hypothesis, defined as a testable statement about the empirical relationship between an independent variable and a dependent variable. The hypothesis tells us exactly how different values of the independent variable are related to different values of the dependent variable. Here is a hypothesis for the relationship between partisanship and gun-control attitudes:

> In a comparison of individuals, those who are Democrats will be more likely to favor gun restrictions than will those who are Republicans.

This hypothesis makes an explicit, testable comparison. It tells us that when we compare units of analysis (individuals) having different values of the independent variable (partisanship), we will observe a difference in the dependent variable (gun-control opinions). The hypothesis also reflects the tendency of the relationship. As the values of the independent variable change, from Democrat to Republican, the dependent variable changes from more support for restrictions to less support. This example suggests a template for writing any hypothesis:

> In a comparison of [units of analysis], those having [one value on the independent variable] will be more likely to have [one value on the dependent variable] than will those having [a different value on the independent variable].

Let's look at a different example, using different units of analysis and different variables, and see if the template works.

Suppose a student of comparative politics observes interesting variation in this dependent variable: the percentage of countries' voting-age populations that turns out in national elections, such as parliamentary or congressional elections. In fact, turnouts range widely—from less than 50 percent in some countries to more than 80 percent in others. The researcher proposes the following explanation:

Potential voters think of their ballots in the same way that potential investors think of their dollars. If people believe that they will get a good return on their money, then they are more likely to invest. If people view a potential investment as having weak prospects, then they will be less likely to commit their money. By the same token, if individuals think that their votes will elect the parties they support and produce the policies they favor, then they will be more likely to turn out. If, by contrast, eligible voters do not believe their ballots will have an impact on the outcome of the election or on the direction of public policy, then they are more likely to stay home. Just as stocks vary in how well they translate dollars into return on investment, countries differ in how well their electoral systems translate voters' preferences into legislative representation. Many countries use some form of proportional representation (PR). In a PR system, a political party is allocated legislative seats based on the percentage of votes it receives in the election. So a party that represents a specific policy or point of view—an environmental party, for example—can gain political power in a PR system. Countries using proportional representation are likely to have multiple parties that afford voters a wide array of choices. Turnout in these countries will be high, because voters will see a strong link between their votes and their impact on government. Other countries use some form of a plurality system. In a plurality system, only the party that receives the most votes wins representation in the legislature. A party receiving less than a plurality of the vote—again, such as an environmental party—ends up with no representation. Countries using plurality systems will have fewer political parties, which offer blunt choices on a large array of issues. Turnout in these countries will be low, because voters will perceive a weak link between their votes and their impact on government.

This explanation describes several causal linkages. Countries vary in how directly their electoral systems convert citizens' votes into legislative power. Proportional representation systems foster multiple parties, giving citizens more choices and greater impact. Plurality systems produce fewer parties, giving individuals fewer choices and lesser impact. The explanation connects a dependent variable, voter turnout, with an independent variable, type of electoral system, and it tells us how the two variables are related. Using the template, we can use this explanation as the basis for a hypothesis:

In a comparison of *countries*, those having *PR electoral systems* will be more likely to have *higher voter turnout* than will those having *plurality electoral systems*.

The independent and dependent variables are easily identified, and the tendency of the hypothetical relationship between them is clear. The hypothesis tells us what to compare—turnout in countries having PR systems with turnout in countries having plurality systems—and what we should find: Turnouts should be higher in countries having PR systems. As we observe a change in the values of the independent variable, from PR systems to plurality systems, the dependent variable should change from higher turnout to lower turnout. So the template works.

The hypothesis-writing template will never steer you in the wrong direction. By following it, you can always draft a testable statement for the relationship between an independent variable and a dependent variable. Most of the time, the template produces a clear and readable hypothesis. The electoral systems-turnout hypothesis is a good example. Other times, you may wish to edit the statement to make it more readable. Using the template, we wrote the following hypothesis for the partisanship–gun opinions relationship: In a comparison of individuals, those who are Democrats will be more likely to favor gun restrictions than will those who are Republicans. Consider this syntax-friendly rewrite: In a comparison of individuals, Democrats will be more likely than Republicans to favor gun restrictions. The first version is acceptable, but the second may be preferred. In any event, by becoming familiar with the template, you can learn to identify—and avoid—some common mistakes in writing hypotheses.[12] Examine the following statements:

A. In a comparison of individuals, some people are more likely to donate money to political candidates than are other people.
B. Highly religious people vote at high rates.
C. In a comparison of individuals, gender and abortion attitudes are related.
D. Because of important cultural changes that began in the 1960s, many current political conflicts are based on generational differences.

Statement A is not a poor hypothesis. It simply is not a hypothesis! It describes one variable—whether or not people donate money to political candidates—but it does not state a relationship between two variables. A hypothesis must compare values on a dependent variable for subjects that differ on an independent variable. Statement B is a poor hypothesis because it does not make an explicit comparison. Highly religious people vote at high rates compared with whom? Such a comparison may be implied, but it is never a good idea to leave the basis of comparison unstated. Statement C is defective because it fails to state the tendency of the relationship. In what way, exactly, are gender and attitudes related? As we have seen, an acceptable explanation does more than simply connect two variables. A good hypothesis tells us exactly how different values of the independent variable are related to different values of the dependent variable. Statement D actually sounds interesting, and it would certainly be a conversation-starter. Stimulating as it is, however, statement D is much too vague to qualify as a testable hypothesis. Vague hypotheses can always be traced to vague explanations. And good explanations are never vague. What is meant by "cultural changes"? What is the dependent variable? What is the independent variable? How would we describe the causal process that connects them?

Intervening Variables

A good explanation always describes a causal process: a causal linkage or series of linkages that connects the independent variable with the dependent variable. These linkages, in turn, suggest the existence of other testable relationships. Let's consider an example that is especially rich in causal linkages. This example helps illustrate an additional, important point about explanations and hypotheses.

Students of presidential elections have long discussed and debated explanations for the different choices that individual voters make between candidates. A typical "Why?" question asked by scholars is: "Why do some voters vote for the incumbent president whereas others vote for the challenger?"

When it comes to making a vote choice, people look at how good a job the incumbent has done in managing the economy. Most voters don't have a great deal of knowledge about political or social problems, but they do know one thing: whether their personal financial situation has improved or gotten worse while the incumbent has been in office. If things have gotten better for them, they attribute this to the good job the incumbent has done in handling the economy and thus are likely to vote for the incumbent candidate. By contrast, if their economic situation has gotten worse, they will blame the incumbent's poor handling of the economy, and they are likely to support the challenging candidate. As one of former president Clinton's advisers memorably put it, it's "the economy, stupid." Differences in voters' economic situations cause differences in their vote choices.[13]

This explanation describes the values of an independent variable, voters' economic situations having either "gotten better" or "gotten worse," and it connects this independent variable to the dependent variable, a vote choice for the incumbent candidate or a vote for the challenger. It suggests the following hypothesis:

> In a comparison of voters, those whose economic situations have gotten better will be more likely to vote for the incumbent candidate than will voters whose economic situations have gotten worse.

The explanation also describes a causal linkage between the independent variable and the dependent variable. Voters evaluate their own situations, thumbs up or thumbs down, tie this evaluation to the incumbent's management skills, and then cast their votes accordingly. Therefore, the independent variable affects another variable, favorable or unfavorable opinions about the incumbent's job performance in handling the economy, which in turn affects the dependent variable. This causal linkage suggests *two* additional hypotheses:

> In a comparison of voters, those whose economic situations have gotten better will be more likely to have favorable opinions about the incumbent's handling of the economy than will voters whose economic situations have gotten worse.

> In a comparison of voters, those who have favorable opinions about the incumbent's handling of the economy will be more likely to vote for the incumbent than will those who have unfavorable opinions.

The first hypothesis says that voters' opinions about the incumbent's handling of the economy depend on voters' economic situations. In the first hypothesis, the job performance variable is the dependent variable. The second hypothesis says that vote choice depends on opinions about job performance. So, in the second hypothesis, the job performance variable is the independent variable. In using an explanation to construct different hypotheses, we might cast the same variable as the dependent variable in one hypothesis and as the independent variable in another hypothesis.

This example is not at all unusual. Many explanations in political science describe an **intervening variable**—a variable that acts as go-between or mediator between an independent variable and a dependent variable. Intervening variables often are of central importance in describing *how* the independent variable is linked to the dependent variable. If someone were to ask how individuals' economic situations and their vote choices are linked, the task of explaining the connection would fall to the intervening variable—voters' opinions about the incumbent's job performance. What is more, if the explanation is correct, then the two additional hypotheses—that voters' economic fortunes affect their opinions about job performance and that opinions about performance affect vote choice—should stand up to empirical scrutiny.

A variable is not inherently an intervening variable. A variable assumes the role of intervener in the context of an explanation that describes a causal process. Indeed, intervening variables often come to mind as we think more carefully about a relationship. Consider once again the hypothesized link between educational attainment (independent variable) and the likelihood of voting (dependent variable). It is easy to become intellectually lazy in thinking about relationships such as this. We might simply say, "Obviously, people who have more years of education will vote at higher rates than will the less educated." So, education magically transforms a nonvoter into a voter? Surely, the educational process must affect an intervening variable (or variables), which in turn affects turnout. Educational life is accompanied by a series of participatory experiences, such as school elections, clubs and other extracurricular activities, and classroom collaborations with peers. What is more, these experiences evolve from symbolic rituals (homeroom elections in fifth grade) into substantive instruments of accountability (student government elections in college). It seems plausible to hypothesize that school provides practical lessons in "civic education"—for example, instilling in people a sense of political efficacy, the belief that participation in collective choices can help to advance desired social goals. According to this explanation, individuals with more education will have more exposure to this process and, thus, evince higher levels of efficacy than the less educated. And those with stronger efficacy will be more likely to participate in politics than will the less efficacious.[14] As this example illustrates, intervening variables often lurk beneath the surface and frequently play a role in carefully constructed explanations.

MAKING COMPARISONS

A hypothesis suggests a comparison. It suggests that if we separate subjects according to their values on the independent variable and compare their values on the dependent variable, we should find a difference. Subjects that differ on the independent variable also should differ on the dependent variable in the hypothesized way. We can now put a finer point on this method. When the dependent and independent variables are categorical variables—that is, they are measured at the nominal or ordinal level—we test the hypothesis using cross-tabulation analysis. When we have a categorical independent variable and an interval-level dependent variable, mean comparison analysis is performed. Let's take a closer look at each of these situations.[15]

Cross-tabulations

To illustrate cross-tabulation analysis, we will use the hypothesis framed by the partisanship explanation of gun control: In a comparison of individuals, Democrats will be more likely than Republicans to favor gun restrictions. The dependent variable, support for gun restrictions, has two possible values: people think gun ownership should be "more difficult" or they think that it should be "the same or easier." The independent variable, partisanship, is an ordinal-level measure, with shadings from "strong Democrat" at one pole to "strong Republican" at the other, and "Independent" in between. To simplify the example, we will collapse partisanship into three categories: Democrat, Independent, and Republican.[16]

Table 3-1 Gun-Control Opinions, by Partisanship (cross-tabulation)

Opinion on gun restrictions	Party identification			Total
	Democrat	*Independent*	*Republican*	
More difficult	73.3%	56.8%	44.8%	58.6%
	(277)	(262)	(155)	(694)
Same/easier	26.7%	43.2%	55.2%	41.4%
	(101)	(199)	(191)	(491)
Total	100.0%	100.0%	100.0%	100.0%
	(378)	(461)	(346)	(1,185)

Source: 2004 American National Election Study.

Note: Question: "Do you think the federal government should make it more difficult for people to buy a gun than it is now, make it easier for people to buy a gun, or keep these rules about the same as they are now?"

Before looking at the data, try to visualize three side-by-side frequency distributions, one for each partisan group. Imagine a group of subjects, all of whom are Democrats, distributed across the two values of the dependent variable, "more difficult" or "same/easier"; a group of Independents distributed across the variable; and a group of Republicans, a percentage of whom favor stronger restrictions and a percentage of whom oppose them. Now ask yourself: If the hypothesis were correct, what would a *comparison* of these three frequency distributions reveal?

This mental exercise is the basis for elemental hypothesis testing in political research. Table 3-1, which displays the relationship between partisanship and gun-control attitudes, introduces the most common vehicle for analyzing the relationship between two variables, the cross-tabulation. A cross-tabulation is a table that shows the distribution of cases across the values of a dependent variable for cases that have different values on an independent variable. So, the first column of numbers in Table 3-1 displays the distribution of the 378 Democrats across the values of the gun-control variable, the second column the distribution of the 461 Independents, and the third the distribution of the 346 Republicans. Each column contains the raw frequency and percentage of cases falling into each category of the dependent variable. Thus, 277 of the 378 Democrats, which is 73.3 percent, support more restrictions, 262 of the 461 Independents (56.8%) are in favor, as are 155 of the 346 Republicans (44.8%). Notice that the right-most column, labeled "Total," which combines all 1,185 respondents, is the frequency distribution for the dependent variable. Of all 1,185 cases, 694 (58.6%) fall into the "more difficult" category of the dependent variable, and 491 (41.4%) fall into the "same/easier" category.

Table 3-1 was constructed by following three rules that should always guide you in setting up a cross-tabulation. The first two rules help you organize the data correctly. The third helps you interpret the relationship.

Rule One: Set up a cross-tabulation so that the categories of the independent variable define the columns of the table, and the values of the dependent variable define the rows. For each value of the independent variable, the raw frequencies falling into each category of the dependent variable are displayed (by convention, in parentheses), totaled at the bottom of each column.

Rule Two: *Always* calculate percentages of categories of the independent variable. *Never* calculate percentages of categories of the dependent variable. This may sound rigid, but Rule

Two is the most essential—and most frequently violated—rule. The partisanship explanation states that the independent variable, party identification, is the cause. The dependent variable, gun opinions, is the effect. As partisanship changes, from Democrat to Independent to Republican, gun-control attitudes should change—from stronger support for restrictions to weaker support for restrictions. Does the balance of opinion change, as the hypothesis suggests, as we move from one level of the causal process to another? To answer this question, we want to know if differences in the causal factor, partisanship, lead to different outcomes, different opinions on gun control. We need to know the percentages of the respondents in each category of the independent variable who favor or oppose gun restrictions. Accordingly, the percentages in each column are based on the column totals, not the row totals. As a visual cue, these percentages are totaled to 100.0 percent at the bottom of each column.

Rule Three: Interpret a cross-tabulation by comparing percentages across columns at the same value of the dependent variable. This is the comparison that sheds light on the hypothesis. We could compare the percentage of Democrats favoring more gun controls with the percentages of Independents and Republicans favoring more controls. Alternatively, we could compare the partisan groups across the other category of the dependent variable, "same/easier." But we could not mix and match.

These simple rules help avoid some common mistakes. The most serious of these errors, calculating percentages of the dependent variable, already has been pointed out. Another mistake is the tendency to interpret a cross-tabulation by referring to the largest percentages in a column. If I were to say, "While 73.3 percent of the Democrats support gun restrictions, 55.2 percent of Republicans think gun ownership rules should stay the same or be easier," I certainly would be describing the numbers, but I would not be directly addressing the hypothesis. The hypothesis says that one value of the dependent variable should be different for cases having different values of the independent variable. Hone in on a comparison of percentages across only one value of the dependent variable.

A related error, one that affects the interpretation of the results, occurs when the analyst gets distracted by the absolute magnitudes of the percentages and neglects to make the comparison suggested by the hypothesis. Suppose, for the sake of illustration, that 80 percent of the Democrats favored tougher gun laws, 70 percent of the Independents were in favor, and 60 percent of the Republicans were in favor. In interpreting percentages such as these—percentages that in this case would show high absolute support for gun control in each partisan group—we might be tempted to say, "The hypothesis is wrong. Large majorities of Democrats, Independents, and Republicans favor stronger gun control." This would be an incorrect interpretation. Remember that a hypothesis is a *comparative* statement. It requires a comparison of *differences* in the dependent variable across values of the independent variable. Always apply a comparative standard. Applying this standard to the illustrative 80–70–60 percentages, we would have to say that, as partisanship changes from Democrat to Independent to Republican, the percentage of people who support gun control declines.

Let's return to the data in Table 3-1. Does the partisanship hypothesis on gun control appear to be correct? Following the rules, we would focus on the percentage of each partisan group favoring restrictions and read across the columns, starting with the Democrats. Democrats are most in favor (73.3%), Independents show somewhat weaker support (56.8%), and Republicans are least likely to favor (44.8%). The pattern is systematic and not inconsistent with the hypothesis. Each time we change the independent variable, from Democrat to Independent to Republican, the distribution of the dependent variable changes, too, and in the hypothesized way.

Table 3-2 Smoking, by Income (cross-tabulation)

Smoker?	Income category					
	$13,999 or less	*$14,000– $24,999*	*$25,000– $39,999*	*$40,000– $59,999*	*$60,000 or more*	Total
Yes	32.5%	27.0%	24.6%	21.8%	16.4%	24.2%
	(90)	(62)	(76)	(58)	(52)	(338)
No	67.5%	73.0%	75.4%	78.2%	83.6%	75.8%
	(187)	(168)	(233)	(208)	(265)	(1,061)
Total	100.0%	100.0%	100.0%	100.0%	100.0%	100.0%
	(277)	(230)	(309)	(266)	(317)	(1,399)

Source: Steven J. Rosenstone, Donald R. Kinder, Warren E. Miller, and the American National Election Studies, American National Election Study, 1996: Pre- and Post-election Survey [Computer file], 4th version. Ann Arbor: University of Michigan, Center for Political Studies [producer], 1999; Inter-university Consortium for Political and Social Research [distributor], 2000.

Note: Question: "Are you a smoker?"

Here is another example. A student of health policy is investigating the relationship between income and cigarette smoking. Are poor people more likely than rich people to smoke cigarettes? The policy researcher's hypothesis is as follows: In a comparison of individuals, people earning lower incomes will be more likely to smoke than will individuals earning higher incomes. Again, before looking at the data, try to visualize a cross-tabulation. You should see groups of people separated into categories of the independent variable, income. Each group will be distributed across values of the dependent variable, smoker or nonsmoker. A certain percentage of the low-income group will be smokers and a certain percentage of higher-income individuals will be smokers. If the hypothesis is incorrect, what will a comparison of these percentages look like? If the hypothesis is correct, what should we find?

Table 3-2 shows the cross-tabulation, constructed according to the rules. The independent variable, a five-value ordinal measure of income, defines the columns. The values of the dependent variable, smoker and nonsmoker, define the rows. The cells of the table contain the raw frequencies and percentages of smokers and nonsmokers for each category of the independent variable, totaled at the bottom of each column. The frequency distribution for the entire sample appears in the right-most column.

How would you interpret the percentages? Obviously, if the hypothesis were incorrect, the percentages of smokers in each category of income should be about the same. But they are not. Comparing percentages across columns at the same value of the dependent variable—we will focus on the percentage of smokers—reveals an unambiguous pattern. Whereas 32.5 percent of the lowest income group are smokers, the percentage of smokers drops systematically as income goes up, to 27.0 percent, then to 24.6 percent, to 21.8 percent, and finally to 16.4 percent among the most affluent respondents. The hypothesis survives.[17]

Mean Comparisons

The same logic of comparison applies in another common hypothesis-testing situation, when the independent variable is nominal or ordinal and the dependent variable is interval. For a change of scene, let's consider a hypothesis using countries as the units of analysis. The dependent variable of interest is a measure of the political rights and freedoms enjoyed by countries' citizens. This index can range from 0 (citizens have very few rights and freedoms) to 12 (citizens have many rights and freedoms). As you might guess, countries vary a lot in this respect. Of the 177 countries being analyzed here, almost 25 percent scored 3 or lower on the 12-point scale, another 25 percent scored 10 or higher, with the remaining 50 percent spread fairly evenly across the middle ranges of the variable.[18] What causes such huge variation in political freedom across countries? We know that characteristics involving economic development, such as economic growth or education levels, create a social and political environment conducive to the expansion of freedoms. One such measure of development, per capita gross domestic product (per capita GDP), becomes the independent variable in this hypothesis: In a comparison of countries, those having lower per capita GDP will have fewer political rights and freedoms than will countries having higher per capita GDP.

Perform the mental exercise. Imagine 177 countries divided into four groups on the independent variable: countries with low GDP per capita, countries with medium-low GDP per capita, countries with medium-high GDP per capita, and countries having high GDP per capita. Now, for each group of countries, "calculate" a mean of the dependent variable—the mean score on the 12-point political freedoms index for low-GDP countries, for medium-low GDP countries, and so on. Your mental exercise, as before, sets up the comparison we are after. If the hypothesis is correct, will these means be different? If so, in what way?

Table 3-3 presents a mean comparison table that shows the results of an analysis using actual data. A **mean comparison table** is a table that shows the mean of a dependent variable for cases that have different values on an independent variable. Table 3-3 introduces a common mean-comparison format, the type of tabular display you are most likely to see in research articles and computer output.[19] In Table 3-3, the means of the dependent variable, along with the number of cases, appear in a single column, next to the values of the independent variable. By simply reading down the column from low to high values of the independent variable, we can see what happens to the average value of the dependent variable. Thus, the mean score on the 12-point political rights and freedoms index is 3.5 for the 45 low-GDP countries, 4.9 for the 44 medium-low countries, 7.0 for the 44 medium-high countries, and 9.3 for the 44 high-GDP countries. The overall mean for all 177 countries, an average score of 6.1, is displayed next to "Total" at the bottom of the table.

How would we interpret Table 3-3? Just as in interpreting a cross-tabulation, we would compare values of the dependent variable between values of the independent variable. If the hypothesis were incorrect, we should find that political rights and freedoms scores are about the same in all categories of the independent variable. But what do the data reveal as we move down the column of numbers, from low GDP to high GDP? Clearly, each time we compare countries with higher values of the independent variable to countries with lower values of the independent variable, we find the hypothesized difference: Higher-GDP countries have higher mean values on the dependent variable than do lower-GDP countries. So the hypothesis passes muster.

GRAPHING RELATIONSHIPS AND DESCRIBING PATTERNS

The above examples illustrate the mechanics of presenting and interpreting comparisons. And these comparisons have fit neatly into one of two classic patterns. The GDP–political rights and freedoms relationship is a **positive relationship**: An increase in the independent variable, per capita GDP, is associated with an increase in the dependent variable, political rights and

Table 3-3 Political Rights and Freedoms, by Country Per Capita GDP

Country per capita GDP	Mean score[a]
Low	3.5
	(45)
Medium-low	4.9
	(44)
Medium-high	7.0
	(44)
High	9.3
	(44)
Total	6.1
	(177)

Source: GDP per capita based on data from United Nations Development Programme, *Human Development Report 2003* (New York: Oxford University Press, 2003), www.undp.org. Political rights and freedoms score based on data from Freedom House, www.freedomhouse.org/. All variables were compiled by Pippa Norris, John F. Kennedy School of Government, Harvard University, and made available on her Web site, pippanorris.com.

[a] Score calculated by summing the Freedom House 7-point political rights index and the 7-point civil liberties index. Combined index was rescaled to range from 0 (fewest rights and freedoms) to 12 (most rights and freedoms).

freedoms. The income-smoking relationship is a **negative relationship**: An increase in the independent variable, income, is associated with a decrease in the dependent variable, the likelihood of smoking. Obviously, if the independent variable is a nominal measure, such as region or race or gender, questions about whether the relationship is positive or negative never come up. It would be nonsensical to talk about an "increase" in region, from Northeast to West, or to describe "decreasing" values of gender, from female to male.[20] If the independent variable is measured at the ordinal or interval level, however, we can ask whether the relationship between the independent and dependent variables is positive or negative.

In Chapter 2 we saw how bar charts can be a great help in describing variables. An adaptation of the bar chart—along with another form of visual display, the **line chart**—also can help to describe relationships between variables. Figure 3-1 displays a bar chart for the income–smoking cross-tabulation (Table 3-2). As simple as the cross-tabulation is, the bar chart is simpler still. The categories of the independent variable, income, appear along the horizontal axis. This placement of the independent variable is just as it was in the bar charts introduced in Chapter 2 that were based on the frequency distributions of single variables. In using this graphic form to depict relationships, however, the vertical axis does not represent the percentage of cases falling into each value of the independent variable. Rather, the vertical axis records the percentage of cases in each value of the independent variable that fall into *one* value of the dependent variable—the percentage who smoke. The height of each bar, therefore, tells us the percentage of subjects at each income level who are smokers. As you move along the horizontal axis from low to high income, notice that the heights of the bars decline

Figure 3-1 Smoking, by Income (bar chart)

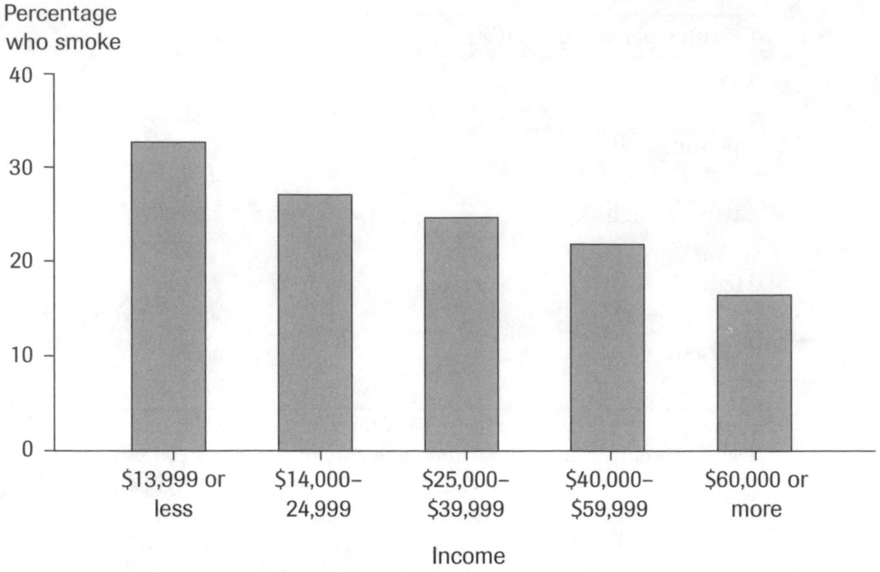

Percentage
who smoke

Source: 1996 American National Election Study.
Note: Question: "Are you a smoker?"

in a systematic pattern. Figure 3-1 shows the visual signature of a negative or inverse relationship. As the values of the independent variable increase, the value of the dependent variable, the percentage of smokers, decreases.

For a hypothesis that has an interval-level dependent variable, such as the per capita GDP–political rights and freedoms relationship (Table 3-3), a line chart is often used (Figure 3-2). The basic format is the same as a bar chart. The values of the independent variable, countries' GDP levels, are along the horizontal axis. The vertical axis records the mean of the dependent variable for each value of the independent variable. This graphic rendition of the mean comparison table reveals the visual signature of a positive or direct relationship. As per capita GDP increases, mean scores on the rights and freedoms index go up.

Both the income–smoking relationship and the GDP–political rights and freedoms relationship are linear. In a **linear relationship**, an increase in the independent variable is associated with a consistent increase or decrease in the dependent variable. Linear relationships can be positive or negative. In a negative linear relationship, any time you compare the dependent variable for subjects having different values on the independent variable, you find a negative relationship: The lower value of the independent variable is associated with a higher value of the dependent variable. In the income-smoking bar chart (Figure 3-1), for example, it doesn't matter which two income groups you compare—people making $13,999 or less with people making $25,000 to $39,999, or people in the $14,000–$24,999 category with people whose income is $60,000 or more, or any other comparison you care to make. Each comparison shows a negative relationship. People with lower incomes are more likely to smoke. The same idea applies to positive linear relationships, such as the GDP–political rights and freedoms example. Each time you compare values of the dependent variable for cases having different values on the independent variable, you find a positive relationship: The lower value of the

Figure 3-2 Political Rights and Freedoms, by Country Per Capita GDP (line graph)

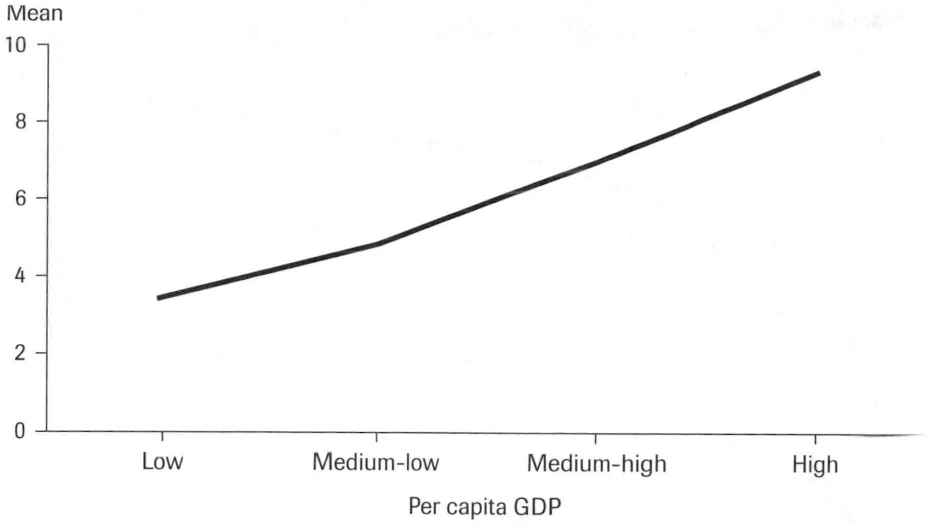

Source: GDP per capita based on data from United Nations Development Programme, *Human Development Report 2003* (New York: Oxford University Press, 2003), www.undp.org. Political rights and freedoms score are based on data from Freedom House, www.freedomhouse.org/. All variables were compiled by Pippa Norris, John F. Kennedy School of Government, Harvard University, and made available on her Web site, pippanorris.com.

independent variable is associated with a lower value of the dependent variable. A comparison between any two groups of countries in Figure 3-2 will show a positive pattern.

Many relationships between political variables do not fit a clear linear pattern. They are nonlinear or curvilinear. In a **curvilinear relationship**, the relationship between the independent variable and the dependent variable depends on which interval or range of the independent variable is being examined. The relationship may change direction, from positive to negative or from negative to positive. Or the relationship might remain positive (or negative) but change in strength or consistency, from strong to weak or from weak to strong. As one moves along the values of the independent variable, comparing values of the dependent variable, one might find a positive relationship; that is, an increase in the independent variable occasions an increase in the dependent variable. But as one continues to move along the independent variable, entering a different range of the independent variable, the relationship may turn negative; that is, as the independent variable continues to increase, the dependent variable decreases. Or a relationship that initially is positive or negative may flatten out, with an increase in the independent variable being associated with no change in the dependent variable. Many other possibilities exist. Let's consider some common patterns.

The first pattern is well within your everyday experience. What is the relationship between the number of hours you spend studying for an exam (independent variable) and the grade you receive (dependent variable)? Generally speaking, of course, the relationship is positive: The more you study, the better you do. But is the amount of improvement gained from, say, the first two hours spent studying the same as the amount of improvement gained from the fifth and sixth hours spent studying? Perhaps not. Figure 3-3 depicts a hypothetical relationship for this example. Notice that, for the lowest values of the independent variable, between 0 hours and 1 hour, the relationship between the independent and dependent variables is steeply

Figure 3-3 Relationship between Exam Score and Hours Spent Studying

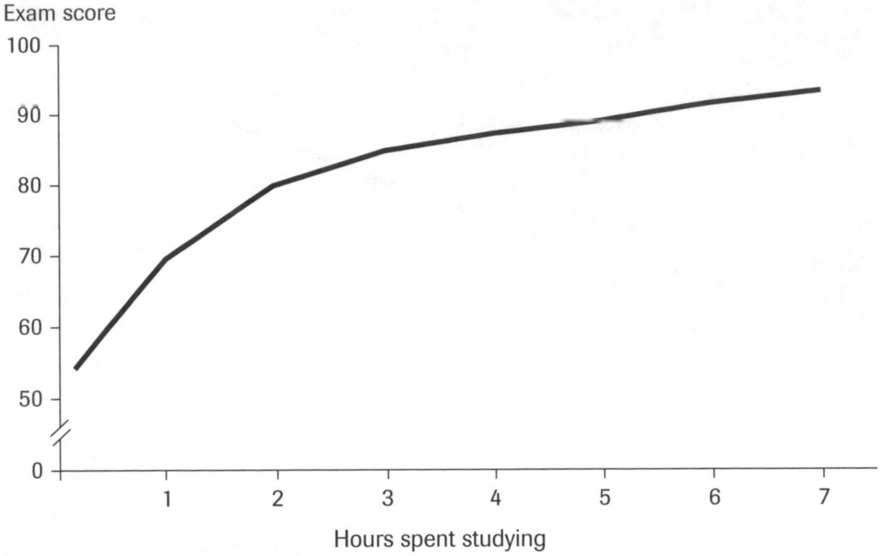

Note: Hypothetical data.

positive. That first hour translates into about a 20-point improvement on the test, from a less-than-lustrous 50 to about a 70. Adding a second hour returns dividends, too, but not as much as the first hour: about 10 additional points. As the hours wear on, the relationship, though still positive, becomes weaker and weaker. The curve begins to flatten out. The common expression, "diminishing rate of return," characterizes this sort of curvilinear relationship.

The nonlinear pattern shown in Figure 3-3 tells a plausible but hypothetical story. Figure 3-4 displays a real world relationship—the relationship between age (independent variable) and voter turnout (dependent variable). As age goes up, does turnout increase? Comparing average turnout among the youngest group (ages 18 through 30) with that of the next-oldest group (31 through 42), one sees the field mark of a positive relationship: turnout is higher among the older group. A comparison between the 31 through 42 age group and the next-oldest group (43 through 52) again reveals a positive relationship. However, as we move into the older age ranges, the positive pattern breaks down and is replaced by a flat, no-change relationship. In its rough outline, then, Figure 3-4 provides an empirical example of the hypothetical pattern shown in Figure 3-3: a strong relationship in the lower range of the independent variable, and then a weakening of the relationship as we move into the higher range.

Another pattern, one often encountered in social research, is the V-shaped or U-shaped relationship. For lower values of the independent variable, the relationship is negative. For higher values, the relationship is positive. The inverted V- or U-shape is based on the same idea: positive in the lower range, negative in the higher range. Consider Figure 3-5, which displays a bar chart of the relationship between the 7-point party identification scale and reported turnout. Strong Democrats turn out at a high rate. But, as you can see, as Democratic identification weakens, reported turnout declines, hitting a low point among pure Independents. As the scale shades through Independent-leaning Republican and weak Republican, the relationship turns positive, finally culminating among respondents who occupy the second tip of the V-shape, strong Republicans.[21]

Figure 3-4   Relationship between Turnout and Age

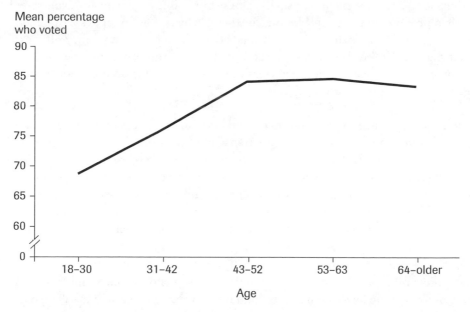

Mean percentage
who voted

Source: 2004 American National Election Study.
   Note: Based on the following percentages (and number of cases): Age 18–30, 68.49 (219); 31–42, 75.59 (213); 43–52, 83.02 (212); 53–63, 83.41 (217); 64–older, 82.44 (205). Reported turnout for the entire sample of 1,066 respondents was 78.52.

Figure 3-5   Relationship between Turnout and Partisanship

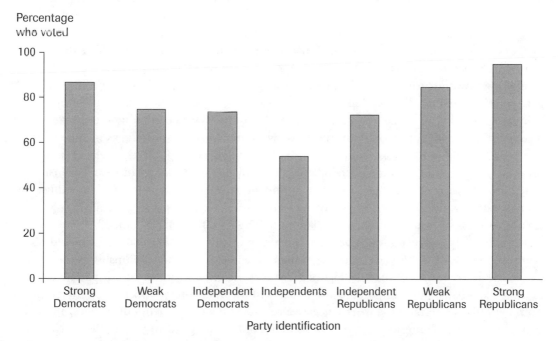

Percentage
who voted

Source: 2004 American National Election Study.
   Note: Based on the following percentages (and number of cases): Strong Democrat, 86.59 (179); Weak Democrat, 74.67 (150); Independent Democrat, 73.60 (178); Independent, 54.00 (100); Independent Republican, 72.36 (123); Weak Republican, 84.62 (143); Strong Republican, 94.92 (177). Reported turnout for the entire sample of 1,050 respondents was 79.05.

Curvilinear relationships are fairly common in political research. And they are always interesting. Sometimes we encounter such relationships while organizing or exploring data, and we then propose and test explanations for them. Why is it, we might ask, that the age–turnout relationship is shaped the way it is? Sometimes, however, we may offer explanations that suggest nonlinear relationships. Let's explore this route, beginning with this fact: Some people belong to labor unions, whereas others do not. What accounts for this fact? Consider a plausible explanation:

> Union organizers need a stable group of workers who can be easily contacted and who can afford to join. Many low-wage jobs have transient or part-time workers whose employment status changes with economic conditions or seasonal cycles. In addition to being difficult for organizers to contact, these low-income workers may consider union dues an unaffordable expense. High-wage workers present union organizers with a different set of challenges. Unions seek to ensure job security and protect wages by bargaining collectively on behalf of all workers. Unions discourage or prevent individual employees from bargaining on their own. For this reason, high-wage workers will have little incentive to join. These employees tend to have skills that they can shop around to the highest bidder. Therefore, union membership will be a middle-class phenomenon. Membership will be uncommon among low-income people and high-income people, and it will be highest in the middle-income ranges. In a comparison of individuals, middle-income earners will be more likely to belong to unions than will low-income or high-income earners.[22]

This explanation describes a relationship between an independent variable, income, and union membership. But it does not describe a linear relationship. Suppose we have a measure that gauges income by ten ordinal categories, from $5,999 or less through $90,000 or more. Imagine that the median value of the income variable—the value that splits the sample into roughly equal halves—occurs at about $35,000. Suppose further that we can calculate, for each of the ten income groups, the percentage belonging to a union. Now visualize the inverted-V pattern suggested by the explanation. As we move along the income variable, comparing lower incomes to higher incomes, the relationship should follow a positive pattern, with membership increasing as income goes up. However, the percentage of union members should peak in the middle-income range, around $35,000. Moving into higher incomes, the relationship should turn negative, with membership dropping as income rises.

Table 3-4 shows the mean comparison that illuminates the hypothesized relationship between income and union membership. Figure 3-6 displays a line chart of the relationship. In the lowest income group, the percentage of individuals belonging to a union is a miniscule 1.9 percent. The percentage rises slightly but remains anemic (5–6 percent) until we encounter the low-middle group at $20,000–$24,999. As we move higher, membership increases sharply, nearly doubling between the group of earners at $25,000–$34,999 and the group at $35,000–$39,999. The percentages then flatten out, and the relationship turns negative in the upper ranges of income. Figure 3-6 shows the graphic profile of this pattern: a modestly positive relationship in the lower values of income, a steeply positive relationship from low-middle to middle, a weakly negative relationship across the middle ranges, and a strongly negative relationship in the upper reaches of the independent variable. Clearly the proposed explanation has merit.

Table 3-4   Relationship between Union Membership and Income (mean comparison)

Respondent income	Mean percent union members
$5,999 or less	1.9
	(209)
$6,000–$14,999	5.7
	(279)
$15,000–$19,999	5.5
	(183)
$20,000–$24,999	7.2
	(195)
$25,000–$34,999	11.8
	(323)
$35,000–$39,999	21.5
	(135)
$40,000–$49,999	21.5
	(233)
$50,000–$59,999	20.1
	(154)
$60,000–$89,999	17.8
	(214)
$90,000 or more	6.3
	(158)
Total	11.5
	(2,083)

Source: James A. Davis, Tom W. Smith, and Peter V. Marsden, General Social Surveys, 1972–2006 [Cumulative File] [Computer file]. ICPSR04697-v2. Chicago, Ill.: National Opinion Research Center [producer], 2007; Storrs, Conn.: Roper Center for Public Opinion Research, University of Connecticut/Ann Arbor, Mich.: Inter-university Consortium for Political and Social Research [distributors], 2007-09-10. Income based on GSS variable RINCOME06 (Respondent income), collapsed into 10 categories. Union membership based on GSS variable UNION (code 1, R belongs).

Note: Displayed data are from 2006.

As this example suggests, you must be alert to the possibility that a political relationship of interest may not follow a classic positive or negative pattern. By carefully examining cross-tabulations and mean comparisons, you can get a clearer idea of what is going on in the relationship. Graphic depictions, such as bar charts and line charts, offer clarity and simplicity. It is recommended that you always sketch an appropriate graphic to help you interpret a relationship. Graphic forms of data presentation will become especially useful when you begin to analyze more complicated relationships —when you begin to interpret controlled comparisons. We turn in the next chapter to a discussion of control.

Figure 3-6 Relationship between Union Membership and Income (line chart)

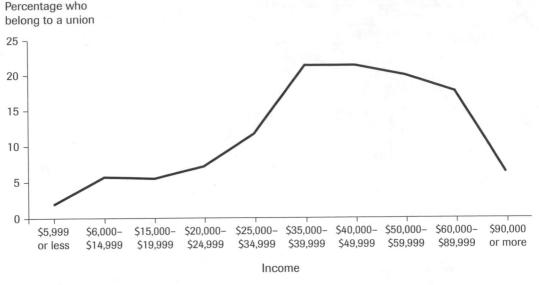

Percentage who
belong to a union

Source: Table 3-4.
Note: Displayed data are from 2006.

SUMMARY

In this chapter we introduced the essential ideas behind proposing explanations, stating hypotheses, and making comparisons. We have seen that political research often begins by observing an interesting variable and then developing a causal explanation for it. An explanation describes how a causal variable, the independent variable, is linked to an effect variable, the dependent variable. Rich explanations often suggest intervening variables, variables that tell us how an independent variable is causally connected to a dependent variable. Remember that an explanation must be plausible. An explanation must be testable. It must suggest a hypothesis, a testable statement about the empirical relationship between an independent variable and a dependent variable.

Cross-tabulation analysis is the most common method for testing a hypothesis about the relationship between nominal or ordinal variables. A cross-tabulation presents side-by-side frequency distributions of the dependent variable for subjects that have different values on the independent variable. To test a hypothesis about the relationship between a nominal or an ordinal independent variable and an interval-level dependent variable, mean comparison analysis is used. In mean comparison analysis, we compare the mean values of the dependent variable for subjects that have different values of the independent variable.

The relationship between two variables may have a linear pattern. In a linear relationship, an increase in the independent variable is associated with a consistent increase or decrease in the dependent variable. A linear relationship can be positive or negative. Some relationships have nonlinear or curvilinear patterns. In a curvilinear relationship, the relationship between the independent variable and the dependent variable is not the same for all values of the independent variable. The relationship may differ in direction or it may differ in strength, depending on which interval or range of the independent variable is being examined. You might discover nonlinear patterns while exploring data and then develop explanations for them. Sometimes you might propose explanations that imply nonlinear relationships.

KEY TERMS

cross-tabulation (p. 55)
curvilinear relationship (p. 61)
dependent variable (p. 46)
hypothesis (p. 45)
independent variable (p. 46)
intervening variable (p. 53)

line chart (p. 59)
linear relationship (p. 60)
mean comparison table (p. 58)
negative relationship (p. 59)
positive relationship (p. 58)

EXERCISES

1. The "early call"—declaring a winner before all the polls have closed—is a controversial staple of the media's election night coverage. Convinced that "projections of the winner in key states may depress voter turnout on the West Coast if it appears that the election is or will be decided before polls close in the West," many reformers advocate measures that would require all polls to close at the same time.[23] These proposed reforms are based on the idea that the media's early declarations depress turnout in areas where the polls are still open.

   A. Think about the relationship between an independent variable, whether or not people have knowledge of an election's predicted outcome (they either "know" or they "don't know" the predicted outcome) and a dependent variable (they either "voted" or they "did not vote"). The reformers' idea links one value of the independent variable, "know the predicted outcome," with one value of the dependent variable, "did not vote." For the reformers' idea to qualify as an acceptable explanation, what else must it describe?

   B. Suppose you believe that knowledge of an election's predicted outcome is causally linked to turnout. Why might differences in knowledge of the outcome cause differences in turnout? Write a paragraph describing the causal linkages between these two variables. Be sure to describe the tendency of the relationship.

   C. Using proper form, state a testable hypothesis for the relationship between the independent variable and the dependent variable.

2. Third-party candidates are something of a puzzle for students of electoral behavior. In presidential elections, most voters cast their ballots for one of the major-party candidates, but many voters support candidates of minor parties, such as the Reform Party (which grew out of Ross Perot's candidacy in 1992), the Green Party (which nominated Ralph Nader in 2000), the Constitution Party, the Libertarian Party, or the Natural Law Party, to name a few.[24] What causes some people to vote for a major-party candidate and some voters to vote for a minor-party candidate? Voters can be measured by one of two values on this dependent variable: major-party voter and minor-party voter.

   A. Think up a plausible independent variable that may explain differences between voters on the dependent variable. Write a paragraph describing an explanation for why some voters support the major parties' candidates and some support the minor parties' candidates. Make sure you connect the causal variable to the dependent variable and be sure to describe the tendency of the relationship.

   B. Using proper form, state a testable hypothesis for the relationship between the independent variable and the dependent variable.

3. Robert Michels's famous "iron law of oligarchy" is familiar to students of large-scale organizations. According to Michels, every large voluntary organization—parties, labor unions, environmental groups, business associations, and so on—eventually develops internal decision-making procedures that severely limit input from ordinary members. Important decisions are made by a small band of organizational oligarchs who tenaciously hold on to their power within the group. Michels claimed that this inexorable organizational law held true for all groups: "It is organization which gives birth to the dominion of the elected over the electors, of the mandataries over the mandators, of the delegates over the delegators. Who says organization, says oligarchy."[25]

Assume that you are skeptical of the iron law of oligarchy. You think that the tendency toward internal oligarchy is shaped by this independent variable: the degree to which an organization is outwardly committed to democracy in society at large. Your hypothesis: In a comparison of organizations, those that are committed to spreading democracy in society are less likely to be oligarchies than are those that are not committed to spreading democracy in society. To test this hypothesis, you gather information on 100 large organizations. You classify 50 as not being committed to spreading democracy in society and the other 50 as being committed to spreading democracy in society. Each organization is further classified on the dependent variable: oligarchy/not an oligarchy.

A. Using the 100 organizations as raw data, fabricate a cross-tabulation analysis that would support your hypothesis. Remember to put the independent variable on the columns and the dependent variable on the rows. Make sure to percentage the independent variable. Write a brief interpretation of your "findings." That is, explain why the fabricated cross-tabulation supports your hypothesis.

B. Fabricate a cross-tabulation analysis that would support Michels's iron law of oligarchy. Again, make sure to follow the rules of cross-tabulation construction. Write a brief interpretation of these fabricated results. That is, explain why the cross-tabulation supports the iron law of oligarchy.

4. Four statements appear below. For each one: (i) Identify and describe at least one reason why it is a poor hypothesis. (ii) Rewrite the statement in proper hypothesis-writing form. (You may have to embellish a bit, using your own intuition.)

Example. Income and partisanship are related.

   (i) This statement does not identify the units of analysis. Also, it does not state the tendency of the relationship because it does not say how income and partisanship are related.

   (ii) In a comparison of individuals, people with lower incomes will be more likely to be Democrats than will people with higher incomes.

A. In a comparison of individuals, some people will be more likely to have served in the military than will other people.

B. Decentralized workplaces have highly satisfied workers.

C. Education and smoking are related.

D. Some people support increased funding for space exploration.

5. Robert Putnam's research, which was discussed briefly in this chapter, has stimulated renewed interest in the role played by voluntary associations in American democracy. Putnam's work seems to suggest that, when people get involved in groups and help make collective decisions for the group, they develop participatory skills. These participatory skills, in turn, cause people to participate more in politics—voting at higher rates than people who are not involved in any groups.

A. This explanation says that the causal relationship between the independent variable, group membership, and the dependent variable, turnout, is mediated by an intervening variable. What is the intervening variable?

B. Based on the explanation, write a hypothesis in which the intervening variable is the dependent variable.

C. Based on the explanation, write a hypothesis in which the intervening variable is the independent variable.

6. Who strongly believes that spanking is an appropriate practice for disciplining children? This has become a controversial question in recent years, as some advocacy organizations have pressed state legislatures to enact tougher child abuse laws. Perhaps such laws would not be welcomed by all groups in society. Consider two hypotheses.

   Generational hypothesis: In a comparison of individuals, older people are more likely to strongly approve of spanking than are younger people.

   Income hypothesis: In a comparison of individuals, lower-income people are more likely to strongly approve of spanking than are higher-income people.

   The raw frequencies for strongly approve/not strongly approve for three age cohorts are as follows: born in 1949 or earlier, 163 strongly approve/431 not strongly approve; born 1950–1965, 165 strongly approve/433 not strongly approve; born 1966 or later, 203 strongly approve/571 not strongly approve. The raw frequencies for strongly approve/not strongly approve for three income groups are as follows: low income, 188 strongly approve/409 not strongly approve; middle income, 177 strongly approve/502 not strongly approve; high income, 89 strongly approve/341 not strongly approve.[26]

   A. Construct two cross-tabulations from this information. One cross-tabulation will test the generational hypothesis. The second cross-tabulation will test the income hypothesis. For each cross-tabulation, write a paragraph interpreting the results. What do you think? Does the generational cross-tabulation support the generational hypothesis? Does the income cross tabulation support the income hypothesis? Be sure to explain your reasoning.

   B. Sketch two bar charts: one for the generation–spanking relationship and one for the income–spanking relationship. For each bar chart, remember to put the values of the independent variable on the horizontal axis. Put the percentage strongly approving of spanking on the vertical axis. To make the bar charts more readable, make 0 percent the lowest value on the vertical axis and make 40 percent the highest value on the vertical axis.

   C. How would you describe the relationship in the income–spanking bar chart? Does it more closely approximate a negative relationship or a positive relationship? Explain your reasoning.

7. It could be that people who frequently read a newspaper know more about public policy than people who read a newspaper less frequently. A plausible hypothesis: In a comparison of individuals, people who read a newspaper less often will have a lower level of knowledge about public policy than will people who read a newspaper more often. Suppose you can measure the frequency of newspaper reading, from less frequently to more frequently, with the following values: never, less than once a week, once a week, a few times a week, and every day. Suppose that policy knowledge is an interval-level scale that begins at 0 for people with the lowest knowledge and takes on positive values for people with higher knowledge.

   A. Does the hypothesis suggest a positive relationship between the independent and dependent variables? Or does the hypothesis suggest a negative relationship between the variables? Explain your reasoning.

   B. The mean public policy knowledge scores for each value of the independent variable are as follows: For people who never read the newspaper, 3.6; less than once a week, 3.8; once a week, 4.0; a few times a week, 4.3; every day, 4.8. The overall mean for the total sample is 4.3. Construct a mean comparison table from these data.[27] (The mean comparison table will display the mean values of the dependent variable for each value of the independent variable. It will also show a mean for "Total." However, because this exercise does not provide the number of cases, your mean comparison table will not contain information about the number of cases.)

C.   Use the information in your mean comparison table to sketch a line chart of the relationship. Remember to put the values of the independent variable on the horizontal axis. Record the mean of the dependent variable on the vertical axis. To make the line chart more readable, make 3.0 the lowest value on the vertical axis, and make 5.0 the highest value on the vertical axis.

D.   Interpret the results. Are the data consistent or inconsistent with the hypothesis? Explain.

8.   Two scholars are hypothesizing about the relationship between educational attainment (independent variable) and economic liberalism (dependent variable). Economic liberalism is defined as the extent to which a person thinks that government spending and services should be expanded and business regulation increased. Economic liberals favor an expansion of government spending and business regulation, whereas economic conservatives oppose these policies.

Scholar 1: "People's beliefs about government spending and regulation depend on how much economic security they have. Less-educated individuals are more vulnerable to economic downturns and thus will favor a larger role for government. As education goes up, economic security goes up—and support for economic liberalism goes down. Therefore, the relationship is linear and negative: In a comparison of individuals, those with lower educational attainment will be more economically liberal than those with higher levels of educational attainment."

Scholar 2: "Yes, generally speaking, you are correct: as education increases, economic liberalism declines. But let's not ignore the role of the liberal intelligentsia in political life. People at the highest level of educational attainment have been the guiding lights of economically liberal policies for many years. I think that people with the highest levels of education will be just as supportive of economic liberalism as people at the lowest educational level. Therefore, I hypothesize a V-shaped pattern."

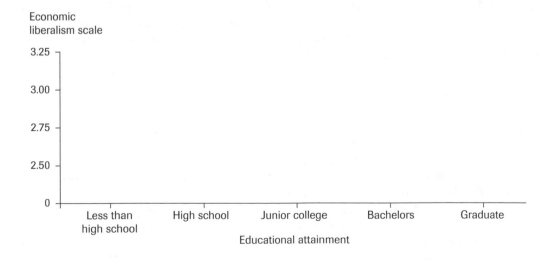

Draw and label three sets of axes like the one at the bottom of page 70. The horizontal axis records five levels of educational attainment, from less than high school to graduate degree. The vertical axis records mean values on an economic liberalism scale. Scores toward the bottom of the scale (around 2.5) denote less support for economic liberalism, whereas scores at the top of the scale (around 3.25) denote more support for economic liberalism.

A. In the first set of axes you constructed, draw a line that would fit Scholar 1's hypothesis about the relationship between educational attainment and economic liberalism. (Do not worry too much about the exact economic liberalism scores. Just draw a line that would be consistent with Scholar 1's hypothesis.)

B. In the second set of axes you constructed, draw a line that would fit Scholar 2's idea that the relationship fits a V-shaped pattern.

C. In the third set of axes you constructed, draw a line chart based on the following real-world information.[28] Mean economic liberalism scores for each value of educational attainment: less than high school, 3.0; high school, 2.9; junior college, 2.7; bachelors, 2.6; graduate, 3.0.

D. Which scholar, Scholar 1 or Scholar 2, is more correct? Explain how you know.

9. You are probably familiar with (and may have used) back belts, which are widely used by workers to protect their lower backs from injuries caused by lifting. A study was conducted to determine the usefulness of this protective gear. Here is a partial description of the study, published in the *Journal of the American Medical Association* and reported by the Associated Press (December 5, 2000):

> New research suggests that back belts, which are widely used in industry to prevent lifting injuries, do not work. The findings by the National Institute for Occupational Safety and Health stem from a study of 160 Wal-Mart stores in 30 states. Researchers [based their findings on] workers' compensation data from 1996 to 1998.

Although you do not know the study's particulars, think about how you would go about investigating the effect of back belt usage on back injuries. Assume that you have data on each of the 160 retail stores in your study. For each store, you know whether back belt usage was low, moderate, or high. You classify 50 stores as having low belt usage by employees, 50 stores as having moderate usage, and 60 stores as having high usage. You also know the number of back-injury workers' compensation claims from each store. This information permits you to calculate the mean number of claims for low-usage, moderate-usage, and high-usage stores.

A. The following hypothesis suggests that back belt usage helps prevent injury: In a comparison of stores, stores with low back belt usage by employees will have more worker injuries than will stores with high back belt usage. What is the independent variable? What is the dependent variable? Does this hypothesis suggest a positive or negative relationship between the independent and dependent variables? Explain.

B. Fabricate a mean comparison table showing a linear pattern that is consistent with the hypothesis. Sketch a line chart from the data you have fabricated. (Because you do not have sufficient information to fabricate a plausible mean for all the cases, you do not need to include a "Total" row in your mean comparison table.)

C. Use your imagination. Suppose the data showed little difference in the worker injury claims for low-usage and moderate-usage stores, but a large effect in the hypothesized direction for high-usage stores. What would this relationship look like? Sketch a line chart for this relationship.

Research Design and the Logic of Control

4

LEARNING OBJECTIVES

In this chapter you will learn:
- The importance of rival explanations in political research
- How experimental studies rule out rival causes
- How observational studies control for rival causes
- Three possible scenarios for the relationship between an independent variable and a dependent variable, controlling for a rival cause

In his book *Data Analysis for Politics and Policy*, Edward R. Tufte recounts the story of a famous surgeon, a pioneer in the technique of vascular reconstruction, who delivered a lecture to medical school students on the large number of patients saved by the surgical procedure:

> At the end of the lecture, a young student at the back of the room timidly asked, "Do you have any controls?" Well, the great surgeon drew himself up to his full height, hit the desk, and said, "Do you mean did I not operate on half of the patients?" The hall grew very quiet then. The voice at the back of the room very hesitantly replied, "Yes, that's what I had in mind." Then the visitor's fist really came down as he thundered, "Of course not. That would have doomed half of them to their death." God, it was quiet then, and one could scarcely hear the small voice ask, "Which half?"[1]

Humiliating as it may have been, the young questioner's skepticism was precisely on point. If the famous surgeon wished to demonstrate the effectiveness of his surgical technique, then he would have to make a comparison. He would have to compare values of the dependent variable, survival rates, for patients who had different values on the independent variable—patients who had received the surgery and patients who had not received the surgery. As it was, the surgeon was reporting values of a dependent variable from observations that he made on a test group. A **test group** is composed of subjects who receive a treatment that the researcher believes is causally linked to the dependent variable. In this case, the test group was composed of patients who had received the surgical procedure. But the surgeon had no basis for comparison. He had no control group. A **control group** is composed of subjects who do not receive the treatment that the researcher believes is causally linked to the dependent variable. In this case, the control group would be composed of patients who had not received the surgical procedure. Without data from a control group, there is simply no way he could have known whether the independent variable had any effect on the dependent variable, survival rates.

To get an accurate assessment of the effectiveness of the surgery on survival rates, the medical researcher would have to make sure that the patients in both the test group and the control group were identical in every other way that could affect the dependent variable. Thus, for example, patients in both groups would need to be identical or very similar in age, medical history, current prognosis, and so on. Why put such stringent conditions on testing the effect of the surgery? Because each of these factors represents a **rival explanation** for surviving or dying, an alternative cause for different values of the dependent variable.

Think about this last point for a moment. Suppose that patients in the test group, those receiving the surgery, were younger and in better health than patients in the control group, those not receiving the surgery. In comparing the surgical test cases with the nonsurgical control cases, we would also be comparing younger and healthier patients (the test group) with older and less healthy patients (the control group). It could be these differences, not the surgery, that explain the test group's better rates of survival. Only by making these other factors equal between the test group and the control group—by making sure that the surgical procedure is the *only* difference between the groups—can we measure the effect of the surgery on survival.

This anecdote is a cautionary tale for political researchers because we face the same problem of controlling for rival explanations. The political world is a complicated place. For every explanation we describe, there is a competing explanation that accounts for the same dependent variable. In Chapter 3 we discussed an explanation proposing that partisanship is causally linked to opinions about gun control. Let's say that we test the hypothesis that Democrats will be more likely than Republicans to favor gun restrictions. And suppose we find that, sure enough, a much larger percentage of Democrats than Republicans support gun control. Imagine that, after reporting these results to a large group of students, one student asks: "Do you have any controls? Democrats are more likely to be women and Republicans are more likely to be men. What is more, women are stronger gun control advocates than men. So a Democrat-Republican comparison is at least partly a female-male comparison. You might be making a big mistake, confusing the effect of party with the effect of gender. Did you control for gender?" This imaginary questioner has proposed a plausible rival explanation, an alternative cause for the same dependent variable, gun-control opinions. For every explanation we propose, and for every hypothesis we test, there are alternative causes, rival explanations, for the same phenomena. These rival explanations undermine our ability to evaluate the effect of the independent variable on the dependent variable. Is the independent variable causing the dependent variable, as our explanation suggests, or is some other variable at work, distorting our results and leading us to erroneous conclusions?

Our ability to rule out worrisome alternative explanations depends on the power of our **research design,** an overall set of procedures for evaluating the effect of an independent variable on a dependent variable. The **experimental study** ensures that the test group and the control group are the same in every way, except one—the independent variable. Because the independent variable is the only way the groups are not alike, any differences on the dependent variable can be attributed to the independent variable. The **observational study** allows the researcher to make controlled comparisons—that is, to observe the effect of the independent variable on the dependent variable while holding constant other plausible causes of the dependent variable. Experimental studies are stronger because they control for the possible effects of all rival explanations, even rivals the investigator has not thought of or cares about. In observational studies, the investigator can make sure to control for known or suspected rivals, but there may be unknown factors that are affecting the dependent variable and contaminating the results.

Although observational studies are the most common approaches in political science, experiments have a long tradition in political research and have been enjoying a renaissance.[2] In the first part of this chapter, we describe the experimental protocol and show how its signature procedure, random assignment, vanquishes rival explanations. By understanding experiments, you can better appreciate the challenges faced in the observational settings where most political research takes place. In the last section of this chapter, we turn to a detailed consideration of the logic of controlled comparison, the main methodology of observational research.

EXPERIMENTAL STUDIES

In all experiments, the investigator manipulates the test group and control group in such a way that, in the beginning, both groups are virtually identical. In many (though not all) experiments, the investigator then measures the dependent variable for both groups. This is called the premeasurement phase. The two groups then receive different values of the independent variable. This is the treatment or intervention phase, in which the test group typically gets some treatment while the control group does not. In the postmeasurement phase, the dependent variable is measured again for both groups. Since, by design, the independent variable is the only way the groups differ, any observed differences in the dependent variable can be attributed directly to the independent variable and cannot be attributed to any other cause.

The preceding paragraph probably conjured the sterile image of a serious-looking technician in a white lab coat assiduously recording numbers on a clipboard. In practice, "experiments are nothing if not amazingly diverse."[3] In a **laboratory experiment,** the control group and the test group are studied in an environment wholly created by the investigator. For example, participants might be asked to leave their homes or workplaces and travel to the research venue. And, although they most likely do not know the exact purpose of the experiment, participants doubtless are aware they are being studied. In a **field experiment,** the control and test groups are studied in their normal surroundings, living their lives as they naturally do, probably unaware that an experiment is taking place. These two approaches may sound different, but they both bear the defining feature of all experimental research: individuals are randomly assigned to the control group or test group. **Random assignment** occurs when every prospective participant—every individual that the investigator wants to study—has an equal chance of ending up in the control group or the test group. Thus, if there are two groups, a control and a test, then each individual has a probability equal to .5 of being selected for the control and a probability equal to .5 of being selected for the test. Let's take a closer look at random assignment. Then we will describe examples of a laboratory experiment and a field experiment.

Random Assignment

To appreciate the elegant power of random assignment, consider how you might address an interesting research question. Assume you want to investigate the role of mass media in setting the political agenda and affecting the issues people care the most about. Suppose the media are covering a political scandal and giving less play to the state of the economy. You reason that this unbalanced emphasis will affect citizens' political priorities, causing people to view the scandal as more important than the economy. Your hypothesis: In a comparison of individuals, those who are exposed to issues covered by the media will be more likely to perceive those issues as important than will those who are not exposed to issues covered by the media.

The comparison seems straightforward. Ask a large number of people how closely they follow media. Based on this information, divide them into two groups: those who closely follow the media (the test group) and those who do not pay close attention to media coverage (the control group). Present a list of political issues to both groups and ask the participants to rank the issues in descending order of importance. After comparing the two lists suppose you discover that, just as your hypothesis suggests, the test group's priorities closely mirror the media's priorities, whereas the control group's list does not.

Here is the pivotal question: Can you conclude that the media's agenda *caused* the difference in priorities between the test and control groups? Think about this question. To impute causation, we must be able to say that the amount of attention paid to media is the only difference between the two groups. We must be able to say that the individuals in the test group (the attention-payers) are identical to the individuals in the control group (the nonattention-payers) in every other way that could affect the sorts of issues that people find important. We must be able to say that if we were to increase the media exposure of the nonattention-payers, then their issue priorities also would change, looking just like the attention-payers' list.[4]

But clearly we cannot say these things. Perhaps attention-payers are better educated than nonattention-payers. Education could affect which issues people find important, quite apart from how much media they consume. The same could be said for a large number of other plausible differences between the test and control groups, such as differences in age, gender, or income. In all these ways—and any others we have not thought of—attention-payers and nonattention-payers may already have been different *before* we compared them. Random assignment takes these prior differences out of play:

> The intellectual beauty of randomized design is this: We do not have to know the names of any of the prior variables that need control. Randomly constituted classes will show only random differences . . . for *any* prior variable whatsoever.[5]

When the investigator cannot control the group assignment process, the participants are liable to bring a lot of rival-explanation baggage to the research setting. Random assignment, by contrast, is the great equalizer of groups and the great neutralizer of rivals: "By randomly assigning subjects to treatments, the experimenter, in one elegant stroke, can be confident that any observed differences [on the dependent variable] must be due to differences in the treatments themselves."[6]

A Laboratory Experiment

Now consider Shanto Iyengar and Donald Kinder's classic series of laboratory experiments aimed at finding out whether agenda setting by the media causes individuals to alter their issue priorities.[7] Iyengar and Kinder used random assignment, the cornerstone of experimental research, the advantages of which we have just discussed. They also premeasured the dependent variable by giving each group a list of eight national problems—defense, inflation, illegal drugs, the environment, and so on—and asking subjects to rank the problems from most important to least important. Why premeasure the dependent variable? Shouldn't the random assignment of subjects be sufficient to eliminate alternative causes? In fact, premeasurement is especially important in laboratory experiments. Suppose that, despite random assignment and other efforts to eliminate rival causes, the control group's value on the dependent variable changes between the premeasurement and the postmeasurement. This could happen for a variety of reasons that are difficult to control—for example, the natural human response to the knowledge of being studied. Since these potentially confounding conditions can be plausibly attributed to the test group as well, the researcher can get a good idea of how much the test group changed, over and above any changes observed in the control group.[8]

During the treatment phase of the experiment, the control group and the test group gathered to watch the news each night for six days. The control group was shown a regular newscast. For the test group, the researchers used state-of-the-art video equipment and editing techniques to insert stories dealing with a specific national problem. At the end of the six days, subjects were again asked to rank the eight issues in terms of importance. The results were stunning. Compared with the control subjects, whose rankings did not change appreciably from beginning to end, individuals who had been shown the doctored newscasts dramatically altered their lists of priorities. For example, in one instance where the investigators had interspliced stories about defense preparedness, test subjects promoted defense from sixth to second place, whereas control subjects reported no change in their assessments of the importance of this national problem.[9]

Iyengar and Kinder's findings are convincing. They followed strict procedures that did not permit rival explanations to account for their results. Neutralizing rival explanations, however, required a lot of creative control. They created and controlled everything in the research environment: the premeasurement instrument, the setting in which subjects watched media newscasts, the type of stories and their placement in the newscasts, and the postmeasurement instrument. Because the laboratory experiment imposes such a large degree of creative control, the results obtained from them have a high degree of internal validity. **Internal validity** means that, within the conditions artificially created by the researcher, the effect of the independent variable on the dependent variable is isolated from other plausible explanations. One would have to say that, within the conditions created by Iyengar and Kinder, the effect of media exposure on the perceived importance of media stories was isolated and precisely measured.

However, whether or not laboratory experiments can be externally valid is more problematic. **External validity** means that the results of a study can be generalized—that is, its findings can be applied to situations in the nonartificial, natural world. When people watch the news each evening, do they travel to a university campus, check in with a professor or graduate student, and then watch television with a group of strangers? To be sure, Iyengar and Kinder went to considerable lengths to recreate natural circumstances—putting subjects at ease, providing a comfortable living-room setting, encouraging casual conversation, and so on. Even so, some inferential distance exists between the experimental finding that "viewers exposed to news devoted to a particular problem become more convinced of its importance" to the real-world conclusion that "[n]etwork news programs seem to possess a powerful capacity to shape the public's agenda."[10]

A Field Experiment

Field experiments, which are conducted in the real world, can sometimes overcome the limitations of the laboratory environment. By way of introducing field experiments, consider a phenomenon of keen interest to political scientists and professional politicians alike: voter mobilization. Political scientists view robust turnouts as a sign of electoral health. And, of course, grassroots organizers and partisan elites are always interested in discovering new ways to woo potential supporters to the polls. You have probably been on the receiving end (and perhaps even the purveyor) of some of these get-out-the-vote (GOTV) methods—door-to-door canvassing, telephone calls, and direct-mail appeals. Do such mobilization efforts work? Does being contacted and urged to vote *cause* people to vote? A working hypothesis: In a comparison of individuals, those who a receive GOTV contact will be more likely to vote than will those who do not receive a GOTV contact.

Again, the test would appear to hinge on a simple comparison: Divide people into two groups on the basis of the independent variable—those who received a contact (the test group) and those who didn't (the control group)—and compare turnout rates. By this point you know that such a comparison cannot tell us whether GOTV contact causes people to vote. Can we say that the contacted and the uncontacted are identical in every other way that could affect turnout? In fact, party operatives and organized canvassers work from targeted lists of partisan supporters and regular voters, individuals who are likely to vote whether they are contacted or not. Thus, in a perverse turn of causal events, the independent variable (the likelihood of contact) is in part caused by the dependent variable (the likelihood of voting).[11]

Alan S. Gerber and Donald P. Green have pioneered the use of field experiments to determine whether nonpartisan GOTV contact causes turnout and whether some methods have bigger causal effects than others.[12] In a study conducted during the 1998 general election, the investigators first obtained a list of about 29,000 registered voters in one metropolitan area. These individuals were then randomly assigned to receive either personal canvassing, direct mail, or telephone calls. A fourth group, the controls, did not receive any GOTV appeals from the researchers. Again, appreciate the importance of random assignment. Each individual had an equal chance of ending up in the canvassing group, the direct-mail group, the phone-call group, or the control group. Thus the four groups were randomized clones, essentially identical on every known (and unknown) characteristic that might affect the dependent variable, turnout.

In addition to the different methods of mobilization, Gerber and Green used three different realistic appeals during the experiment's treatment phase. One evoked citizen responsibility ("We want to encourage everyone to do their civic duty . . ."), another the potential closeness of the election ("Each year some election is decided by a handful of votes . . ."), and another neighborhood solidarity ("Politicians sometimes ignore a neighborhood's problems if the people in that neighborhood don't vote . . ."). These different appeals were delivered face to face, by mail, by phone, or not at all. Gerber and Green were thus able to gauge the effects of three mobilization methods and three mobilizing messages—all received by identical treatment and control groups, and all delivered in a way that appeared perfectly natural to the people being studied.

Following the election, the investigators obtained postmeasurements by consulting voting records to see who had voted and who had stayed home. Their findings might give pause to professional campaigners, especially those who specialize in phone-bank mobilization. Not surprising, personal canvassing produced the biggest turnout boost. People contacted face to face had a turnout rate almost 9 percentage points higher than the control group. Mail recipients were somewhat more likely to vote, although this effect was modest. Individuals in the phone-call group were no more likely to turn out than were individuals in the control group. Interestingly, whereas the method of contact mattered a lot, the content of the message did not matter at all.[13]

Field experiments have powerful methodological appeal. Even so, they can be hampered by problems of internal and external validity. For example, Gerber and Green were not able to contact all the people in the personal canvassing and telephone treatment groups. This is an internal validity problem because it involves the design of the protocol itself. Each member of a randomly assigned treatment group needs to receive the treatment meant for that group, because those who receive the treatment may differ in some important way from those who are missed. Although field experiments generally enjoy greater external validity than laboratory experiments, one can still raise questions about their generalizability. Gerber and Green's study tells us a great deal about nonpartisan voter mobilization in one election and in one city. What about partisan efforts in different sorts of elections? Do negative messages increase turnout, or might they have the opposite effect? Like all experiments, field experiments are carefully created with a specific dependent variable in mind. It can be hazardous to apply their findings to different but related questions.

OBSERVATIONAL STUDIES

Drawbacks aside, experimental approaches represent the gold standard in research procedure. For many research questions in political science, however, experimental methods are not feasible—or they are simply impossible. Usually we study units of analysis as we find them, as they naturally occur in society and politics. Obviously, many naturally occurring independent variables of potential interest, such as age or sex or education level, cannot be manipulated. More generally, the top entry on the experimenter's checklist, random assignment, is missing from the repertoire of the observational study. Because of this shared omission, all observational studies suffer from selection bias, or selection for short. **Selection** occurs when the subjects who find their way into the test group differ from subjects who find their way into the control group. These differences, in turn, affect the dependent variable. You are likely to encounter two observational approaches: the natural experiment and the controlled comparison. The natural experiment is not uncommon, but the controlled comparison is prevalent. Accordingly, we will take a brief look at natural experiments and a much longer look at controlled comparisons.

A Natural Experiment

In a **natural experiment,** the researcher studies two groups—a test group and a control group, premeasures the dependent variable for both groups, applies a treatment to the test group, and then measures the dependent variable again for both groups. In each of these ways, a natural experiment is the same as an experiment. The treatment phase can be completely controlled, and the investigator seeks to equalize all other aspects of the research setting. Plus, by comparing pre- and postmeasurements of the treatment and control groups, the investigator can distinguish the treatment's effect from other changes that might occur during the study. But selection is a central concern. Because subjects themselves select the group to which they will belong, test or control, rival causes are onboard.

Natural experiments are frequently used in pedagogical research—studies aimed at determining whether one teaching method is more effective than another. Is Internet-enhanced instruction better than traditional lecture-based teaching? One study compared two sections of a U.S. politics course. The control group was taught in the conventional manner. The test group met less frequently and used Web-based learning modules, quizzes, and online discussions.[14] Both sections had the same instructor, the same textbook, the same assigned readings, and the same exams. Both groups were given a pretest measuring basic knowledge of political concepts and facts, and both were tested again at the end of the term. So the investigators controlled everything they could control. But because they could not control the assignment of subjects, the test and control groups were two quite different groups of students. Compared with their counterparts in the Internet section, students who enrolled in the conventional lecture setting had more academic experience, were more likely to be political science majors, and professed greater interest in and attentiveness to politics. Thus, from the beginning of the study, the control group was predisposed to perform better than the test group.

Selection issues should be viewed as a challenge, not as a cause for surrender. Although subjects find their own ways into the research setting, the investigator can exercise some methodological leverage over the problems this creates. Since the natural experiment requires premeasurement of the dependent variable, the researcher can compare these measurements for the test and control groups. A large pretest difference alerts the researcher to the selection problem. And, armed with postmeasurements of the dependent variable, the researcher can gauge changes in the dependent variable for both groups. These are the special advantages that natural experiments enjoy over other observational approaches.

Yet the investigator must also search for naturally occurring differences between the test and control groups, find out how these uncontrolled variables are affecting the dependent variable, and then neutralize their effects. This is accomplished by matching test and control subjects as closely as possible on these alternative causal variables and then comparing their values on the dependent variable. In this way, the natural experiment is the same as other observational approaches. In other words, the researcher tackles selection bias by applying the methodology of controlled comparison.

Controlled Comparisons

"In what ways, other than the independent variable," the researcher always must ask, "do the test group and the control group differ? How else are they not the same?" In those uncommon instances when we are doing a laboratory experiment or a field experiment, this question has a clear answer: Except for the independent variable, the test group and the control group are indistinguishable. For most of us most of the time, however, the "How else?" question defines life in the real world of political research. Ordinarily, in testing the effect of an independent variable on a dependent variable, we do two things. First, we make the comparison suggested by our hypothesis. Second, we make a controlled comparison. A **controlled comparison** is accomplished by examining the relationship between an independent and a dependent variable, while holding constant other variables suggested by rival explanations and hypotheses. In Chapter 3 you learned how to perform the first task. We now turn to the logic behind the second.

Return to the explanation of gun-control opinions introduced in Chapter 3. That explanation, which we will call the partisanship explanation, proposed that individuals' party loyalties predisposed them to adopt the policy positions of partisan opinion leaders. Gun-control opinions are one example. The hypothesis: In a comparison of individuals, those who are Democrats will be more likely to favor gun restrictions than will those who are Republicans. You may remember the results from the cross-tabulation that tested the hypothesis. More than 70 percent of Democrats favor stronger restrictions, compared with only 45 percent of Republicans—a 25-point difference.[15] As the methodological tenor of this chapter has made clear, however, we cannot regard these findings as confirming the correctness of the partisanship explanation.

The "How else?" question is the unofficial mantra of social research. And it needs to be repeated every time we test a hypothesis. How else, other than partisanship, are the two groups—Democrats and Republicans—not the same? Might these other, unexamined differences account for differences in gun-control attitudes? Consider a rival explanation, which we will call the gender explanation. This explanation also accounts for differences between people in their attitudes toward gun control:

> Attitudes toward political issues like gun control are products of gender-specific socialization practices in American culture. The socialization process is applied differently to males and females. Males engage in competitive play, often involving make-believe weapons, and they are taught that solving problems often requires the use of force. For females, cooperative play and care-giving activities are the norm in our society. Females are taught that problems are best resolved through dialogue and discussion, not force. As a result of these two different types of childhood socialization, males and females enter adulthood with different predispositions on many political issues that represent the use of force: fighting wars versus engaging in diplomacy, punishing criminals versus addressing the social problems that cause crime, or opposing gun control versus favoring stronger restrictions on guns. Therefore, females will be more likely than males to favor gun control.

The gender explanation describes a general causal process and suggests the existence of several empirical relationships. It links a dependent variable (opinions on gun control) to an independent variable (gender). The test of the hypothesis centers on a simple comparison—the percentage of women favoring gun restrictions compared with the percentage of men. This comparison, again using data from the 2004 American National Election Study, reveals a large difference: 68 percent of women, but only 48 percent of men, support stronger restrictions. How could this simple finding represent a potential rival to the partisanship explanation?

Think about it this way. Democrats and Republicans are not randomized twins, created by experimenters using random assignment. Just as with students who choose to enroll in either a lecture section or a Web-enhanced format, nonrandom processes are at work, causing some individuals to end up as Democrats and not as Republicans. Here is the potential rival. If Democrats are more likely to be female than are Republicans, then a Democrat-Republican comparison is, at least in part, also a female-male comparison. Just as in a comparison between the lecture section, which contained many political science majors, and the Internet section, which contained only a few, a comparison of the gun opinions of Democrats and

Republicans runs the risk of saying that partisanship produced the difference, when in fact gender may be the true causal agent.

How would we go about neutralizing this rival? Suppose we observed a group of Democrats and Republicans, all of whom were women. If the partisanship explanation is correct, then female Democrats should be more likely to favor gun control than female Republicans. If the partisanship explanation is incorrect, then there will be no difference between partisans on the dependent variable. Now imagine another group of Democrats and Republicans, all of whom were men. Again, if the partisanship explanation is correct, then male Democrats should also be more likely to favor gun control than male Republicans. If the partisanship explanation is off base, we will not observe this partisan difference.

That is the elemental logic of controlled comparison. We neutralize the effects of a rival cause by holding it constant—by not permitting it to operate. The variable that represents the rival cause is called the control variable. A **control variable** is a variable that is held constant while the researcher examines the relationship between an independent and a dependent variable. By convention, a control variable is represented by the letter Z. An independent variable is represented by the letter X, and a dependent variable is represented by the letter Y. When we control for Z, we can say that any relationship between X and Y cannot be attributed to Z. Obviously, any difference in gun-control opinions between female Democrats and female Republicans cannot be caused by gender differences—everybody's a female. By the same token, any difference in gun-control opinions between male Democrats and male Republicans cannot be caused by gender—everybody's a male.

THREE SCENARIOS: X→Y, CONTROLLING FOR Z

The causal connection between an independent variable and a dependent variable is symbolized by the expression X→Y, which can be read, "X is a cause of Y." So the partisanship explanation says, partisanship → gun opinions. Similarly, the causal connection between a control variable and a dependent variable is symbolized by the expression Z→Y, meaning "Z is a cause of Y." The gender explanation says, gender → gun opinions. Now consider the following expression:

> Z→X, which means "Z is a cause of X." Gender could be a cause of partisanship: gender → partisanship.[16]

The causal Z→X link might be strong. If so, then when we compare different values of X we are also (perhaps unwittingly) comparing different values of Z. If gender is a strong cause of partisanship, then Democrats will be more likely to be female than will Republicans. So by comparing different values of partisanship, Democrat with Republican, we would also be comparing different values of gender, female with male. On the other hand, the Z→X link might be much weaker, or even nonexistent. In that case, gender and party will not have anything to do with one another. As we move across the values of partisanship, from Democrat to Republican, we are not also moving across the different values of gender, from female to male. Thus, an uncontrolled Z can have a big impact on the X→Y relationship, or it can have a small impact on the X →Y relationship. It depends on whether Z is a strong cause of X or a weak cause of X.

Depending on the status of the Z–X relationship, one of three things can happen to the relationship between X and Y after Z is held constant. First, in a **spurious relationship** the

Figure 4-1 Spurious Relationship between X and Y (arrow diagram)

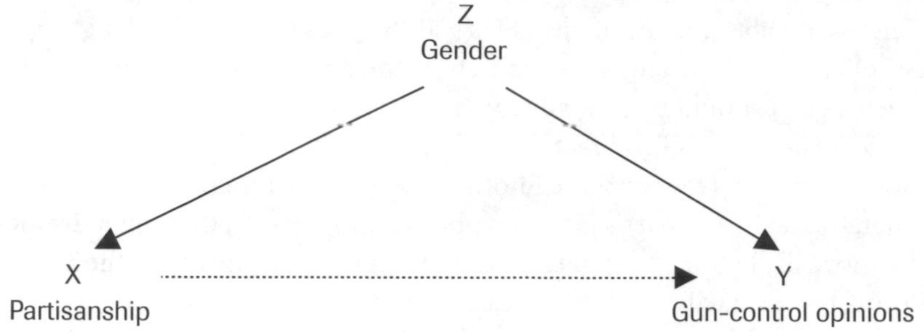

control variable has strong relationships with the independent variable and the dependent variable. After holding Z constant, the causal connection between X and Y turns out to be completely coincidental—not causal at all. Second, in an **additive relationship,** the control variable has a weak relationship with the independent variable and a strong relationship with the dependent variable. Because the Z–X relationship is weak, X retains a causal relationship with Y after controlling for Z. The control variable, Z, also helps to explain the dependent variable. A third possible scenario is more complex. In an **interaction relationship,** the relationship between the independent variable and the dependent variable depends on the *value* of the control variable. For one value of Z, the X–Y relationship might be stronger than for another value of Z.

We will now consider spurious relationships, additive relationships, and interaction relationships in greater detail. In the sections that follow, all three possibilities are illustrated with hypothetical data, using the gun-control example. But because these possibilities tell us what *can happen* to the relationship between X and Y, controlling for Z, they give us the interpretive tools we need for describing what *does happen* when, in the next chapter, we perform analyses using real data.

Spurious Relationships

There are two useful ways to represent a spurious relationship. Figure 4-1 shows an **arrow diagram**, which is a schematic representation that depicts the causal relationships between the variables. The solid arrow between Z and Y says, "Z is a cause of Y." Change Z, and Y changes, too. The solid arrow between Z and X says, "Z is a cause of X." Change Z, and X changes, too. Much like a puppeteer, who manipulates X with one hand and Y with the other, Z causes changes in the both the independent and the dependent variables. That is the puppeteer's illusion: X *appears* to cause Y. This nonrelationship between X and Y is depicted by the dashed arrow between X and Y.

Applied to the gun-control example, Figure 4-1 says that gender is a cause of gun opinions in that women are more pro-control than are men. Furthermore, Figure 4-1 tells us that gender is a cause of partisanship in that Democrats are more likely to be women and Republicans more likely to be men. Now consider how these two sets of relationships work together to create spuriousness. When we examine Democrats, we are looking at a group more heavily populated by women (who tend to be pro-gun control) than by men (who tend to be anti-control). This female bias gives the entire group of Democrats a pro-control opinion. As we move across the values of partisanship, from Democrat to Republican, the female-male

Figure 4-2 Spurious Relationship between Partisanship and Gun-Control Opinion (line chart)

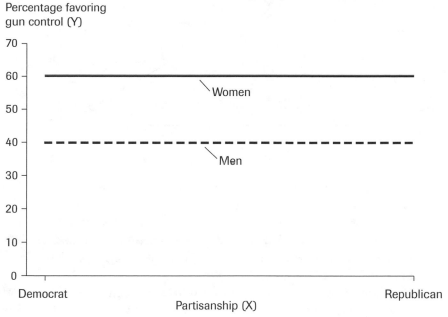

balance changes, with women (still strongly pro-control) being replaced by men (still strongly anti-control). This male bias gives the entire group of Republicans an anti-control opinion. There is nothing about party *per se* that causes the dependent variable. The party–opinion relationship is completely explained by the distribution of women and men within each value of partisanship. The X–Y relationship is an illusory artifact created by the different gender compositions of the two partisan groups.[17] As gender varies, partisanship varies and gun-control opinions vary, but no causal link exists between party and opinions.

Figure 4-2 presents an idealized line chart of a spurious relationship between party (X) and gun opinions (Y), controlling for gender (Z). The categories of X are represented along the horizontal axis, Democrats on the left and Republicans on the right. The vertical axis reports the percentage of people falling into one value of the dependent variable (Y), the percentage favoring stronger gun restrictions. Two lines inside the graph, the solid line for women and the dashed line for men, represent values of Z. By reading along each line, from one value of X (Democrat) to the other value of X (Republican), you get a visual feel for the effect of X on Y. As you can see, tracing the line for women, nothing happens to the line. It remains flat. Ditto for men. Therefore, a comparison of Democrats and Republicans reveals no difference. By looking at the distance between the lines, you can gauge the effect of Z on Y. For Democrats and Republicans alike, gender is the main determinant of gun-control opinions. This is the visual signature of a spurious relationship. If the empirical data were to produce a line chart such as the graphic depicted in Figure 4-2, one would have to conclude that partisanship bears a spurious relationship with gun-control opinions.

Mundane examples are sometimes used to illustrate spuriousness. The age-old legend that storks "cause" babies to appear may have been based on the abundance of storks in

Figure 4-3 Additive Relationships between X, Y, and Z (arrow diagram)

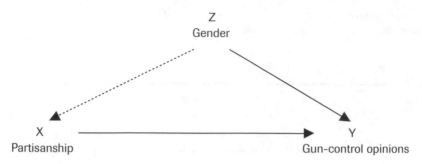

geographic areas with a high number of births and their rarity in areas where births were fewer. So the independent variable, the number of storks (X), is linked to the dependent variable, the number of births (Y). But, of course, storks prefer to roost in the nooks and crannies of buildings, so as the numbers of buildings and people (Z) increase, so too will the number of storks and the number of babies. The storks–babies relationship is purely an artifact of the number of people.[18]

In serious research, however, spurious relationships are not always so easily identified. We might find, for example, that adolescents who play violent video games (X) are more prone to antisocial behavior (Y) than adolescents who do not play violent video games. But can we say that the game playing *causes* the behavior? Or, to consider another hypothetical example, an education researcher may discover that school districts with higher per-student spending (X) have higher student achievement (Y) than do districts with lower per-student spending. Does spending cause achievement? Parents with higher socioeconomic status (Z) may be more likely to encourage their children to achieve (Z→Y) and to live in districts that spend more per student (Z→X), creating a false relationship between spending and achievement.

The specter of an unknown Z making spurious mischief with an X→Y relationship is a constant worry in observational research. The good news is that, in controlling for potentially troublesome variables, we almost always learn something new about the phenomenon being studied. For example, the relationship between smoking and lung cancer, for years vulnerable to skeptics touting an array of uncontrolled variables, is now more firmly established. This enhanced understanding has occurred in large measure because of—not despite—the suggestion that the relationship could be spurious.

Additive Relationships

These examples edge us closer to the second possible scenario for the interrelationships of X, Y, and Z. Consider the arrow diagram of Figure 4-3. A major difference exists between Figure 4-1, which diagrammed a spurious relationship, and Figure 4-3. In a spurious relationship, a strong link exists between Z and X. But in Figure 4-3, which diagrams a set of additive relationships, Z and X are weakly related, or not related at all. The dashed arrow between Z and X communicates the weak Z–X relationship. Changes in the values of Z are not associated with changes in the values of X.

This weak Z→X link is the key to additive relationships. To return to the gun control example, suppose that gender (Z) affects gun opinions (Y) in that women are more likely to

Figure 4-4 Additive Relationships between Partisanship, Gender, and Gun-Control Opinion (line chart)

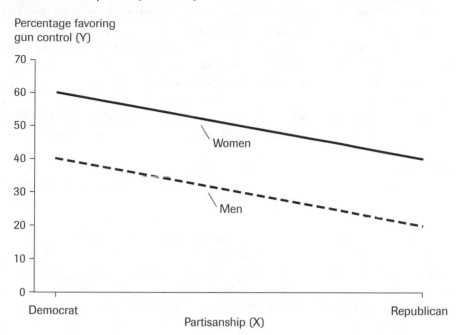

favor restrictions than are men. Suppose further that partisanship (X) causes attitudes (Y) such that Democrats are more supportive of restrictions than Republicans. Now imagine that as we move across the values of party, from Democrat to Republican, the gender composition does not change. The percentage of women in the group of Democrats is equal to the percentage of women in the group of Republicans. Thus, the fact that Democrats tend to be pro-control and Republicans anti-control cannot be attributed to a bias toward female composition among the Democrats and bias toward male composition among the Republicans. The gender makeup is the same in both partisan groups. Therefore, we cannot rule out the possibility that party *per se* is exerting an independent effect on gun attitudes.

The term *additive* describes this situation, in that knowledge of both sets of relationships, X→Y and Z→Y, adds to or strengthens the explanation of the dependent variable. Suppose the researcher finds that partisanship "works"—a larger percentage of Democrats than Republicans favor gun control. Then, by knowing partisanship, the researcher can explain some differences in gun-control opinions between people. But suppose that the gun-control opinions of some subjects cannot be explained by party differences. Some Democrats are anti-control and some Republicans are pro-control. What might account for these "unexplained" individuals? If gender also explains gun-control opinions, then women will be more likely than men to fall into the pro-control camps of both parties, Democrats *and* Republicans. Similarly, men will be more likely than women to be anti-control, regardless of party. By adding gender to the explanation, the researcher can account for additional differences in gun opinions, over and above differences explained by partisanship alone.

Figure 4-4 provides an idealized line chart of additive relationships. Again, by reading along each line, from Democrat to Republican, you can see what happens to gun-control attitudes separately for women (solid line) and for men (dashed line). Each line, considered by itself, drops predictably: Democrats are more supportive of control than are Republicans. Notice that the lines are parallel, communicating that the effect of party is the same for both genders. In the hypothetical depiction of Figure 4-4, 60 percent of Democratic women favor restrictions, compared with 40 percent for Republican women, a 20-percentage-point "party effect." This 20-point party effect is the same for males: 40 percent of male Democrats are pro-control, compared with only 20 percent of male Republicans. Notice, too, the sizeable "gender effect" within each partisan group. The female-male difference is equal to 20 percentage points among Democrats (60 percent for women versus 40 percent for men) and 20 points among Republicans (40 percent versus 20 percent). In sum, knowledge of both variables, party and gender, allows an enhanced explanation of gun-control attitudes.[19]

Interaction Relationships

In the case of spuriousness, holding Z constant has a devastating effect on the relationship between X and Y, reducing it to zero. In the case of additive relationships, controlling for Z has little impact on the relationship between X and Y, leaving it as strong as it was before Z was controlled. As incompatible as these two options seem, the third possibility can be understood by combining what you have learned thus far.

Try to imagine the following set of relationships between partisanship, gender, and attitudes toward gun control. Suppose that, for women, there is no difference between Democrats and Republicans. When we control for gender (Z) and look at the relationship between partisanship (X) and gun opinions (Y), we find no difference among women between the categories of X: Democratic women and Republican women are equally supportive of gun control. For women, therefore, the effect of X on Y is zero. But suppose that, for men, there is a huge difference between Democrats and Republicans. That is, when we control for gender and look at the relationship between partisanship and gun opinions, we find, among men, a large difference between categories of X: Democratic men are much more supportive of gun control than are Republican men. Thus, for men, the effect of X on Y is large. Now, if someone were to ask you, "What is the relationship between partisanship and gun opinions, controlling for gender?" you might reply: "It depends on whether you're talking about the opinions of women or men. To be specific, for women partisanship has no effect on gun-control attitudes, but for men partisanship has a big effect."

Political researchers use two interchangeable terms, *interaction relationships* or *specification relationships*, to describe this situation. As depicted in Figure 4-5, the arrow from X to Y

Figure 4-5 Interaction Relationships between X, Y, and Z (arrow diagram)

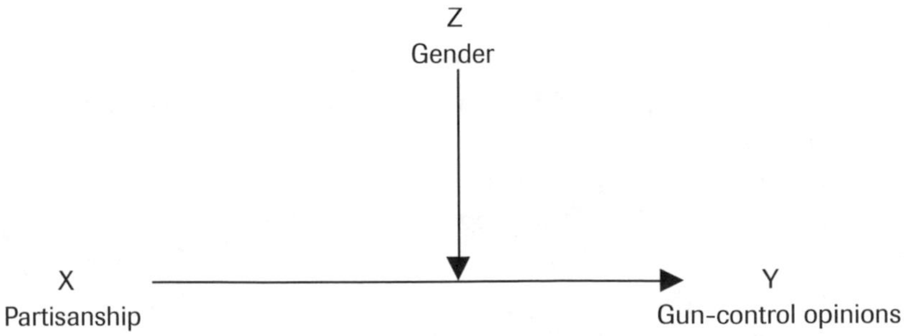

Z
Gender

X
Partisanship

Y
Gun-control opinions

Figure 4-6   Interaction Relationship between Partisanship and
Gun-Control Opinion, Controlling for Gender (line chart)

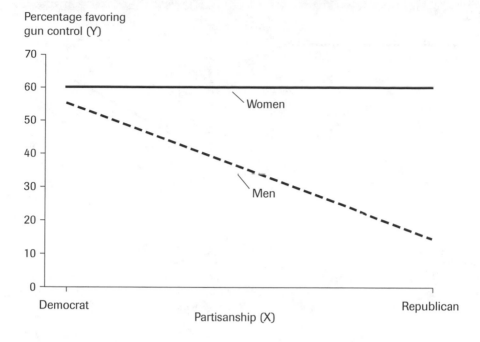

says that, yes, a causal relationship exists. But notice that the arrow from Z is not pointed directly at X or Y; rather, it is pointed at the *relationship* between X and Y. This says that the specific relationship between X and Y will depend on the value of Z. For one value of Z, when gender takes on the value "women," the relationship between partisanship (X) and gun opinion (Y) is not the same as it is for a different value of Z, when gender takes on the value "men." [20]

Figure 4-6 illustrates an idealized line chart of interaction for the gun-control example. The visual profile of interaction is in the tracks of the lines, which are different for each gender. Fill a room with women—Democrats and Republicans—and most of them would favor a gun ban. Their party allegiances would not matter. But fill a room with men, and their partisanship would make a great deal of difference. A proposal to restrict guns would be widely supported by the male Democrats and widely rejected by the male Republicans. And notice, too, the varied distance between the lines. Although there is virtually no gender gap among Democrats, a big gender gap is seen among Republicans.

The Many Faces of Interaction

The hypothetical example of interaction depicted in Figure 4-6 shows one particular profile: For one value of Z, the effect of X on Y is zero, but for another value of Z, the effect of X on Y is large. This is a common pattern of interaction. However, this is not the only way that interaction can work. The idea behind interaction is that the relationship between X and Y depends on the value of Z, and this can take many forms. Consider Figure 4-7, which uses hypothetical relationships to depict three basic patterns of interaction. (Two possible variants of each basic pattern, labeled A and B, are shown.) By now, pattern 1A is familiar to you.

Figure 4-7 Patterns of Interaction

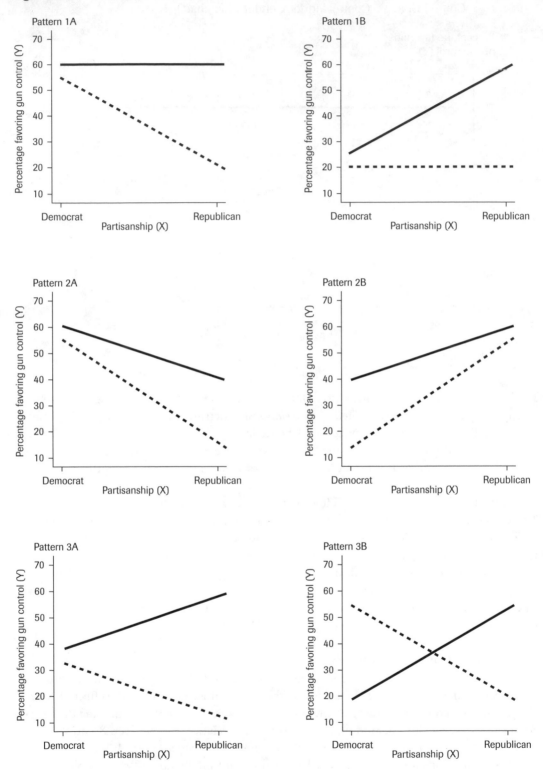

Note: Hypothetical data. For illustrative purposes, women's opinions are represented by the solid line in each graph, and men's opinions are represented by the dashed line.

This is the same pattern shown in Figure 4-6: For women, the relationship is zero; for men, the percentage of pro-control Democrats is much higher than the percentage of pro-control Republicans, resulting in a downward-sloping line as one moves from left to right along the horizontal axis. Of course, in your own analyses you may encounter one of several upward sloping variants of this same pattern (as in 1B). But profiles 1A and 1B share a basic similarity: For one value of the control variable, the X–Y relationship is zero; for another value of the control variable, the relationship is strongly positive or negative.

Often the relationship between X and Y has the same tendency for all values of Z, but the tendency is stronger for some values of Z. For example, tracing along the solid line in pattern 2A, we see that female Democrats (60 percent in favor) are more likely than female Republicans (40 percent in favor) to support controls—a 20-point difference. So, among women, partisanship works as expected, with Republicans being less supportive than Democrats. Yet the relationship is clearly stronger for men, with 55 percent of Democrats in favor, compared with only 15 percent of Republicans—a 40-point difference. Again, pattern 2B displays the same thing with tendencies reversed. For patterns 2A and 2B, the relationship between X and Y has the same tendency for both values of the control variable, but it is stronger for one value of Z than for the other value of Z.

Occasionally you will find the interesting profile depicted in patterns 3A and 3B, in which the relationship between X and Y has a different tendency for each value of Z. According to the 3A profile, for example, Republican women are more supportive of gun control than are Democratic women, producing an upward-sloping line, as partisanship shades from Democrat to Republican. For men, however, party identification has just the opposite effect, resulting in lower levels of support among Republican males than among Democratic males. An extreme variant of this form of interaction is depicted in pattern 3B. Consider this pattern for a moment. Notice that, just as in 3A, the effect of X on Y is strongly positive for one value of Z and strongly negative for the other value of Z. Unlike 3A, however, Z also has large and very different effects on Y, controlling for X. So, in pattern 3B's mocked-up scenario, Democrats are deeply divided on gender, with this qualification: Men are much more pro-control than women. The Republicans show a large gender gap, too—with women much more supportive than men. Should you encounter pattern 3B in your research, you can label it with one of these commonly used descriptors: *crossover interaction* or *disordinal interaction*.

SUMMARY

In testing a hypothesis to see if an independent variable affects a dependent variable, the researcher tries to control for rival or alternative explanations—other plausible causes of the dependent variable. The ability to rule out or neutralize rival causes depends on the power of the research design being used. In experimental studies, such as laboratory experiments and field experiments, the small miracle of random assignment creates a treatment group and a control group that are stochastic duplicates, identical in every respect that matters and in every respect that does not. By design, the experimenter-administered treatment is the only difference between the groups, guaranteeing that any observed differences in the dependent variable can be attributed directly to the independent variable and cannot be attributed to any other cause. Laboratory experiments generally have high internal validity, but they may lack external validity or generalizability. Because they are conducted in natural settings, field experiments get high marks for external validity, but uncontrolled events or measurement issues may degrade internal validity.

Observational studies do not use random assignment and so must contend with selection bias, naturally occurring differences between the treatment group and the control group that can affect the dependent variable. One type of observational study, the natural experiment, mimics some features of experimental approaches—pre- and postmeasurement and treatment intervention—but ultimately must rely on controlled comparison methodology to distinguish treatment effects from selection effects. Controlled comparison aficionados have a credo—more like a catchphrase framed as a question: "How else, other than the independent variable, are the test group and control group not the same?" In a controlled comparison, the researcher examines the relationship between an independent variable and a dependent variable while holding a rival causal variable constant, controlling its effect on the dependent variable. Any observed relationship between the independent and dependent variables may not be attributed to the variable being controlled. However, there may be other, uncontrolled causes for the observed relationship.

After controlling for a rival causal variable, there are three things that can happen to the relationship between the independent and dependent variables. In a spurious relationship, the relationship between the independent and dependent variables weakens, perhaps dropping to zero. In additive relationships, both variables—the independent variable and the control variable—make meaningful contributions to the explanation of the dependent variable. A third pattern, specification or interaction, is quite common in political research. In an interaction relationship, the relationship between the independent and the dependent variable is not the same for all values of the control variable. The relationship may differ in strength, or it may differ in tendency, depending upon which value of the control variable is being examined. The three scenarios are logical possibilities, illustrated in this chapter using idealized data. Even so, every real-world controlled comparison comes close to one of the three patterns. In the next chapter you will learn how to set up controlled comparisons and how to apply the logical scenarios.

KEY TERMS

additive relationship (p. 82)

arrow diagram (p. 82)

control group (p. 73)

control variable (p. 81)

controlled comparison (p. 79)

experimental study (p. 74)

external validity (p. 76)

field experiment (p. 74)

interaction relationship (p. 82)

internal validity (p. 76)

laboratory experiment (p. 74)

natural experiment (p. 78)

observational study (p. 74)

random assignment (p. 74)

research design (p. 74)

rival explanation (p. 73)

selection (p. 78)

spurious relationship (p. 81)

test group (p. 73)

EXERCISES

1. This chapter discussed laboratory experiments, field experiments, and natural experiments. The names of these different procedures appear in the columns of the table that follows. A number of features appear on the rows. Draw the table on a sheet of paper. In the table you have created, match each procedure with its features by making checkmarks in the appropriate cells. *Note:* The same feature may apply to more than one approach.

Feature	Laboratory experiment	Field experiment	Natural experiment
Controlled comparison method			
Control group			
Premeasurement			
Test group			
Random assignment			
Selection			
Postmeasurement			
Natural setting			

2. The enthusiastic instructor of a large class wants to determine how he can help his students review for exams. One possible method is a video-based review that can be accessed online. A few days prior to an examination, the instructor announces that the video review is available on the course Web site, and he encourages (although he does not require) students to take advantage of it. After the exam, the instructor records students' scores. Using instructional software, he then determines which students had accessed the video review and which students had not. "Yippee!," he exclaims. "The group of students who accessed my online video review scored much higher on the exam than the students who did not access the online review. My video review causes higher scores!"

   A. The instructor's enthusiasm obviously outweighs his methodological savvy. (i) Write down two rival causes that could explain why students who accessed the video review scored higher on the exam than students who did not access the video review. (ii) Describe how each rival cause could account for the video review–exam score relationship.

   B. Suppose the instructor would like to improve his procedure for determining whether the video review causes students to score higher on exams. Suppose further that you are asked to recommend three procedures that would greatly improve the instructor's study. (i) What three procedures would you recommend? (ii) Explain why these procedures would help isolate the effect of video review on exam scores.

3. Each of the following conclusions is based on a relationship between X and Y that could be spurious. For each one: (i) Identify a plausible control variable, Z. (ii) Briefly describe how Z might be affecting the relationship between X and Y.

   A. The level of ice cream sales (X) and crime rate (Y) are strongly related: As sales go up, so does the crime rate. Conclusion: To reduce the crime rate, ice cream sales should be prohibited.

   B. Car color (X) and accident rates (Y) are linked: Red cars are more likely to be involved in accidents than are nonred cars. Conclusion: If red cars are banned, the accident rate will decline.

   C. When one looks at the relationship between marital status (X) and party identification (Y), one finds: Married people are more likely to be Republicans than people that are not married. Conclusion: Getting married causes people to become Republicans.

   D. Individuals' attendance at religious services (X) is related to the number of children they have (Y). Conclusion: Declining religious attendance causes declining birthrates.

4. In observational studies, the observed relationship between an independent variable (X) and a dependent variable (Y) may be caused by some other, uncontrolled difference (Z). So we always ask, "How else, besides the independent variable, do the subjects differ?" For each of the three hypotheses below: (i) Think up a plausible alternative causal variable on which the subjects might differ. (ii) Describe how this variable might affect the relationship between the independent variable and the dependent variable.

Example. In a comparison of countries, those with decentralized governments tend to have higher percentages of citizens who trust the government than do countries with centralized governments.

(i) Plausible alternative variable: size of country's population
(ii) Countries with decentralized governments might also have smaller populations than countries with centralized governments. If smaller countries also have higher percentages of trusting citizens than do larger countries, then differences in size, not differences in decentralization-centralization, could explain the relationship between the independent and dependent variables.

A. In a comparison of congressional candidates, candidates who raise large amounts of money during their campaigns are more likely to win than are candidates who raise small amounts of money.

B. In a comparison of students, those who arrive late to class will be more likely to receive poor grades than will those who arrive on time.

C. In a comparison of individuals, southerners are more likely than are nonsoutherners to approve of spanking as a way to discipline children.

5. Here is a variable: Some people support a government-backed health insurance plan that would cover all citizens, whereas others oppose such a plan. This variable has two values: "support government plan" and "oppose government plan." Let's say that one explanation, the "income explanation," suggests this hypothesis: In a comparison of individuals, those who have low incomes will be more likely to support a government plan than will those who have high incomes. A rival explanation, the "age explanation," suggests this hypothesis: In a comparison of individuals, those who are older will be more likely to support a government health plan than will those who are younger. For the purposes of this exercise, income will be X, age will be Z, and support for government-backed health insurance will be Y.

A. Suppose that after controlling for age (Z), the relationship between income (X) and health insurance opinions (Y) turned out to be spurious. (i) Using this chapter's discussion of spuriousness as a guide, write 4–5 sentences explaining how the Z–Y relationship and the Z–X relationship would produce a spurious relationship between X and Y. (ii) Sketch a line graph depicting a spurious relationship between X and Y, controlling for Z. The vertical axis will show the percentage supporting a government plan. Invent plausible percentages for the values of Y. The horizontal axis will show the two values of income. There will be two lines inside the graph: one for younger people and one for older people.

B. Suppose that after controlling for age (Z), the set of relationships between X, Z, and Y turned out to be additive. (i) Using this chapter's discussion of additive relationships as a guide, write 4–5 sentences explaining how this set of relationships would fit an additive pattern. (ii) Sketch a line graph depicting the additive relationships between X, Z, and Y. The vertical axis will show the percentage supporting a government plan. Just as you did in part A, invent plausible percentages for the values of Y. The horizontal axis will show the two values of income. There will be two lines inside the graph: one for younger people and one for older people.

C. Suppose that a set of interaction relationships exists between X, Z, and Y. Suppose further that the interaction takes this form: Among younger people, income has no effect on the dependent variable; among older people, those with lower incomes are more likely to support a government plan than are those with high incomes. Sketch a line graph depicting this set of interaction relationships. Remember that the vertical axis shows the percentage supporting a government plan. As before, invent plausible percentages for the values of Y. The horizontal axis will show the two values of income. There will be two lines inside the graph: one for younger people and one for older people.

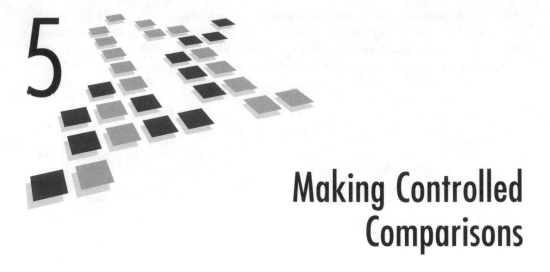

5

Making Controlled Comparisons

LEARNING OBJECTIVES

In this chapter you will learn:
- How to set up controlled comparisons using cross-tabulation analysis and mean comparison analysis
- How to identify spurious relationships, additive relationships, or interaction relationships in empirical data
- How to construct line charts of the relationship between an independent variable and a dependent variable, while holding a control variable constant

Describing explanations, stating hypotheses, making comparisons, and looking for interesting patterns in the data—these are all genuinely creative activities, and sometimes they are even accompanied by the joy of discovering something new. But, as we have seen in Chapter 4, discovery is an ongoing process. For every explanation we describe, a plausible alternative explanation exists for the same phenomenon. Each time we test the relationship between an independent and a dependent variable, we must ask, "How else, besides the independent variable, are the units of analysis not the same?" Is the dependent variable being caused by the independent variable, or is some other cause, some other independent variable, affecting the dependent variable?

Chapter 4 introduced the methodology of controlled comparison, the most common method for dealing with the "How else?" question in observational research. You now have an idea of what *can happen* to the relationship between an independent variable and a dependent variable, controlling for a rival cause. The three logical scenarios—a spurious relationship, a set of additive relationships, or interaction—provide the tools you need to interpret the relationships you will find when performing controlled comparison analysis.

In this chapter we use empirical data to illustrate the practice of controlled comparison in the real world of political research. After controlling for a rival cause of the dependent variable, what *does happen* to the relationship between the independent variable and the dependent variable? The procedures for setting up controlled comparisons are natural extensions of procedures you learned in Chapter 3 for setting up cross-tabulations and mean comparisons. You also will find that the three possible scenarios are valuable interpretive tools in helping you to understand and describe complex empirical relationships. We turn first to cross-tabulation analysis and then consider mean comparison analysis.

CROSS-TABULATION ANALYSIS

Chapter 3 laid out the ground rules for using cross-tabulation analysis to make comparisons. Let's begin by getting reacquainted with these procedures. Table 5-1 displays a cross-tabulation of the relationship between party identification and gun-control opinions.[1] The values of the independent variable—party identification—define the columns of the table, with the 378 Democrats in the column on the left and the 346 Republicans in the column on the right. (The "Total" column shows the distribution of all 724 respondents across the dependent variable.) The values of the dependent variable—"high" or "low" support for gun control—define the rows. It is clear that a substantial partisan divide exists on this issue. The percentage of Democrats that expressed a high level of support for stronger gun restrictions, 60.6 percent, is more than 30 points higher than the percentage of high gun-control supporters among Republicans, 26.9 percent. A difference obtained from a simple comparison is called a **zero-order relationship**. Also known as a *gross relationship* or an *uncontrolled relationship*, a zero-order relationship is an overall association between two variables that does not take into account other possible differences between the cases being studied.

Zero-order relationships always invite the "How else?" question. How else do Democrats and Republicans differ that may account for the difference in gun-control opinions? In Chapter 4 we described a plausible rival cause, gender. Gender could be related to gun attitudes because women may be more pro-control than men. Furthermore, if Democrats are more likely than Republicans to be women, then the party–opinion relationship could be a spurious artifact of the gender–party relationship. To address this rival, we need to reexamine the relationship between party and gun-control opinions while holding gender constant, thereby controlling for its effects.

Control Tables

Table 5-2, a **controlled comparison table**, demonstrates how to accomplish this goal. A controlled comparison table, or *control table* for short, presents a cross-tabulation between

Table 5-1 Relationship between Partisanship and Gun-Control Opinion

Support for gun control	Partisanship		Total
	Democrat	*Republican*	
High	60.6%	26.9%	44.5%
	(229)	(93)	(322)
Low	39.4%	73.1%	55.5%
	(149)	(253)	(402)
Total	100.0%	100.0%	100.0%
	(378)	(346)	(724)

Source: 2004 American National Election Study.

Table 5-2   Relationship between Partisanship and Gun-Control Opinion, Controlling for Gender

Support for gun control	Gender					
	Female			Male		
	Partisanship			Partisanship		
	Democrat	*Republican*	Total	*Democrat*	*Republican*	Total
High	64.4%	33.9%	51.3%	54.2%	19.5%	35.4%
	(152)	(60)	(212)	(77)	(33)	(110)
Low	35.6%	66.1%	48.7%	45.8%	80.5%	64.6%
	(84)	(117)	(201)	(65)	(136)	(201)
Total	100.0%	100.0%	100.0%	100.0%	100.0%	100.0%
	(236)	(177)	(413)	(142)	(169)	(311)

Source: 2004 American National Election Study.

an independent variable and a dependent variable for each value of the control variable. Cases are first divided into groups according to their values on the control variable. So, in constructing Table 5-2, respondents were separated into two groups on the basis of gender (the control). A separate cross-tabulation analysis between the independent variable and the dependent variable is then performed for each value of the control. The left-hand cross-tabulation of Table 5-2 shows the relationship between partisanship (the independent variable) and gun-control opinions (the dependent value) for the 413 respondents sharing the same value of gender, female. The right-hand cross-tabulation shows the same party–opinion relationship for the 311 respondents sharing the other value of gender, male. (These totals, 413 women and 311 men, appear at the bottom of the "Total" column of each cross-tabulation.)

A control table illuminates two sets of relationships. First, it reveals the controlled effects of the independent variable on the dependent variable. A **controlled effect** is a relationship between a causal variable and a dependent variable within one value of another causal variable. In Table 5-2, we can assess the relationship between party affiliation and gun-control opinions separately for females and for males. Thus, we can obtain two controlled effects of partisanship, one for women and one for men. Second, a control table permits us to describe the effects of the control variable on the dependent variable at each value of the independent variable. By comparing the gun-control opinions of female Democrats with the opinions of male Democrats—and comparing the opinions of female Republicans with those of male Republicans—we can obtain the controlled effects of gender on opinions, with one controlled effect for Democrats and one controlled effect for Republicans. Let's take a close look at both sets of relationships.

First consider the relationship between party and gun-control opinions, holding gender constant. Clearly, sizable partisan differences exist for women and for men. Focusing first on females, 64.4 percent of female Democrats are high gun-control supporters, compared with 33.9 percent of female Republicans. Calculating the controlled effect of partisanship among women, we have: 64.4 − 33.9 = 30.5 percentage points. For males we find a similar partisan divide. According to the control table, 54.2 percent of male Democrats are high gun-control supporters, compared with only 19.5 percent of male Republicans. For males the controlled effect of partisanship is: 54.2 − 19.5 = 34.7 points. Pause for a moment and think about what these two differences—a controlled effect of 30.5 points for females and 34.7 points for males—tell us about the relationship between party and opinion, controlling for gender.

Partial Effect

Controlled effects are summarized by a **partial relationship** or **partial effect.** Just as a zero-order relationship summarizes an overall relationship between variables, a partial relationship or partial effect summarizes a relationship between two variables after taking rival variables into account.[2] Suppose someone asked, "What is the partial effect of partisanship on gun-control opinions, controlling for gender?" In asking for a partial effect, this questioner would like a concise, one-number summary that conveys the partial relationship between partisanship and opinion, controlling for gender. We have two numbers to choose from: 30.5, the controlled effect of partisanship among females, and 34.7, the controlled effect of partisanship among males. These numbers are quite close. A handful of calculations will produce a more precise summary,[3] but it seems reasonable and conservative to say that, controlling for gender, the partial effect of partisanship on gun-control opinions is about 30 points. Democrats are about 30 points more likely than Republicans to favor stronger gun restrictions. So the partial effect of partisanship is 30.

Now turn your attention to the second set of relationships, the relationship between gender and gun-control opinions among Democrats and among Republicans. To evaluate these relationships, we jump between cross-tabulations, comparing respondents who share the same partisanship (for example, all Democratic Party identifiers) but differ on gender (females compared with males). How do we accomplish these comparisons? As you can see from Table 5-2, if we subtract the percentage of gun control supporters among males from the percentage of gun control supporters among females, we will obtain a positive (+) number. Subtracting the female percentage from the male percentage yields a negative (-) number. Since gender is a nominal variable, it is silly to ask whether an "increase" in gender is associated with an increase or decrease in support for gun control. So, why worry about positive or negative signs? Because we need a way to identify interaction relationships with nominal variables. This book follows the **rule of direction for nominal relationships.** The value of the variable that defines the left-most column of a cross-tabulation is the base category. Determine the direction of a nominal relationship by subtracting other values from the base category's value. (See Box 5-1.)

In Table 5-2, females appear on the left-hand side of the control table and males appear on the right. By the rule, "female" defines the base category. The controlled effects of gender on opinion are determined by subtracting the male percentages from the female percentages. Among female Democrats, 64.4 percent are strong supporters. Among male Democrats, 54.2 percent are strong supporters. Thus, for Democrats the controlled effect of gender is: 64.4 − 54.2 = 10.2 percentage points. Similarly, female Republicans, at 33.9 percent, are more likely to be strong supporters than are male Republicans, at 19.5 percent. For Republicans, then, the controlled effect of gender is: 33.9 − 19.5 = 14.4 points. The gender-opinion relationship "runs in the same direction" for both partisan groups—when we subtract the male percentage from the female percentage we get a positive number—but the gender difference is somewhat larger among Democrats.

Our imaginary questioner returns: "What is the partial effect of gender on gun-control opinions, controlling for partisanship?" As before, we have two numbers to choose from: 10.2 percentage points, the controlled effect of gender among Democrats, and 14.4 points, the gender difference among Republicans. Inasmuch as these numbers are not wildly different, it seems reasonable and conservative to say that, controlling for partisanship, the partial effect

---

Box 5-1   The Rule of Direction for Nominal Relationships

In a positive relationship, signified by a plus sign (+), increasing values of the independent variable are associated with increasing values of the dependent variable. In a negative relationship, signified by a minus sign (−), increasing values of the independent variable are associated with decreasing values of the dependent variable. These definitions of positive and negative relationships make intuitive sense. This intuitiveness evaporates, however, for nominal-level independent variables. In summarizing the gender–gun-control opinions relationship displayed in Table 5-2, for example, we subtracted the percentage of male supporters from the percentage of female supporters, resulting in a positive number. Because the values of gender cannot be ranked, it would be just as correct to subtract the female percentage from the male percentage and end up with a negative number. However, if interaction is in play, the independent variable–dependent variable relationship might "run in different directions," depending on the value of the control. The female percentage may be higher than the male percentage for one value of the control and lower than the male percentage for another value of the control. We need to establish a rule that allows us to use plus signs and minus signs to communicate the presence of interaction in nominal relationships.

In this book we follow the **rule of direction for nominal relationships.** The value of the variable that defines the left-most column of a cross-tabulation is the base category. Figure out the direction of a nominal relationship by subtracting other values from the base category's value. Thus, a relationship is positive if the base category's value is higher than the value in the category to which it is being compared, and negative if the base category's value is lower than the value in the category to which it is being compared. In Table 5-2, "Democrat" is the left-most value of the independent variable, and "Female" is the left-most value of the control. In determining the direction of the partisanship–gun-control opinions relationship, we subtract the percentage of Republican supporters from the base percentage, Democratic supporters. In determining the direction of the gender–gun-control opinions relationship, we subtract the percentage of male supporters from the base percentage, female supporters. This rule is consistent with the algorithm computer packages follow in calculating the direction of relationships in cross-tabulation analysis.

---

of gender on gun-control opinions is about 10 points. Women are about 10 points more likely than men to favor stronger gun restrictions. So the partial effect of gender is 10.[4]

Identifying the Pattern

Which scenario—spurious, additive, or interaction—best characterizes the relationships between partisanship, gender, and gun-control opinions? Let's review the definitions of spurious, additive, and interaction relationships.

- In a spurious relationship, after holding the control variable constant, the relationship between the independent variable and the dependent variable weakens or disappears.
- In a set of additive relationships, the tendency and strength of the relationship between the independent variable and the dependent variable are the same or very similar at all values of the control variable.

- In a set of interaction relationships, the tendency or strength of the relationship between the independent variable and the dependent variable is different, depending on the value of the control variable.

These definitions suggest three questions that will help you determine which pattern comes closest to the relationships you are analyzing. After setting up a control cross-tabulation—or a mean comparison control table, covered later in this chapter—examine the data and answer the following questions:

1. After holding the control variable constant, does a relationship exist between the independent variable and the dependent variable within at least one value of the control variable?
   - If the answer is no, then the relationship is spurious. If the answer is yes, then go to question 2.

2. Is the tendency of the relationship between the independent variable and the dependent variable the same at all values of the control variable?
   - If the answer is no, then interaction is taking place. If the answer is yes, then go to question 3.

3. Is the strength of the relationship between the independent variable and the dependent variable the same or very similar at all values of the control variable?
   - If the answer is yes, then the relationships are additive. If the answer is no, then interaction best characterizes the relationships.

For the partisanship–gender–gun-control opinions relationships, we can answer question 1 in the affirmative. After controlling for gender, the party–opinion relationship is a robust 30 points. Democrats are 30 percentage points more likely than Republicans to support gun restrictions. If spuriousness were in play, then the partial relationship would be weak or nonexistent. Because the partial effect of the independent variable on the dependent variable persists after control, we can rule out spuriousness. The dynamic behind the persistence of the partisanship–opinion relationship can be gleaned from Table 5-2's raw numbers: Democrats are only somewhat more likely to be female than are Republicans.[5] Thus, when we compare the two values of the independent variable, Democrat with Republican, we are not also comparing a group heavily composed of females with a group heavily composed of males.

Having ruled out spuriousness, let's consider question 2. As we have seen, the tendency of the partisanship–gun-control opinions relationship is the same for both women and men. For both genders, Democrats are more pro-control than are Republicans. Suppose for the sake of illustration that Democratic women were less likely to be pro-control than Republican women, but Democratic men were *more* likely to be pro-control than Republican men. In such situations, in which the relationships run in different directions within categories of the control, question 2 will alert you to the presence of interaction. In the current example, however, the relationships have the same tendency, so we can't rule interaction in, and we can't rule additive out.

Question 3 permits a clear choice: the relationships are additive. We can say this because the strength of the controlled effect of partisanship on gun-control opinions is the same or very similar at both values of gender. Controlling for gender, the partial effect of party is 30 points. For both women and men, going from Democrat to Republican decreases gun-control support by 30 percentage points. Going from Republican to Democrat increases support by 30 points. Controlling for party, the partial effect of gender is 10 points. For both Democrats and Republicans, going from female to male decreases gun-control support by 10 percentage points. Going from male to female increases support by 10 points.

All additive relationships have a straightforward, symmetrical quality. Considered separately, each causal variable—the independent variable and the control variable—helps to explain the dependent variable. Both variables, considered together, enhance the power of the explanation. To illustrate, suppose we wanted to compare respondents having the least supportive gun-control combination of attributes, Republican males, with respondents having the most supportive combination, Democratic females. Because the partial effect of partisanship is equal to 30, the partisanship "part" of the comparison translates into a 30-point difference between Republican males and Democratic females. Based on partisan differences alone, we would expect the opinions of Democratic females to be 30 points more supportive than the opinions of Republican males. But we still need to add in the female-male comparison. Because the partial effect of gender is equal to 10, the gender "part" of the comparison translates into an additional 10-point difference between Republican males and Democratic females. Thus we would expect a 40-point difference between Republican males and Democratic females, a difference that fits the data pretty well (19.5 percent versus 64.4 percent).

How about a Republican female–Democratic male comparison? The Republican-Democratic part adds 30 points. So, based on partisanship alone, we would expect Democratic males to be 30 points more supportive than Republican females. But we still need to add—or, in this comparison, subtract—the female-male part. Because males are 10 points less supportive than females, we must subtract 10 points from the 30-point partisan difference, yielding an estimated 20-point difference between Republican females and Democratic males. Again, this combination of additive building blocks reproduces the data: 33.9 percent for Republican females and 54.2 percent for Democratic males.

GRAPHING CONTROLLED COMPARISONS

Chapter 3 described how to construct a line chart based on the cross-tabulation of an independent variable and a dependent variable. Line charts are especially useful for lending clarity and simplicity to controlled comparison relationships. How would one construct a line chart for the relationships in Table 5-2? Consider Figure 5-1. As in the line charts you constructed for simple comparisons, the values of the independent variable appear on the figure's horizontal axis—with Democrats on the left and Republicans on the right. And, as before, the percentage of cases falling into one value of the dependent variable—the percentage with a high level of gun-control support—are represented on the vertical axis. For controlled comparisons, the lines inside the graph depict the relationship between the independent and dependent variables for each value of the control. The "female" line (solid line) connects two percentages from Table 5-2: 64.4 percent (the percentage of female Democrats with high support) and 33.9 percent (the percentage of female Republicans with high support). The "male" line (dashed line) connects 54.2 percent (the percentage of male Democrats with high support) and 19.5 percent (the percentage of male Republicans with high support).

Whereas control tables can be tedious to contemplate, graphic depictions often assist interpretation. It is plain from Figure 5-1 that the independent variable, partisanship, does substantial explanatory work within each value of the control. Each line drops about 30 points from left to right. The effect of the control variable on the dependent variable is

Figure 5-1   Relationship between Partisanship and Gun-Control Opinions, Controlling for Gender (line chart)

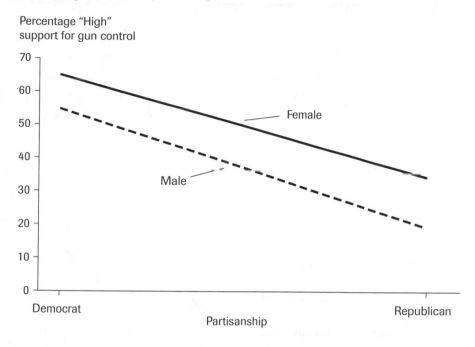

depicted in the distance between the lines. The female line is about 10 points higher than the male line. Is the linear symmetry a geometric perfection? No, but real-world relationships rarely are. One would have to say that Figure 5-1 closely approximates the visual profile of an additive relationship.

An Example of Interaction

To bring the additive character of the party–gender–gun-control opinions relationships into stark relief, let's examine a set of interaction relationships. As discussed in Chapter 4, interaction relationships are protean shape-shifters, assuming a variety of forms. However, all interaction relationships share one characteristic: *The tendency or strength of the relationship between the independent variable and the dependent variable is different, depending on the value of the control variable.* Sometimes we discover an interaction pattern where we expected to find something else, such as a set of additive relationships. Many times, however, we have an explanation that implies interaction. If our explanation is correct, not only do we expect to find interaction, but we also know which form the interaction should take.

Consider the plausible causal link between individuals' positions on issues and their political behavior. For example, we might hypothesize that individuals' abortion opinions (independent variable) are causally linked to their vote choices (dependent variable). In a comparison of individuals, those who hold liberal opinions on abortion will be more likely to vote Democratic than those who hold conservative opinions on abortion. Notice that this proposed linkage assumes that abortion is *salient* to people—that they personally care about the issue and consider it extremely important. Think of two eligible voters, both of whom say that abortion "should always be permitted." One thinks that the abortion issue is "extremely important," whereas the other thinks it is only "somewhat important." According to the salience idea, even though both individuals hold the same position on the abortion issue, the

Table 5-3 Relationship between Abortion Opinion and Vote Choice, Controlling for Issue Salience

Vote choice	Issue salience					
	Low			*High*		
	Abortion opinion			Abortion opinion		
	Always permit	*Not always permit*	Total	*Always permit*	*Not always permit*	Total
Democratic	58.5%	43.4%	49.1%	81.8%	27.3%	50.0%
	(117)	(144)	(261)	(90)	(42)	(132)
Republican	41.5%	56.6%	50.9%	18.2%	72.7%	50.0%
	(83)	(188)	(271)	(20)	(112)	(132)
Total	100.0%	100.0%	100.0%	100.0%	100.0%	100.0%
	(200)	(332)	(532)	(110)	(154)	(264)

Source: 2004 American National Election Study.

"extremely important" voter will be more likely than the "somewhat important" voter to cast a ballot for the Democratic presidential candidate. Of course, the same idea applies to abortion conservatives, those who do not think that abortion should always be permitted. An abortion opponent who places great importance on the issue will be less likely to vote Democratic than will an abortion opponent who considers the issue to have less salience, even though both voters share the same issue position.

Salience implies interaction. Salience says that the strength of the relationship between the independent variable (abortion opinions) and the dependent variable (vote choice) will depend on the value of a control variable (whether people consider the issue important or salient). Under one value of the control variable—when the issue is salient to people—the opinion–vote relationship will be strong. Under another value of the control variable—when the issue is *not* salient—the opinion–vote relationship will be weak.

Table 5-3 displays a control table for the relationship between abortion opinions and vote choice, controlling for issue salience.[6] Table 5-3 was constructed according to protocol. Respondents were first divided into two groups on the basis of the professed importance of the abortion issue: 532 low-salience respondents and 264 high-salience respondents. The cross-tabulation on the left displays the abortion opinion–vote choice relationship for people who place low importance on the issue, and the cross-tabulation on the right shows the relationship for people who place high importance on the issue.

First consider the opinion–vote choice relationship, controlling for salience. For low-salience respondents, issue position obviously is related to vote. Among those who think abortion should always be permitted, 58.5 percent voted Democratic, compared with 43.4 percent of respondents who do not think abortion should always be permitted. The controlled effect is: 58.5 – 43.4 = 15.1 percentage points. Thus, among low-salience respondents, abortion liberals were 15 points more likely to vote Democratic than were abortion conservatives. Not exactly an eye-popping difference, but interesting nonetheless. Now shift your attention to the high-salience cross-tabulation, where you will find a much stronger relationship. Among "always permit" respondents who place high importance on the issue, 81.8 percent voted Democratic. By contrast, only 27.3 percent of the high-salience "not always permit" respondents voted Democratic. When salience switches from low to high, some sort of relationship amplifier gets switched on as well: 81.8 − 27.3 = 54.5 percentage points. Among high-salience respondents, abortion liberals were nearly 55 points more likely to vote Democratic than were abortion conservatives.

Our imaginary questioner now asks: "What is the partial effect of abortion opinions on vote choice, controlling for salience?" There are, of course, two controlled effects from which to choose: 15 percentage points, obtained from the low-salience cross-tabulation, and 55 points, from the high-salience cross-tabulation. Would either of these numbers suffice as a fair summary of the partial effect of the independent variable on the dependent variable? No. Let's be clear. Both effects have the same tendency. For low-salience and high-salience respondents alike, abortion liberals are more likely to vote Democratic than are abortion conservatives. But the relationships differ dramatically in strength: a 15-point difference versus a 55-point difference. This is a sure sign of interaction. When it makes little sense to settle upon a one-number summary of a partial effect—when empirical reality requires that you describe each controlled effect separately—then interaction is at work.

We also see the interaction profile in the controlled effects of salience on vote choice. Focus on abortion liberals, who think abortion should always be permitted. Among low-salience abortion liberals, 58.5 percent voted Democratic, compared with 81.8 percent of their high-salience counterparts. Therefore, the controlled effect of salience among abortion liberals is: 58.5 − 81.8 = −23.3. The minus sign tells us that low-salience abortion liberals were 23 points *less likely* to vote Democratic than were high-salience liberals. Among low-salience abortion conservatives, 43.4 percent voted Democratic, compared with 27.3 percent of the high-salience abortion conservatives. The controlled effect is: 43.4 − 27.3 = 16.1. Here the difference is a positive number, telling us that low-salience conservatives were 16 points *more likely* to vote Democratic than were high-salience conservatives. Thus, the relationship between issue salience and vote choice has different tendencies, depending on the value of the independent variable. Issue salience heightens the Democratic proclivities of abortion liberals and depresses the Democratic proclivities of abortion conservatives. This is another sure sign of interaction. If you hold one variable constant—the independent variable or the control variable, it does not matter which one—and you find that the controlled effects run in different directions, then interaction is at work.

Although interaction is clearly taking place here, we will ensure that the three rhetorical questions also lead us to the correct conclusion:

1. After holding the control variable constant, does a relationship exist between the independent variable and the dependent variable within at least one value of the control variable?
   - Yes. So the relationship is not spurious.

2. Is the tendency of the relationship between the independent variable and the dependent variable the same at all values of the control variable?
   - Yes. For both low-salience and high-salience respondents, abortion liberals are more likely to vote Democratic than are abortion conservatives. Interestingly, if we had framed the research problem differently—making salience the independent variable and abortion position the control—then question 2 would lead us directly to interaction. As we have seen, the salience–vote relationship has opposite tendencies for abortion liberals (salience increases the likelihood of a Democratic vote) and abortion conservatives (salience decreases the likelihood of a Democratic vote).

Figure 5-2 Relationship between Abortion Opinion and Vote Choice, Controlling for Issue Salience (line chart)

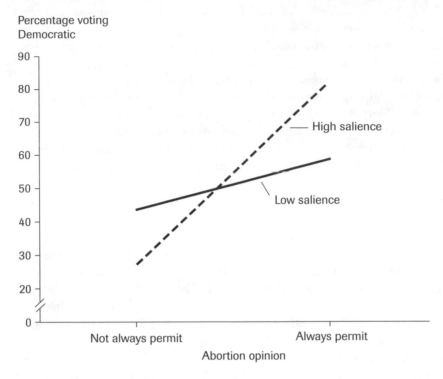

3. Is the strength of the relationship between the independent variable and the dependent variable the same or very similar at all values of the control variable?
   • No, clearly. At low salience, the controlled effect of abortion opinion is 15 points. At high salience, the controlled effect is 55 points.

Rest assured that, no matter how you set up the analysis, the three-question sequence will keep you on track.

Figure 5-2 displays the interaction relationships just discussed. Both the low-salience line (solid line) and the high-salience line (dashed line) rise from left to right. As the independent variable changes from "not always permit" to "always permit," there is an increase in the percentage voting Democratic. But the gradient of the low-salience line is much more modest than the increase in the high-salience line, communicating the independent variable's stronger effect under the high-salience condition. One way to spot interaction, then, is to create a graphic rendition and look for lines that are clearly not parallel. Notice, too, the different effects of the control variable among "not always permit" respondents and "always permit" respondents. Here we can plainly see that low-salience conservatives are more likely to vote Democratic than high-salience conservatives, and low-salience liberals are less likely to vote Democratic than high-salience liberals.

To label a set of relationships as interaction, do the lines need to cross, as shown in Figure 5-2? No. The only requirement for interaction is this: *The tendency or strength of the relationship between the independent variable and the dependent variable is different, depending on the value of the control variable.* Crossed lines—indicative of disordinal or crossover interaction— provide dead-giveaway certainty and add visual flair, but they are not required.

MEAN COMPARISON ANALYSIS

The same mechanics of control—and the same graphic representations—apply when the dependent variable is summarized by means. In many ways, mean comparison control tables are easier to interpret than cross-tabulation control tables. Cross-tabulations display the distribution of cases across the values of the dependent variable. Even in simple controlled relationships between variables having the bare minimum of two values, cross-tabulations produce a fair amount of information to sort out and understand. As you learned in Chapter 3, mean comparisons condense relationships to a single, easy-to-interpret measure of central tendency. In a perfect world, in which everything is measured at the interval level, everyone is doing mean comparison analysis all the time. In this section, we will present two examples of controlled mean comparisons. The first illustrates additive relationships. The second is an example of interaction.

An Example of Additive Relationships

Here is an interesting dependent variable: individuals' ratings of "gay men and lesbians, that is, homosexuals," as measured by an American National Election Study (ANES) feeling thermometer scale. The ANES routinely includes a series of questions that ask individuals to rate institutions, groups, and contemporary political figures on a scale from 0 to 100. The scale gauges the range from highly unfavorable (ratings close to 0) to highly favorable (ratings close to 100).[7] What independent variable might be causally linked to ratings of homosexuals? In Chapter 2 we described a Likert-scale measure of egalitarianism, the extent to which individuals believe that government and society should reduce differences between people.[8] It seems plausible to suggest that egalitarianism is causally connected to ratings of homosexuals: In a comparison of individuals, those with low levels of egalitarianism will give homosexuals lower ratings than will individuals with high levels of egalitarianism.

In explaining attitudes toward homosexuality, there are several potential control variables that the analyst might take into account—gender, race, income, religiosity, and age, among others.[9] For example, the evidence suggests that younger people have more positive attitudes toward homosexuals than do older people. Furthermore, if age and egalitarianism are related, then when we compare the ratings of homosexuals by low egalitarians with those of high egalitarians we are also comparing the ratings of older people with the ratings of younger people. We will analyze the egalitarianism–homosexual ratings relationship, controlling for age. Table 5-4 introduces the format for a mean comparison control table. The categories of the control variable, age, appear across the top of the table and define the columns. The values of the independent variable, egalitarianism, appear along the side of the table and define the rows. By reading down each column, we can see what happens to the dependent variable (mean ratings of homosexuals), as the independent variable (egalitarianism) changes from "low" to "high" within each category of the control variable (age). By reading across each row, we can see the effect of the control variable, age, on the dependent variable, at each value of the independent value, egalitarianism. This efficient little table contains a wealth of information. By reading down the right-most "total" column, we can gauge the zero-order effect of the independent variable on the dependent variable. By reading across the bottom-most "total" row, we see the zero-order effect of the control variable on the dependent variable.

Examine Table 5-4 for a few minutes. Retrace the three steps we followed in diagnosing the partisanship–gender–gun-control opinion relationships and the abortion opinion–salience–vote choice relationships. The egalitarianism–homosexual ratings relationship

Table 5-4 Mean Feeling Thermometer Ratings of Homosexuals, by Egalitarianism, Controlling for Age

Egalitarianism	Age			Total
	18–35	*36–55*	*56–older*	
Low	49.9	44.4	41.2	44.5
	(143)	(204)	(231)	(578)
High	58.0	54.0	48.0	53.8
	(162)	(172)	(122)	(456)
Total	54.2	48.8	43.6	48.6
	(305)	(376)	(353)	(1,034)

Source: 2004 American National Election Study.

persists within all values of the control. As we move from "low" to "high" egalitarianism, average homosexual ratings increase by 8.1 points among the youngest cohort, 9.6 points among the middle age group, and 6.8 points among the oldest cohort. So the egalitarianism–homosexual ratings relationship is not spurious.

Each controlled relationship has the same tendency: higher values of egalitarianism are associated with higher mean ratings of homosexuals. If, say, egalitarianism had no effect among older people but a big effect among younger people—or, perversely, if egalitarianism and homosexual ratings were negatively related in one age group but positively related in another—then the data would be making an open-and-shut case for interaction. But nothing even remotely exotic is going on here. Notice also that the relationship between the control variable and the dependent variable has the same tendency at both values of egalitarianism: as age goes up, homosexual ratings decline. As is often the case, question 3—Is the strength of the relationship between the independent variable and the dependent variable the same or very similar at all values of the control variable?—serves as the tie-breaker between interaction and additive relationships. Although the controlled effects of egalitarianism on homosexual ratings are not identical within all age groups, they are quite similar. Moving from "low" to "high" occasions an average increase in ratings of between 7 and 9 points. Erring on the conservative side, we can say that the partial effect of egalitarianism is about 7. Controlling for age, going from low egalitarianism to high egalitarianism adds 7 points, or going from high to low subtracts 7 points. Also note that the controlled effect of age is much the same for low egalitarians and high egalitarians. On average, the oldest cohort rates homosexuals between 9 and 10 points lower than does the youngest cohort. So the partial effect of age is about 9. Controlling for egalitarianism, going from young to old subtracts 9 points, or going from old to young adds 9 points. These two partial effects, 7 points and 9 points, are the additive building blocks of the relationships.

The building blocks fit the data reasonably well. For example, people with supportive homosexual values on both causal variables—high egalitarians who are 18–35 years old—average 58.0 on the scale, while people without supportive values on either causal variable—low egalitarians who are 56 or older—average 41.2 on the scale. Based on differences in egalitarianism alone, 18–35-year-old egalitarians will rate gays 7 points higher than will the older group. Based on age alone, the younger group will rate gays an additional 9 points higher. Thus, the younger egalitarians should rate homosexuals 7 + 9 = 16 points higher than older,

Figure 5-3 Mean Feeling Thermometer Ratings of Homosexuals, by Egalitarianism, Controlling for Age (line chart)

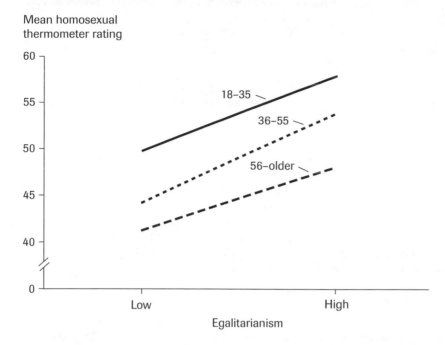

less egalitarian respondents. The data correspond pretty closely to this expectation: 58.0 − 41.2 = 16.8. Compare low egalitarian younger people (49.9) with high egalitarian older people (48.0). Based on the age comparison alone, the younger cohort will be 9 points higher than the older cohort. But make a 7-point downward adjustment to account for the lower egalitarianism of the younger cohort: 9 − 7 = 2. The observed difference is: 49.9 − 48.0 = 1.9.

Figure 5-3 provides a line chart of the egalitarianism–age–homosexual ratings relationships. As with the line charts discussed earlier, the values of the independent variable, egalitarianism, appear along the horizontal axis. Mean values of the dependent variable, mean feeling thermometer ratings, are recorded on the vertical axis. The three lines are graphic summaries of the relationship for each age group. Although the symmetry is imperfect—the line representing 36–55-year-olds rises a bit more steeply than the others—this configuration comes close to an additive pattern.

Another Example of Interaction

So that you can become more comfortable setting up and interpreting controlled comparisons, particularly those involving interaction relationships, we will work through one more example. We start with a hypothesis that links countries' electoral systems (independent variable) to the percentage of women in their legislatures (dependent variable): In a comparison of countries, countries that have proportional representation (PR) electoral systems will have higher percentages of women in their legislatures than will countries that do not have PR electoral systems. Proportional representation was briefly described in Chapter 3. As you may recall, PR is based on the idea that the size of a party's legislative delegation ought to reflect the size of its electoral support. Because it permits smaller parties to win legislative seats, PR fosters an electoral environment with diverse ideological and demographic choices.

Table 5-5 Mean Percentage of Women in Parliament, by Type of Electoral System, Controlling for the Cultural Acceptance of Women in Political Leadership

	Acceptance of women in political leadership		
Electoral system	Low	High	Total
Non-PR	12.0	16.9	13.2
	(32)	(11)	(43)
PR	17.0	28.3	20.4
	(24)	(10)	(34)
Total	14.1	22.3	16.4
	(56)	(21)	(77)

Sources: Democracy Crossnational Data (revised Spring 2008), Pippa Norris, John F. Kennedy School of Government, Harvard University. Data on the percentage of women in parliament are from 2005 and available from Inter-parliamentary Union, *Women in Parliament,* www.ipu.org. The variable that measures cultural acceptance of women in politics was compiled by Norris from the World Values Survey, 1995–2000 waves, www.worldvaluessurvey.org.

Some scholars argue that PR is quite effective in increasing the legislative presence of women.[10] Others point out that, quite apart from their electoral systems, countries differ in their cultural acceptance of women in positions of political leadership. Whereas the citizens of some countries believe that men make better political leaders than women do, other cultures reject this idea and consider women and men to be equal in their leadership abilities.[11] Alternatively, it may be that the cultural acceptance of women in leadership roles is a *precondition* for PR to have an effect. In cultures with a low level of acceptance of women, this precondition is absent. In these countries, the type of electoral system may make little difference in the percentage of women in parliament. In more accepting cultures, in which the precondition is met, the type of electoral system might have a big effect, channeling relatively more women into leadership positions in PR countries than in non-PR countries.[12] From a methodological standpoint, a precondition implies interaction. When the precondition is absent, the electoral system–women legislator relationship will be weak. When the precondition is present, the relationship will be strong.

Table 5-5 presents a mean comparison control table for the relationship between the percentage of women legislators and the type of electoral system, controlling for the cultural acceptance of women in leadership roles.[13] In keeping with the rules for setting up control tables, the 77 countries were divided into two groups on the basis of their level of cultural acceptance—the 56 "low" acceptance countries in the left-hand column and the 21 "high" acceptance countries in the right-hand column. By reading down each column, we evaluate the "electoral system effect," controlling for culture. By reading across each row, we evaluate the "culture effect," controlling for the type of electoral system. Another concise table is at hand.

Examine the numbers. Follow the interpretive steps: (1) Does the relationship between electoral system and the percentage of female parliamentarians persist within at least one category of the control? Yes. The relationship is not spurious. (2) Does the electoral system–women legislators relationship run in the same direction at both values of cultural acceptance? Yes. In both "low" and "high" countries, those with PR systems have higher

Figure 5-4 Relationship between Electoral System and the Percentage of Women in the Legislature, Controlling for the Cultural Acceptance of Women in Political Leadership (line chart)

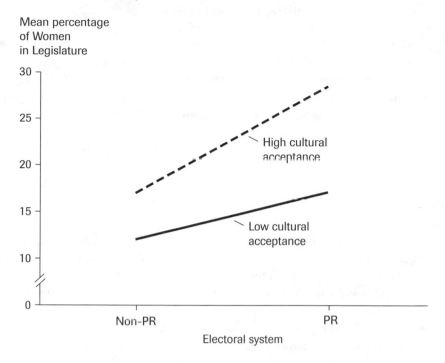

percentages of women legislators than those without PR. So once again we need the tie-breaker. (3) Is the strength of the relationship the same or very similar at both values of cultural acceptance? No. Among less accepting countries, the mean percentage of women legislators increases from 12.0 for those with non-PR systems to 17.0 for those with PR—a 5-point controlled effect. For culturally accepting countries, the mean percentage of women legislators increases from 16.9 percent for non-PR to 28.3 for PR—an 11.4-point controlled effect. As the precondition idea implies, the type of electoral system has a much larger effect when the culture is more accepting of women in politics than when the culture is less accepting. Figure 5-4 shows the interaction relationships quite clearly. Although both lines slope upward from left to right, imparting the positive effect of PR in both cultural settings, the effect is clearly stronger when the cultural precondition is present.

SUMMARY

In this chapter we explored the practice of making controlled comparisons in political research. Whereas an uncontrolled comparison reveals the zero-order relationship between two variables, a controlled comparison reveals the controlled effects of each causal variable on the dependent variable, holding constant another causal variable. Controlled effects are summarized by a partial relationship or partial effect. In an additive relationship, the independent variable's controlled effects can be faithfully summarized by a one-number partial effect. Also, holding the independent variable constant, the relationships between the control variable and the dependent variable can be described with a single partial effect. These partial effects become additive building blocks for estimating values of the dependent variable for cases having different combinations of values on the independent variable and the control variable. In interaction relationships, the controlled effects of the independent variable on the dependent variable cannot be accurately summarized by a single partial effect and must be described separately for each value of the control variable.

This chapter introduced a three-step heuristic for helping you identify the pattern—spuriousness, additive relationships, or interaction—that best describes a set of relationships. First, ask whether the relationship between the independent variable and the dependent variable persists when a rival cause is controlled. If the relationship survives, then you can rule out spuriousness. Second, ask whether the relationship has the same tendency at all values of the control variable. If the relationship runs in different directions for different values of the control, then interaction is at work. If the tendencies are the same, then ask a third question—whether the strength of the relationship between the independent variable and dependent variable is similar at all values of the control variable. If the relationships are similar in strength, then additive is the best description. If the relationships clearly differ in strength, then interaction is a better characterization.

We presented four examples of empirical analysis, two of which were additive patterns and two of which were patterns of interaction. These examples, of course, were chosen for their clarity. Even so, many times you will find a similar clarity in your own analyses. After setting up a control table and constructing a line chart of the relationships, you will find that your results come pretty close to one of the scenarios. Yet you often will need additional assistance in figuring out what is going on in the data. The distinction between additive and interaction relationships, in particular, can be a tough call to make. Just how far off parallel do the graphic lines need to be before we abandon an interpretation of "additive" in favor of "interaction"? It is here that statistical inference enters the picture. Inferential statistics is a key resource for the political researcher and an important interpretive tool for evaluating relationships between variables. In the chapters that follow, you will come to appreciate the central importance of statistical inference in political analysis.

KEY TERMS

controlled comparison table (p. 95)
controlled effect (p. 96)
partial effect (p. 97)

partial relationship (p. 97)
rule of direction for nominal relationships (p. 97)
zero-order relationship (p. 95)

EXERCISES

1. Authoritarianism, the extent to which people scorn unconventional behaviors and obediently defer to strong leaders, is a venerable concept in social psychology. It stands to reason that authoritarianism and opinions about pornography might be causally linked: In a comparison of individuals, people having stronger authoritarian beliefs will be more likely to think that pornography should be outlawed than will people having weaker authoritarian beliefs. Yet we also know that authoritarians tend to have lower levels of education than do non-authoritarians. Education is related to pornography opinions too, in that more educated people have more permissive attitudes than do the less educated.

   A. The control table below will allow you to analyze the authoritarianism–pornography opinions relationship, controlling for level of education.[14] The table displays raw numbers only.

For example, in the low-education group, 104 individuals have low authoritarianism. Of these, 34 said pornography should be illegal to all, and 70 said that pornography should not be illegal to all. Construct a complete control table from this information.

Pornography opinion	Education							
	Low				High			
	Authoritarianism				Authoritarianism			
	Low	Middle	High	Total	Low	Middle	High	Total
Illegal to all	34	61	105	200	60	63	53	176
Not illegal to all	70	70	86	226	190	122	69	381
Total	104	131	191	426	250	185	122	557

B. Decide which pattern—spuriousness, additive, or interaction—best describes the set of relationships. (i) Write out each of the rhetorical questions, introduced in this chapter, that help you determine the pattern. (ii) Write a complete sentence answering each question.

C. Draw a line chart of the authoritarianism–pornography opinions relationship, controlling for education. The values of authoritarianism will appear along the horizontal axis. The vertical axis will record the percentages of respondents saying "illegal to all." Enhance the readability of the chart by making 20 percent the lowest vertical axis value and 60 percent the highest vertical axis value.

2. In one of the examples in this chapter, we looked at the effect of electoral systems on the percentage of women in national legislatures, controlling for the level of cultural acceptance of women in leadership positions. In this exercise you will test a hypothesis that electoral systems affect turnout: In a comparison of countries, countries with PR systems will have higher turnouts than will countries that do not have PR systems. The literacy rate is an important control variable, because literacy is known to be an alternative cause of voter participation. Therefore, evaluate the electoral system–turnout relationship, controlling for literacy rate.[15]

Among low literacy countries, the mean levels of turnout are as follows: non-PR countries, 59.4 percent; PR countries, 71.6 percent. Among high literacy countries: non-PR countries, 68.8 percent; PR countries, 74.9 percent. Among all countries: low literacy countries, 66.8 percent; high literacy countries, 72.0 percent. Among all countries: non-PR countries, 64.7; PR countries, 73.2 percent.

A. Construct a mean comparison control table from the information provided.

B. Decide which pattern—spuriousness, additive, or interaction—best describes the set of relationships. (i) Write out each of the rhetorical questions, introduced in this chapter, that help you determine the pattern. (ii) Write a complete sentence answering each question.

C. Draw a line chart of the electoral system–turnout relationship, controlling for literacy rate. Enhance the readability of the chart by making 55 percent the lowest vertical axis value and 75 percent the highest vertical axis value.

D. Suppose someone made the following statement: "The results show that high literacy is a precondition for the effect of electoral systems. Instituting PR in low-literacy countries will have, at best, modest effects on turnout." (i) Based on your analysis, is this statement correct or incorrect? (ii) Explain how you know.

3. In one of the exercises in Chapter 3 you were asked to consider a study showing that back belts, widely used in industry to prevent injury, do not work. For present purposes, assume that the researchers gathered data on a large number of individual workers. Each worker was measured on a two-category independent variable, labeled "Back belt use." The independent variable's values are: Use belt/Do not use belt. Each worker also was measured on a dependent variable, labeled "Back injury reported." The dependent variable's values are: Reported injury/Did not report injury.[16]

   A. Draw an empty cross-tabulation shell, putting the values of the independent variable on the columns and the values of the dependent variable on the rows. Inside the cross-tabulation, write in fabricated percentages showing that back belt use is not related to back injuries. (Just fabricate percentages. You do not need to fabricate raw cell frequencies.)

   B. According to a report by the Associated Press (December 5, 2000), "The [back belt study's] findings were questioned by a spokesman for the International Mass Retail Association, an industry group whose members include 200 retail chains. The researchers did not directly compare workers doing the same jobs, the spokesman said." The spokesman is suggesting that the original research is flawed because it did not control for the different types of jobs that workers perform. Describe the values of a plausible two-category control variable, labeled "Job type," that measures the types of jobs that workers perform.

   C. The spokesman's claim presents a challenging methodological problem. He is saying that the zero-order relationship, which shows no relationship between back belt use and back injuries, masks a true causal relationship between back belt use and back injuries: Workers who use belts are less likely to report back injuries than workers who do not use belts. The spokesman claims that after controlling for job type, this causal relationship will become evident. (i) Is the spokesman saying that, controlling for job type, the back belt–back injury relationship is spurious? Or is he saying that the back belt–job type–back injury relationships are additive? Or is he saying that interaction is occurring in the back belt–job type–back injury relationships? (ii) Explain how it is possible for the zero-order relationship to show no relationship between belt use and back injuries while at least one of the controlled relationships shows that workers who use belts are less likely to be injured than are workers who do not use back belts.

   D. Draw two cross-tabulation shells, one for each value of job type. Just as you did in part A, put the independent variable on the columns and dependent variable on the rows. Inside the cross-tabulations, write in fabricated percentages that are consistent with your answers in part C.

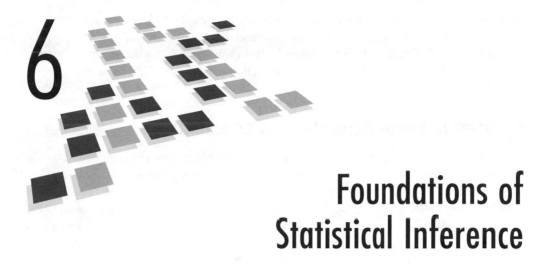

Foundations of Statistical Inference

LEARNING OBJECTIVES

In this chapter you will learn:
- Why random sampling is of cardinal importance in political research
- Why samples that seem small can yield accurate information about much larger groups
- How to figure out the margin of error for the information in a sample
- How to use the normal curve to make inferences about the information in a sample

By this point in the book, you have become comfortable with the essential techniques of political analysis. You know how to think clearly and critically about concepts. You can measure variables, construct explanations, and set up cross-tabulations and mean comparisons. You can interpret complex relationships. As we saw in Chapter 5, however, real-world relationships can present interpretive challenges. For example, suppose that in one of our analyses of the American National Election Study we find that men give the Republican Party an average thermometer rating of 55, compared with a mean rating of 52 among women. Is this 3-point difference "big enough" to support the conclusion that males have higher regard for the Republican Party than do females? Or should we instead decide that the difference is "too small" to warrant that conclusion? Suppose we are investigating the electoral mobilization of military veterans. One of our cross-tabulation analyses shows that 84 percent of veterans reported voting in the presidential election, compared with 77 percent of nonveterans. Does a 7 percentage-point difference allow us to say that veterans are more likely to vote than are nonveterans, or is the difference too fragile to support this implication?

Inferential statistics was invented to help the investigator make the correct interpretations about empirical relationships. **Inferential statistics** refers to a set of procedures for deciding how closely a relationship we observe in a sample corresponds to the unobserved relationship in the population from which the sample was drawn. Inferential statistics can help us decide whether the 3-point feeling thermometer difference between men and women represents a real gender difference in the population or whether the difference occurred by happenstance when the sample was taken. Inferential statistics will tell us how often a random sample will produce a 7 percentage-point difference between veterans and nonveterans if, in fact, no difference exists in the population. In this chapter we cover the essential foundations of inferential statistics. In Chapter 7 we apply these foundational skills to the analysis of empirical relationships.

POPULATION PARAMETERS AND SAMPLE STATISTICS

Anyone who is interested in politics, society, or the economy wants to understand the attitudes, beliefs, or behavior of very large groups. These large aggregations of units are populations. A **population** may be generically defined as the universe of cases the researcher wants to describe. If I were studying the financial activity of political action committees (PACs) in the most recent congressional election, for example, my population would include all PAC contributions in the most recent election. Students analyzing vote choice in the most recent congressional elections, by contrast, would define their population as all voting-age adults. A characteristic of a population—the dollar amount of the average PAC contribution or the percentage of voting age adults who voted—is called a **population parameter.** Figuring out a population's characteristics, its parameters, is a main goal of the social science investigator. Researchers who enjoy complete access to their populations of interest—they can observe and measure every PAC, eligible voter, every member of Congress, Supreme Court decision, or whatever—are working with a **census.** A census allows the researcher to obtain measurements for all members of a population. Thus, the researcher does not need to infer or estimate any population parameters when describing the cases.[1]

More often, however, researchers are unable to examine a population directly and must rely, instead, on a sample. A **sample** is a number of cases or observations drawn from a population. Samples, like death and taxes, are fixtures of life in social research. Because population characteristics are frequently hidden from direct view, we turn to samples, which yield observable sample statistics. A **sample statistic** is an estimate of a population parameter, based on a sample drawn from the population. Public opinion polls, for example, never survey every person in the population of interest (for example, all voting-age adults). The pollster takes a sample, elicits an opinion, and then infers or estimates a population characteristic from this sample statistic. Sometimes such samples—samples of 1,000 to 1,500 are typical—seem too small to faithfully represent their population parameters. Just how accurately does a sample statistic estimate a population parameter? The answer to this question lies at the heart of inferential statistics.

In the sections that follow we will discuss three factors that determine how closely a sample statistic reflects a population parameter. The first two factors have to do with the sample itself: the procedure that we use to choose the sample and the sample's size (the number of cases in the sample). The third factor has to do with the population parameter we want to estimate: the amount of variation in the population characteristic. First we turn to a discussion of the nature and central importance of random sampling. We then consider how a sample statistic, computed from a random sample, is affected by the size of the sample and the amount of variation in the population. Finally, we show how the normal distribution comes into play in helping researchers determine the margin of error of a sample estimate and how this information is used for making inferences.

RANDOM SAMPLING

The procedure we use in picking the sample is of cardinal importance. For a sample statistic to yield an accurate estimate of a population parameter, the researcher must use a **random sample,** that is, a sample that has been randomly drawn from the population. In taking a random sample, the researcher ensures that every member of the population has an equal chance of being chosen for the sample. To appreciate the importance of random sampling, consider a well-known sample that was taken during the 1936 presidential election campaign. Then-president Franklin Roosevelt, a Democrat whose policies were widely viewed as benefiting the lower and working classes, was seeking reelection against Republican candidate Alf Landon, who represented policies more to the liking of higher-income individuals and business interests. In a well-intentioned effort to predict the outcome (and boost circulation), the magazine *Literary Digest* conducted perhaps the largest poll ever undertaken in the history of electoral politics. Using lists of names and addresses obtained from phone records, automobile registrations, and the ranks of its own subscribers, the *Digest* mailed out a staggering 10 million sample ballots, over 2.4 million of which were filled out and returned. Basing its inferences on responses from this enormous sample, the *Digest* predicted a Landon landslide, estimating that 57 percent of the two-party vote would go to Landon and 43 percent to Roosevelt. The election, indeed, produced a landslide—but not for Landon. Roosevelt ended up with more than 60 percent of the vote. (And the *Literary Digest* ended up going out of business.)

What went wrong? In what ways did the magazine's sampling procedure doom its predictions? As you have no doubt surmised, people who owned cars and had telephones (and could afford magazine subscriptions) during the Great Depression may have been representative of Landon supporters, but they decidedly were not a valid reflection of the electorate at large. Certainly, the *Digest* wanted to make a valid inference about the population of likely voters. But it used the wrong **sampling frame,** the wrong method for defining the population it wanted to study. Poor sampling frames lead directly to selection bias, or sampling bias. **Selection bias** occurs when some members of the population are more likely to be included in the sample than are other members of the population. Because people without telephones or cars were systematically excluded from the sample, selection bias was at work. The poll also suffered from **response bias,** which occurs when some cases in the sample are more likely than others to be measured. Because only a portion of the *Digest*'s sample returned their ballots, response bias was at work. People who are sufficiently motivated to fill out and return a sample ballot—or any sort of voluntary-response questionnaire—may hold opinions that are systematically different from the opinions of people who receive a ballot but fail to return it.[2] Samples drawn in this manner are guaranteed to produce sample statistics that are meaningless. Garbage in. Garbage out.

Fortunately, thanks in part to lessons learned from legendary mistakes like the *Literary Digest* poll, social science has figured out how to construct sampling frames that virtually eliminate selection bias and has devised sampling procedures that minimize response bias. A valid sample is based on **random selection.** Random selection occurs when every member of the population has an equal chance of being included in the sample. So, if there are 1,000 members of the population, then the probability that any one member would be chosen is 1 out of 1,000. Thus, the *Literary Digest* should have defined the population they wanted to make inferences about—the entire voting-age population in 1936—and then taken a random sample from this population. By using random selection, every eligible voter, not just those who owned cars or had telephones, would have had an equal chance of being included. But the *Digest*, probably believing that a huge sample size would do the trick, ignored the essential principle of random selection: If a sample is not randomly selected, then the size of the sample simply does not matter.

Let's explore these points, using a plausible example. Suppose that a student organization wants to gauge a variety of student political opinions: how students rate the political parties and the institutions of government, whether they have ever volunteered in a political campaign, their ideological leanings, and so on. As a practical matter, the student researchers cannot survey all 20,000 students enrolled at the university, so they decide to take a sample of 100 students. How might the student pollsters obtain a sample that avoids the infamous pitfalls of the *Literary Digest* poll? The group would first define the sampling frame by assigning a unique sequential number to each student in the population, from 00001 for the first student listed in administration records to 20000 for the last listed. No problem so far. But how do the pollsters guarantee that each student has exactly one chance in 20,000 of being sampled? A systematic approach, such as picking every two-hundredth student, would result in the desired sample size (since 20,000/200 = 100), but it would not produce a truly random sample. Why not? Because two students appearing next to each other in the sampling frame would not have an equal chance of being selected.

To obtain a random sample, the researchers would need a list of five-digit random numbers, created by many computer programs. A random number has a certain chaotic beauty. The first digit is randomly generated from the numbers 0–9. The second is randomly generated from 0–9 as well, and so its value is not connected in any way to the first digit. The third digit is completely independent of the first two, and so on, for each of the five digits. Since there is no rhyme or reason to these numbers, the pollsters can begin anywhere on the list, adding to their sample the student having the same number as the first random number, using the second random number to identify the second student, and continuing until a sample of 100 students is reached. (Any random number higher than 20,000 can be safely skipped, since the list has no systematic pattern.) Variants of this basic procedure are used regularly by commercial polling firms, like Gallup, and academically oriented survey centers, such as the University of Michigan's Institute for Social Research.[3]

The essential methodological goodness of random processes has been previously discussed. In Chapter 1 we saw that random error introduces haphazard static into the measurement process. To be sure, random measurement error is not a welcome sight, but it is a mere annoyance compared with the fundamental distortion introduced by systematic measurement error. In Chapter 4 we found that random assignment is the great neutralizer of selection bias in experimental research design. Random assignment ensures that the test group and the control group will not be systematically different in any way, known or unknown, that could affect the dependent variable. If human choice is allowed to enter the assignment process—the investigators choose one subject over another for the test group or a prospective participant chooses the control group instead of the test—then selection bias is onboard. The rationale for random sampling in observational research is identical to its rationale in experimental design. Because the population is beyond empirical view, we take a random sample, which ensures that each population member has an equal chance of being included. Just as random assignment in experimental research eliminates biased differences between the test group and the control group, so does random sampling eliminate biased differences between the population and the sample.

It is important to point out, however, that in eliminating bias we do not eliminate error. In fact, in drawing a random sample, we are consciously introducing **random sampling error.** Random sampling error is defined as the extent to which a sample statistic differs, *by chance*, from a population parameter. Trading one kind of error for another may seem like a bad bargain, but random sampling error is vastly better because we know how it affects a sample statistic, and we fully understand how to estimate its magnitude. Assuming that we are working with a random sample, the population parameter will be equal to the statistic we obtain from the sample, plus any random error that was introduced by taking the sample:

Population parameter = Sample statistic + Random sampling error.

The student researchers want a sample statistic that provides an unbiased estimate of a true population parameter, a characteristic of all students at the university. They eliminate selection bias by taking a random sample. But they know that random sampling error is affecting their estimate of the population parameter. Assume that the student researchers use a feeling thermometer scale to measure the sample's attitudes toward the Democratic Party. Having collected this information on each member of the sample, they calculate the mean rating of the Democratic Party. Because they are working with a random sample, the student pollsters know that the sample's mean Democratic rating is the same as the population's mean Democratic rating, plus the random error introduced by taking the sample. What makes random sampling error a "better" kind of error is that we have the statistical tools for figuring out how much a sample statistic is affected by random sampling error.

The magnitude of random sampling error depends on two components: (1) the size of the sample and (2) the amount of variation in the population characteristic being measured. Sample size has an inverse relationship with random sampling error: As the sample size goes up, random sampling error goes down. Variation in the population characteristic has a direct relationship with random sampling error: As variation goes up, random sampling error goes up. These two components—the variation component and the sample size component—are not separate and independent. Rather, they work together, in a partnership of sorts, in determining the size of random sampling error. This partnership can be defined by using ideas and terminology that we have already discussed:

Random sampling error = (Variation component) / (Sample size component).

Before exploring the exact properties of this conceptual formula for random sampling error, consider its intuitive appeal. Notice that "Variation component" is the numerator. This reflects its direct relationship with random sampling error. "Sample size component" is the denominator, depicting its inverse relationship with random sampling error. Return to the student organization example and consider an illustration of how these two components work together. Suppose that in the population of 20,000 students there is a great deal of variation in ratings of the Democratic Party. Large numbers of students dislike the Democrats and give them ratings between 0 and 40. Many students like the Democrats and give them ratings between 60 and 100. Still others give ratings in the middle range, between 40 and 60. So the population parameter the student researchers wish to estimate, Democratic Party thermometer ratings, would have a large variation component. Suppose further that the campus group is working with a small-sized random sample. Thus, the variation component is relatively large and the sample size component is relatively small. Dividing the large variation component by the small sample size component would yield a large amount of random sampling error. Under these circumstances, the organization could not be very confident that their sample statistic provides an accurate picture of the true population mean, because their estimate contains so much random sampling error. But notice that if the campus group were to take a larger sample, or if student ratings of the Democratic Party were not so spread out, random sampling error would diminish, and the student pollsters would gain confidence in their sample statistic.

Both components, the variation component and the sample size component, have known properties that give the researcher a good idea of just how much random sampling error is contained in a sample statistic.

Sample Size and Random Sampling Error

As previously noted, the basic effect of sample size on random sampling error is: As the sample size increases, error decreases. Adopting conventional notation—in which sample size is denoted by a lowercase *n*—we would have to say that a sample of $n = 400$ is preferable to a sample of $n = 100$, since the larger sample would provide a more accurate picture of what we are after. However, the inverse relationship between sample size and sampling error is non-linear. Even though the larger sample is four times the size of the smaller one, going from $n = 100$ to $n = 400$ delivers only a twofold reduction in random sampling error. In ordinary language, if you wish to cut random error in half, you must quadruple the sample size. In mathematical language, the sample size component of random sampling error is equal to the square root of the sample size, *n*:

$$\text{Sample size component of random sampling error} = \sqrt{n}.$$

Plugging this into our conceptual formula for random sampling error:

$$\text{Random sampling error} = (\text{Variation component}) / \sqrt{n}.$$

Because of the nonlinear relationship between sample size and random sampling error, samples that seem rather small nonetheless carry an acceptable amount of random error. Consider three samples: $n = 400$, $n = 1,600$, and $n = 2,500$. The sample size component of the smallest sample size is the square root of 400, which is equal to 20. So, for a sample of this size, we would calculate random sampling error by dividing the variation component by 20. Random sampling error for the next sample would be the variation component divided by the square root of 1,600, which is equal to 40. So, by going from a sample size of 400 to a sample size of 1,600, we can increase the sample size component of random sampling error from 20 to 40. Notice that by increasing the sample size component from 20 to 40, we double the denominator, $\sqrt{n}$. This has a beneficial effect on random sampling error, effectively cutting it by half. Thus, if resources permit, obtaining a sample of $n = 1,600$ would be a smart move. Random sampling error for the largest sample would be equal to the variation component divided by the square root of 2,500, which is equal to 50. Boosting the sample size by 900 cases—from 1,600 to 2,500—occasions a modest increase in the sample size component, from 40 to 50. Sophisticated sampling is an expensive undertaking, and survey designers must balance the cost of drawing larger samples against the payoff in precision. For this reason, most of the surveys you see and read about have sample sizes in the 1,500 to 2,000 range, an acceptable comfort range for estimating a population parameter.

Sample size is an important factor in the accuracy of a sample statistic. You now have a better idea of how random sampling error is affected by n. Suppose the campus organization successfully collects its sample ($n = 100$) and computes a sample statistic, mean Democratic rating. Let's say that the sample rates the Democrats at 59, on average. The group wants to know how much random sampling error is contained in this estimate. As we have just seen, part of the sampling error will depend on the sample size. In this case, the sample size component is equal to $\sqrt{100} = 10$. Now what? What does a sample size error component of 10 have to do with the accuracy of the sample mean of 59, the campus group's estimate of the true rating of the Democratic Party in the student population? The answer depends on the second component of random sampling error, the amount of variation in the population characteristic being measured. As we have seen, this connection is direct: As variation in the population characteristic goes up, random sampling error goes up.

To better appreciate how variation in the population parameter affects random sampling error, consider Figure 6-1, which depicts two possible ways that student ratings of the Democrats might be distributed within the student population. First, suppose that Democratic Party ratings are widely dispersed across the student population, as in Panel A of Figure 6-1. There are appreciable numbers of students in every range of the rating scale, from lower to higher, with only a slight amount of clustering around the center of the distribution. Since variation in the population characteristic is high, the variation component of random sampling error is high. A random sample taken from the population would produce a sample mean that may or may not be close to the population mean—it all depends on which cases were randomly selected. Because each student has an equal chance of being chosen for the sample, one sample might pick up a few more students who reside in the upper range of the distribution. Another sample from the same population may randomly choose a few more students from the lower range. In fact, one might draw a very large number of random samples, each one producing a different sample estimate of the population mean. Now, visualize a population like the one depicted in Panel B of Figure 6-1. Notice that the ratings are clustered around a well-defined center, with fewer cases at the extremes of the scale. Since variation in the population characteristic is low, the variation component of random sampling error is low. A random sample taken from the population would produce a sample mean that is close to the population mean. What is more, repeated sampling from the same population would produce sample mean after sample mean that are close to the population mean—and close to each other.

The variation component of random sampling error is statistically defined by a measure you may have encountered before: the standard deviation. After discussing the standard deviation, we return to the question of how this foundational measure affects random sampling error.

Variation Revisited: The Standard Deviation

The amount of variation in a variable is determined by the dispersion of cases across the values of the variable. If the cases tend to fall into one value of a variable, or into a handful of similar values, then the variable has low dispersion. If the cases are more spread out across the variable's values, then the variable has high dispersion. As discussed in Chapter 2, describing the degree of dispersion in nominal and ordinal variables sometimes requires a judgment call.

For interval-level variables, a more precise measure of variation is used. The **standard deviation** summarizes the extent to which the cases in an interval-level distribution fall on or close to the mean of the distribution. Although more precise, the standard deviation is based

Figure 6-1 High Variation and Low Variation in a Population Parameter

A. High Variation in Democratic Thermometer Ratings in the Student Population

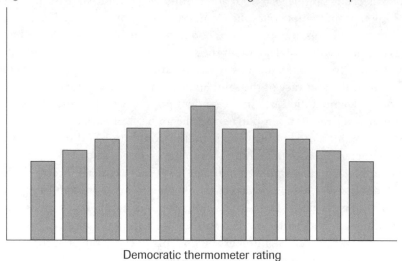

Democratic thermometer rating

B. Low Variation in Democratic Thermometer Ratings in the Student Population

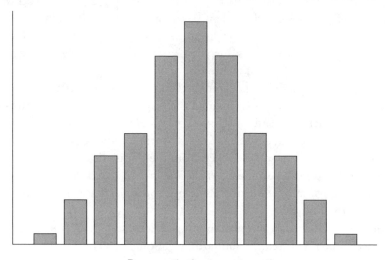

Democratic thermometer rating

on the same intuition as the less precise judgment calls applied to nominal and ordinal variables. If, on the whole, the individual cases in the distribution do not deviate very much from the distribution's mean, then the standard deviation is a small number. If, by contrast, the individual cases tend to deviate a great deal from the mean—that is, large differences exist between the values of individual cases and the mean of the distribution—then the standard deviation is a large number.

To demonstrate the central importance of the standard deviation in determining random sampling error, we will present two hypothetical possibilities for the distribution of Democratic thermometer ratings in the student population. The population means are the same

Table 6-1 Central Tendency and Variation in Democratic Thermometer Ratings: Hypothetical Scenario A

Student	Democratic rating	Deviation from the mean	Squared deviation from the mean
1	20	−38	1,444
2	22	−36	1,296
3	34	−24	576
4	50	−8	64
5	56	−2	4
6	58	0	0
7	60	2	4
8	66	8	64
9	82	24	576
10	94	36	1,296
11	96	38	1,444

Summary information

Central tendency	*Dispersion*
Summation of ratings = 638	Summation of squared deviations = 6,768
	Average of squared deviations (variance) = 615.3
$N = 11$	
$\mu = 58$	$\sigma = 24.8$

in both scenarios, a mean rating of 58. However, in population A student ratings are more spread out—the distribution has a higher standard deviation—than in population B. In discussing population A, we will provide a step-by-step guide for arriving at the standard deviation. Populations A and B are unrealistic in two respects. First, both show calculations that have been performed on a population. Researchers perform calculations on samples, not populations. But here we bend reality so that we can introduce appropriate terminology and lay necessary groundwork. Second, both scenarios depict the student population as having 11 members, not the more realistic 20,000 students we have been using as an example. This is done to make the math easier to follow. After these simplifications have served their purposes, we will restore plausibility to the populations.

Table 6-1 presents student population A. As discussed earlier, a sample size is denoted by a lowercase *n*. In contrast, a population size is denoted by an uppercase *N*. In Table 6-1, then, $N = 11$. The thermometer ratings given by each member of the population are in the "Democratic rating" column, from 20 for the student with the coolest response to the Democrats to 96 for the most pro-Democratic student. To arrive at the mean rating for the population, we divide the summation of all ratings (20 + 22 + 34 + . . .), 638, by the population size, 11, which yields 58. Unlike sample statistics, which (as we will see) are represented by ordinary letters, population parameters are always symbolized by Greek letters. A population mean is symbolized by the Greek letter μ (pronounced "mew"). Thus, in Table 6-1, $\mu = 58$. This is a familiar measure of central tendency for interval-level variables.

How might we summarize variation in ratings among the students in this population? A rough-and-ready measure is provided by the **range,** defined as the maximum actual value minus the minimum actual value. So, in this example, the range would be the highest rating, 96, minus the lowest rating, 20, a range of 76. In gauging variation in interval-level variables, however, the measure of choice is the standard deviation. The standard deviation of a population is symbolized by the Greek letter σ ("sigma"). As its name implies, the standard deviation measures variation as a function of deviations from the mean of a distribution. The first step in finding the standard deviation, then, is to express each value as a deviation from the mean or, more specifically, to subtract the mean from each value.

Step 1. Calculate each value's deviation from the mean:

$$(\text{Individual value} - \mu) = \text{Deviation from the mean.}$$

A student whose rating is below the population mean will have a negative deviation, and a student who gave a rating that is above the mean will have a positive deviation. An individual with a rating equal to the population mean will have a deviation of 0. In Table 6-1, the deviations for each member of the population are shown in the column labeled "Deviation from the mean." These deviations tell us the locations of each population member relative to the population mean. So, for example, Student 1, who rated the Democrats at 20 on the scale, has a deviation of −38, 38 units below the population mean of 58. Student 7, who rated the Democratic Party at 60, is slightly above the population mean, scoring the Democrats 2 points higher than the population mean of 58. Deviations from the population mean provide the starting point for figuring out the standard deviation.

Step 2. Square each deviation. All measures of variation in interval-level variables, including the standard deviation, are based on the square of the deviations from the mean of the distribution. In Table 6-1, these calculations for each student in the population appear in the column labeled "Squared deviation from the mean." Squaring each individual deviation, of course, removes the minus signs on the negative deviations, those members of the population who gave ratings below the population mean. Notice, for example, that the square of Student 1's deviation, −38, is the same as the square of Students 11's deviation, 38. Both square to 1,444. Why perform a calculation that treats Student 1 and Student 11 as equal, when they clearly are not equal at all? Because, in the logic of the standard deviation, both of these students make equal contributions to the *variation* in ratings. Both lie an equal distance from the population mean of 58, and so both deviations figure equally in determining the dispersion of ratings around the mean.

Step 3. Sum the squared deviations. If we add all the squared deviations in the "Squared deviation from the mean" column, we arrive at the sum 6,768. The summation of the squared deviations, often called the *total sum of squares,* can be thought of as an overall summary of the variation in a distribution. When calculated on real-world data with many units of analysis, the total sum of squares is always a large and seemingly meaningless number. However, the summation of the squared deviations becomes important in its own right when we discuss correlation and regression analysis (see Chapter 8).

Step 4. Calculate the average of the sum of the squared deviations. The average of the squared deviations is known by a statistical name, the **variance.** The population variance is equal to the sum of the squared deviations divided by *N*. (*Special note:* To calculate the variance for a sample, you would divide the sum of the squared deviations by $n - 1$. This is discussed below.) For the population depicted in Table 6-1, the variance is the summation of

Table 6-2 Central Tendency and Variation in Democratic Thermometer Ratings: Hypothetical Scenario B

Student	Democratic rating	Deviation from the mean	Squared deviation from the mean
1	25	−33	1,089
2	34	−24	576
3	50	−8	64
4	55	−3	9
5	56	−2	4
6	58	0	0
7	60	2	4
8	61	3	9
9	66	8	64
10	82	24	576
11	91	33	1,089

Summary information

	Central tendency		*Dispersion*
	Summation of ratings = 638		Summation of squared deviations = 3,484
			Average of squared deviations (variance) = 316.7
	$N = 11$		
	$\mu = 58$		$\sigma = 17.8$

the squared deviations (6,768) divided by the population size ($N = 11$), which yields an average of 615.3. Notice that, as with any mean, the variance is sensitive to values that lie far away from the mean. Students toward the tails of the distribution—Students 1 and 2 on the low end and Students 10 and 11 on the high end—make greater contributions to the variance than students who gave ratings that were closer to the population mean. That's the beauty of the variance. If a population's values cluster close to the mean, then the average of the squared deviations will record the closer clustering. As deviations from the mean increase, then the variance increases, too.

Step 5. Take the square root of the variance. The population parameter of current concern, the standard deviation, is based on the variance. In fact, the standard deviation is the square root of the variance. The standard deviation (σ) for the population of students in scenario A, then, is the square root of 615.3, or $\sqrt{615.3} = 24.8$.

Turn your attention to Table 6-2, which depicts a second possibility for the distribution of Democratic ratings in the student population. The mean Democratic rating for population B is the same as population A, $\mu = 58$, but the scores are not as spread out. Evidence of this lower dispersion can be found in every column of Table 6-2. Notice that the range is equal to $91 − 25 = 66$ (compared with 76 for population A) and that there are fewer double-digit deviations from the population mean. Most noticeable are the lower magnitudes of the squared deviations, which sum to 3,484 (compared with 6,678 for population A), and the variance, which is equal to 316.7, substantially less than the value we calculated for the more dispersed population (615.3). Taking the square root of the variance, we arrive at a σ equal to 17.8, which is 7 points lower than the standard deviation of population A ($\sigma = 24.8$). As we will demonstrate, a statistic computed on a random sample from population A will have a higher amount of random sampling error than will a statistic computed on a random sample drawn from population B.

n *and* σ

Let's pause and review the statistical components discussed thus far.

Sample size component: As the sample size goes up, random sampling error declines as a function of the square root of the sample size.

Variation component: As variation goes up, random sampling error increases in direct relation to the population's standard deviation.

Now, we will take a firsthand look at how these components work together. Again consider population A and population B—only this time think of them in a much more realistic light. Instead of a mere 11 members, each population now has 20,000 students. Just as its counterpart in Table 6-1, the distribution of the 20,000 students in population A has a mean equal to 58 and a standard deviation equal to 24.8. And, just as in Table 6-2, the distribution of the 20,000 students in population B has a mean equal to 58 and a standard deviation equal to 17.8. Having artificially created these realistic populations, we can ask the computer to draw random samples of different sizes from each population.[4] We can then calculate and record the mean Democratic rating obtained from each sample.

The results are presented in Figure 6-2. All the sample means displayed in panel A are based on the same student population—a population in which $\mu = 58$ and $\sigma = 24.8$. All the sample means displayed in panel B were drawn from a student population in which $\mu = 58$ and $\sigma = 17.8$. The dashed horizontal line in each panel shows the location of the true population mean, the parameter being estimated by the sample means. For each population, the computer drew ten random samples of $n = 25$, ten random samples of $n = 100$, and ten random samples of $n = 400$. So, by scanning from left to right within each panel, you can see the effect of sample size on random sampling error. By moving between panel A and panel B, you can see the effect of the standard deviation on random sampling error. (So that we don't lose track of our student researchers, their sample's mean of 59 appears as a solid dot in the $n = 100$ group in panel A. We return to this example below.)

Consider the set of sample means with the largest error component, the samples of $n = 25$ in panel A. Even though three or four of these sample means come fairly close to the population mean of 58, most are wide of the mark, ranging in value from the chilly (a mean Democratic rating of 50) to the balmy (a mean rating of 65). A small sample size, combined with a dispersed population parameter, equals a lot of random error. As we move across panel A to the ten sample means based on $n = 100$, we get a tighter grouping and less wildness, but even here the means range from about 53 to 62. The samples of $n = 400$ return much better precision. Four of the ten sample means hit the population mean almost exactly. Plainly enough, as sample size increases, error declines. By comparing panel A with panel B, we can see the effect of the population standard deviation on random sampling error. For example, notice that the ten samples of $n = 25$ in panel B generate sample statistics that are about as accurate as those produced by the samples of $n = 100$ in panel A. When less dispersion exists in the population parameter, a smaller sample can sometimes yield relatively accurate statistics. Naturally, just as in panel A, increases in sample size bring the true population mean into

Figure 6-2   Sample Means from Population with $\mu = 58$ and $\sigma = 24.8$ (Panel A) and $\sigma = 17.8$ (Panel B)

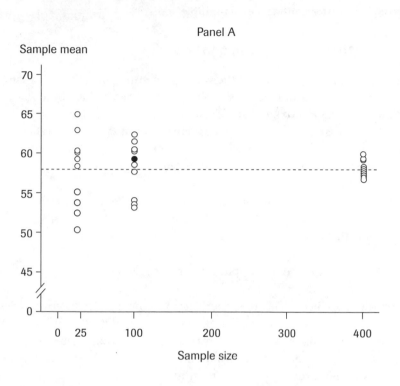

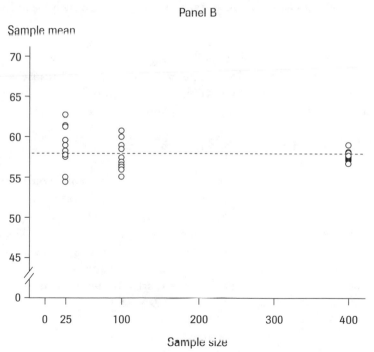

Note: Hypothetical data. Hypothetical student group's sample mean is represented by the solid dot in the $n = 100$ group in panel A. Dashed horizontal line shows location of true population mean ($\mu = 58$).

clearer focus. At $n = 400$ in panel B, six of the ten sample means are within a few tenths of a point of the true population mean. A larger sample, combined with lower dispersion, equals less random error and greater confidence in a sample statistic.

THE STANDARD ERROR OF A SAMPLE MEAN

We can now supply the missing ingredient, the variation component of random sampling error. The variation component of random sampling error is determined by a now familiar measure, the standard deviation (σ). For any given sample size of n, as the population standard deviation increases, random sampling error increases (and vice versa). When we combine this principle with the sample size error component, we can again represent the partnership between the two components of random sampling error:

Random sampling error = Standard deviation / Square root of the sample size.

Or, using symbols that have been discussed:

Random sampling error = $\sigma / \sqrt{n}$.

In this chapter we have been using the generic term *random sampling error* to describe the error introduced when a random sample is drawn. The size of this error, as we have just seen, is determined by dividing the variation component by the sample size component. However, when researchers are describing the random sampling error associated with a sample statistic, they do not ordinarily use the term *random sampling error*. Rather, they refer to the **standard error** of the mean. Computer analysis programs routinely calculate standard errors for mean estimates, and political researchers always report the standard errors for the sample estimates they publish in quantitative research articles. Let's be clear. The terms *standard error of a sample mean* and *random sampling error of a sample mean* are synonymous. Both terms refer to the bedrock foundation of inferential statistics. But because you will often encounter the term *standard error*, this book will use the term, too. Let's reemphasize the point. The standard error of a sample mean is synonymous with random sampling error:

Standard error of a sample mean = $\sigma / \sqrt{n}$.

How can we use the standard error to figure out how closely a sample mean matches the population mean? Let's return to the student-researcher example. For present purposes, assume that we are omniscient observers. We have watched the pollsters take a random sample of $n = 100$ from a population whose parameters are known to us: $\mu = 58$ and $\sigma = 24.8$. We know that the student group's sample mean will be equal to the population mean of 58, plus the associated standard error. According to the formula for the standard error of a sample mean, the magnitude of the standard error of the students' sample mean is equal to:

$$24.8 / \sqrt{100} = 2.48 \approx 2.5.$$

Even before the students drew their random sample of $n = 100$, we could be pretty sure that it would yield a sample mean of 58 plus or minus 2.5 or so—between 55.5 (58 minus

2.5) and 60.5 (58 plus 2.5). For reasons we will demonstrate, there is a fairly good chance, about a 68 percent chance, that any random sample will produce a mean that falls within one standard error of μ. (The student researchers' sample mean, 59, falls in this high-probability give-or-take interval.) We could be almost, but not quite, certain that the students' sample mean would fall within the long bandwidth between two standard errors below the population mean and two standard errors above the population mean. In fact, there is better than a 95 percent chance the sample mean would have a value between 58 minus 2(2.5) and 58 plus 2(2.5)—between 53 at the low end and 63 at the high end. There is a small but real probability, a 5 percent chance, that the students will randomly pick a bona fide fluke, a sample with a mean that falls more than 2 standard errors below the true population mean of 58 (below 53) or more than 2 standard errors above the true mean (above 63).

The Central Limit Theorem and the Normal Distribution

Where do the percentages just discussed—68 percent, 95 percent, and 5 percent—originate? The answer requires an acquaintance with two related topics: the central limit theorem and the normal distribution. As a way of introducing both topics, again consider panel A of Figure 6-2. In creating panel A, the computer drew just ten random samples of $n = 100$. That handful of samples gave us only a fair idea of the true location of the population mean. But imagine drawing an extremely large number of random samples of $n = 100$ students from a population with a mean of 58 and a standard deviation equal to 24.8. Draw a sample, record the mean Democratic rating, and return the cases to the population. Draw another sample, record the mean Democratic rating, and return the cases to the population. Draw another, and another, and another—until hundreds of thousands of means from hundreds of thousands of samples of $n = 100$ are calculated and recorded. What would the distribution of all those sample means look like? Consider Figure 6-3, which shows the distribution of the means of 100,000 random samples drawn from student population A, the same population that was used to generate the sample means in panel A of Figure 6-2.

Notice two aspects of the distribution depicted in Figure 6-3. First, it is centered on the true population mean of 58. In fact, if we were to calculate the overall mean of the 100,000 sample means represented in Figure 6-3, we would arrive at the population mean. The bulk of the 100,000 sample means reside in the thick part of the distribution around 58, between about 56 and 60. Again, the student group's sample mean, 59, would be a typical sample mean representing the mainstream of the distribution. Even so, a few samples serve as very poor representations of μ, returning values of less than about 53 or greater than 63. Second, despite some random "notchiness" here and there, the distribution has a symmetrical bell-like shape. This bell-like symmetry is the signature of the normal distribution.

In both its mean and its shape, Figure 6-3 illustrates the central limit theorem. The **central limit theorem** is an established statistical rule that tells us that, if we were to take an infinite number of samples of size n from a population of N members, the means of these samples would be normally distributed. This distribution of sample means, furthermore, would have a mean equal to the true population mean and have random sampling error equal to σ, the population standard deviation, divided by the square root of n. Thus, most random samples of $n = 100$ that are drawn from a population with a mean of 58 and a standard deviation of 24.8 will yield means that are equal to 58, give or take 2.5 or so. In fact, the normal distribution allows us to make precise inferences about the percentage of sample means that will fall within any given number of standard errors of the true population mean.

Figure 6-3   Distribution of Means from 100,000 Random Samples

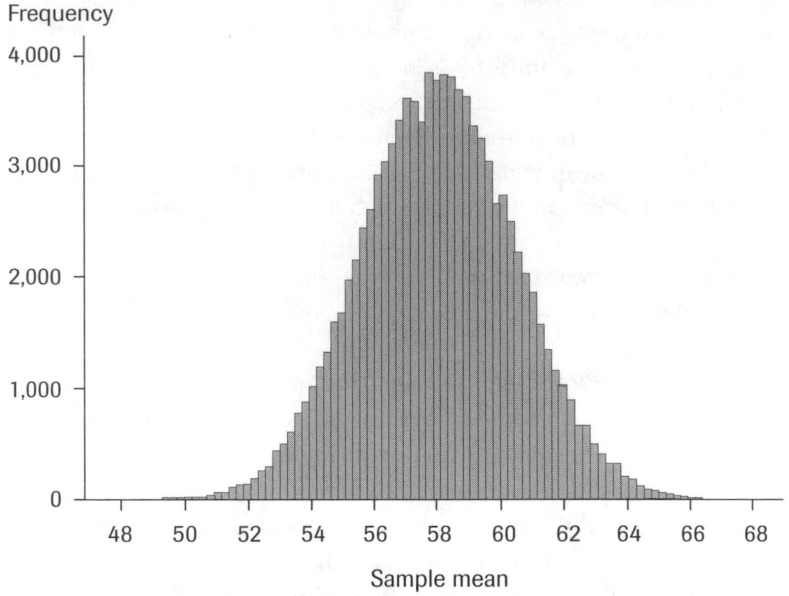

Note: Displayed data are means from 100,000 samples of *n* = 100. Population parameters: μ = 58 and σ = 24.8.

The **normal distribution** is a distribution used to describe interval-level variables. Examine Figure 6-4, which again displays the distribution of the 100,000 sample means. Figure 6-4 differs from Figure 6-3 in two ways. First, a line representing the normal curve has been drawn around the distribution, summarizing its shape. Second, the horizontal axis is simply labeled "*Z*," and Figure 6-3's units of measurement (points on the Democratic feeling thermometer) are replaced in Figure 6-4 with a scale of single-digit numbers. The mean, which appeared as the raw value "58" in Figure 6-3, is labeled "0" in Figure 6-4. Values below the mean are negative, and values above the mean are positive.

The horizontal axis in Figure 6-4 is a standardized transformation of the axis in Figure 6-3. **Standardization** occurs when the numbers in a distribution are converted into standard units of deviation from the mean of the distribution. A value that is expressed in original form—points on a thermometer scale, years of age, dollars of income—is called a raw score or unstandardized score. A standardized value is called a *Z* value or **Z score.** To transform an unstandardized score into a *Z* score, you would divide the score's deviation from the mean by the standard unit you are using:

$$Z = (\text{Deviation from the mean}) / (\text{Standard unit}).$$

The standard deviation, as its name suggests, is a standard unit of deviation from the mean. Take a few minutes to refer back to Table 6-1, which displayed the Democratic ratings for an 11-member population. If you wanted to standardize this list of 11 Democratic thermometer ratings, you would divide each value in the "Deviation from the mean" column by the standard deviation, 24.8. Scores below the mean will have negative values of *Z*, and scores above the mean will have positive values of *Z*. For example, Student 1, who gave the Democrats a rating of 20, has a *Z* score equal to −38/24.8 = −1.53. This value of *Z*, −1.53, locates this

Figure 6-4   Raw Values Converted to Z Scores

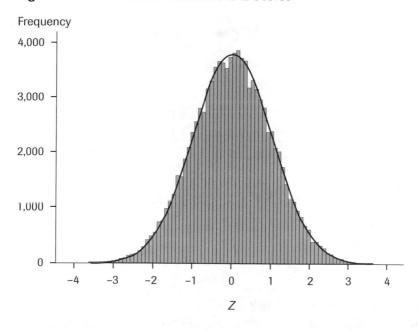

student's rating at one and a half standard deviations below the mean. Student 9, who rated the Democrats at 82, has a Z score equal to 24/24.8 = .97, or about one standard deviation above the mean. Of course, any untransformed value that is equal to the mean has a Z score equal to 0.

The standard error, as its name suggests, is also a standard unit of deviation from the mean. Consider three sample means from Figure 6-3: 54, 58, and 60. Given a standard error equal to 2.5, a sample mean of 54 has a Z score equal to (54 − 58)/2.5 = −1.6, or 1.6 standard errors below the true population mean of 58. So, a sample mean of 54 in Figure 6-3 appears as a Z score of −1.6 in Figure 6-4. A sample mean of 58 in Figure 6-3 hits the population mean right on the money, so in Figure 6-4 it has a Z score equal to 0: (58 − 58)/2.5 = 0. The sample mean of 60 converts into a Z score of (60 − 58)/2.5 = .8, very close to a Z score of 1 in Figure 6-4.

Why go through the ritualistic procedure of converting an untransformed score into a Z score? Because Z is the key to the inferential leverage of the normal distribution. Figure 6-5 again presents the bell-shaped curve, this time displaying percentages inside the curve. The arrow stretching between $Z = −1$ and $Z = +1$ bears the label "68%." What does this mean? It means this: If a list of numbers is normally distributed, then 68 percent of the cases in the distribution will have Z scores between −1 (one standard unit below the mean) and +1 (one standard unit above the mean). Moreover, since the curve is perfectly symmetrical, half of that 68 percent—34 percent of the cases—will fall between the mean (Z equal to 0) and a Z score of +1, and the other 34 percent will fall between the mean and a Z score of −1. So the range between $Z = −1$ and $Z = +1$ is the fattest and tallest part of the curve, containing over two-thirds of the cases. Notice the arrow labeled "95%," the one stretching between $Z = −1.96$ and $Z = +1.96$. These numbers tell us that, in a normal distribution, 95 percent of the cases will have Z scores in the long interval between 1.96 standard units below the mean and 1.96 standard units above the mean. This long interval, in other words, will contain just about all the cases. But of course 5 percent of the cases—those with Z scores of less than −1.96 or greater than +1.96—will lie outside this interval, in the sparsely populated tails of the distribution.

Figure 6-5 Areas under the Normal Curve

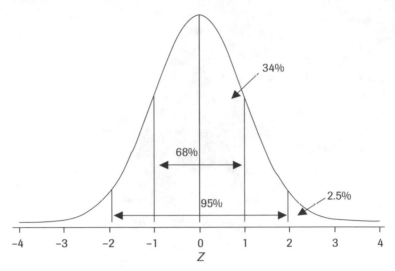

Again, since the curve is symmetrical, half of this 5 percent, or 2.5 percent, will fall in the region below $Z = -1.96$, and the other 2.5 percent will fall in the region above $Z = +1.96$.

Now consider the normal distribution's essential role in making probabilistic inferences. A **probability** is defined as the likelihood of the occurrence of an event or set of events. Imagine being blindfolded and randomly picking a mean from Figure 6-3's distribution of sample means. What is the probability that your randomly chosen mean will have a value between 55.5 and 60.5? Use the standard error, 2.5, to convert each raw value into a Z score. The lower value has a Z score equal to: $(55.5 - 58)/2.5 = -1$. The higher value has a Z score equal to: $(60.5 - 58)/2.5 = +1$. Thanks to the inferential leverage of the normal distribution, you know that 68 percent of all the cases fall in this interval. There is a 68 percent probability that any number drawn at random from a list of normally distributed numbers will have a Z score in the interval between $Z = -1$ and $Z = +1$. But because random processes are at work, there is a chance that the number you choose will have a Z score outside the -1 to $+1$ region. How much of a chance? Since 68 percent of the cases are in the $Z = \pm 1$ area, then the rest of the cases, 32 percent, must lie in the area above $Z = +1$ or below $Z = -1$. Thus, there is a 32 percent probability—about one chance in three—that any number drawn at random from a list of normally distributed numbers will have a Z score outside the fat and tall part of the curve. Suppose you were to bet on the probability of picking a number with a Z score of greater than $+1.96$. Are the odds with you? Hardly. According to the normal distribution, 95 percent of all the cases fall in the long interval between $Z = -1.96$ and $Z = +1.96$, only 5 percent will fall outside these boundaries: 2.5 percent falling below $Z = -1.96$ and 2.5 percent falling above $Z = +1.96$. Thus, you would be betting on an unlikely event, an occurrence that would happen, by chance, fewer than 3 times out of 100. Not impossible, to be sure. Just not very likely.

INFERENCE USING THE NORMAL DISTRIBUTION

Now that we have glimpsed the larger, probabilistic world in which our student pollsters are operating, let's begin to restore realism to the example. We will assume (realistically) that the researchers do not know the true population mean, $\mu = 58$. However, we will also assume (unrealistically) that they know the population standard deviation, $\sigma = 24.8$. (In the next section, we relax this assumption.) Recall that the student researchers took a random sample of $n = 100$ and calculated a sample statistic, the mean Democratic thermometer in their sample, equal to 59. As we have seen, a population mean is symbolized by the Greek letter μ. To distinguish this population parameter from an estimate based on a sample, the mean of a sample is represented by the symbol $\bar{x}$ (pronounced "x bar"), the ordinary letter x with a bar across the top. So in this example, $\bar{x} = 59$, or "x bar" equals 59. Of course, the group does not know the value of μ. But they do know that their estimate of μ, the sample mean of 59, will be equal to the population mean, within the normally distributed boundaries of the standard error: $24.8/\sqrt{100} = 2.48 \approx 2.5$.

Given this information, what is the location of the population mean? Because of the random probability of obtaining a sample statistic that departs dramatically from the population parameter, statisticians never talk about certainty. They talk, instead, about confidence and probability. The most common standard is the **95 percent confidence interval**, defined as the interval within which 95 percent of all possible sample estimates will fall by chance. Thanks to the central limit theorem, we know that the boundaries of the 95 percent confidence interval are defined by the sample mean minus 1.96 standard errors at the lower end, and the sample mean plus 1.96 standard errors at the upper end. Let's use this knowledge to find the 95 percent confidence interval for mean Democratic ratings in the student population.

$$
\begin{aligned}
\text{Lower confidence boundary} \quad &= \quad \bar{x} - 1.96 \text{ standard errors} \\
&= \quad 59 - 1.96\,(2.5) \\
&= \quad 54.1 \\
\text{Upper confidence boundary} \quad &= \quad \bar{x} + 1.96 \text{ standard errors} \\
&= \quad 59 + 1.96\,(2.5) \\
&= \quad 63.9.
\end{aligned}
$$

Conclusion: Ninety-five percent of all possible random samples of $n = 100$ will yield sample means between 54.1 and 63.9.

The 95 percent standard of confidence is widely applied, but its precise boundaries can be a bit tedious to calculate quickly. Therefore, it is customary to round off 1.96 to 2.0, resulting in a useful rule of thumb. To find the 95 percent confidence interval for a sample mean, multiply the standard error by 2. Subtract this number from the sample mean to find the lower confidence boundary. Add this number to the sample mean to find the upper confidence boundary.[5] Applying this rule of thumb, the student organization can be confident, 95 percent confident at least, that the unobserved population mean (μ) has a value between $59 \pm 2(2.5)$, that is, between 54 and 64.

Suppose for the sake of argument (and to demonstrate inference) that the president of the College Democrats, upon hearing the results of the student survey, expresses skepticism, declaring the pollsters' mean "off the mark." "There must be some mistake," the club president suggests. "I would hypothesize that the mean Democratic rating of all students at the university is at least 66, not the rating of 59 you found in your sample." Now, the club president has offered a counterfactual—a hypothetical population mean, a value of μ, that is 7 rating points higher than the researchers' observed sample mean. Is the club president's mean correct or incorrect? Remember, the student pollsters are operating in an environment of uncertainty. They don't know the true population mean, but they can frame and answer this probabilistic question: Assuming that the club president is correct, that the true value of μ is 66, *how often will a random sample yield $\bar{x}$ equal to 59 by chance*? This question illustrates the general logic of inferential reasoning. Assume that the hypothetical population mean is correct. Then determine the probability of obtaining the observed sample statistic, if in fact the hypothetical challenge is true. Assemble all the numbers:

Club president's population mean (the assumed value of μ) =	66
Observed sample mean ($\bar{x}$) =	59
Standard error of $\bar{x}$ =	2.5

Just how far apart are the club president's hypothetical population mean, 66, and the observed sample mean, 59? To find this difference, we subtract the sample mean ($\bar{x}$) from the hypothetical population mean (μ):

Hypothetical mean minus sample mean	=	$\mu - \bar{x}$
	=	66 − 59
	=	7

As we saw earlier, to unlock the inferential power of the normal distribution, we must use a standard unit to convert an untransformed number—in this case, a difference of 7 rating points—into a standardized score, a Z score. The standard unit, of course, is the standard error of the mean. Convert the disputed difference, 7, to a Z score:

$$
\begin{aligned}
Z &= \text{(Hypothetical mean minus sample mean) / standard error} \\
&= (66 - 59) / 2.5 \\
&= 7 / 2.5 \\
&= 2.8
\end{aligned}
$$

How much statistical distance lies between the hypothetical population mean of 66 and the student researchers' observed sample mean of 59? After converting the unstandardized difference (7 points) into standardized units, we obtain a Z score of 2.8. Thus, the club president's proposed mean is 2.8 standard errors higher than the student group's sample mean. If the president is right, what are the chances that the student group's random sample produced a mean that is so far below the true value?

Take a moment and once again consider Figure 6-4, which shows the distribution of 100,000 sample means converted to units of Z. Let's suppose that the student researchers are pondering Figure 6-4, just as we are. Unlike us, however, the students are in the dark about the true location of the population mean. They have a hypothetical value, 66, and an observed sample mean, 59, and they know that the hypothetical value is 2.8 standard units higher than the observed value. To set up the inferential problem, the researchers would assume that the club president's mean *is* the true population mean. Under this working assumption, the researchers would locate the president's mean at $Z = 0$, the true center of the distribution of all possible sample means. Because the researchers' sample statistic is 2.8 units below the hypothetical population mean, the researchers would locate their sample mean at −2.8, close to the tick mark at $Z = -3$. Now imagine standing at $Z = -2.8$ and gazing up toward $Z = 0$. If the club president is correct—the true population mean really is way up there at $Z = 0$—how often will random processes yield a sample mean that is 2.8 standard units below the true mean? What percentage of possible samples would produce such a result?

Table 6-3 Proportions of the Normal Curve above the Absolute Value of Z

First digit and first decimal of Z	Second decimal of Z									
	.00	.01	.02	.03	.04	.05	.06	.07	.08	.09
0.0	.5000	.4960	.4920	.4880	.4840	.4801	.4761	.4721	.4681	.4641
0.1	.4602	.4562	.4522	.4483	.4443	.4404	.4364	.4325	.4286	.4247
0.2	.4207	.4168	.4129	.4090	.4052	.4013	.3974	.3936	.3897	.3859
0.3	.3821	.3783	.3745	.3707	.3669	.3632	.3594	.3557	.3520	.3483
0.4	.3446	.3409	.3372	.3336	.3300	.3264	.3228	.3192	.3156	.3121
0.5	.3085	.3050	.3015	.2981	.2946	.2912	.2877	.2843	.2810	.2776
0.6	.2743	.2709	.2676	.2643	.2611	.2578	.2546	.2514	.2483	.2451
0.7	.2420	.2389	.2358	.2327	.2296	.2266	.2236	.2206	.2177	.2148
0.8	.2119	.2090	.2061	.2033	.2005	.1977	.1949	.1922	.1894	.1867
0.9	.1841	.1814	.1788	.1762	.1736	.1711	.1685	.1660	.1635	.1611
1.0	.1587	.1562	.1539	.1515	.1492	.1469	.1446	.1423	.1401	.1379
1.1	.1357	.1335	.1314	.1292	.1271	.1251	.1230	.1210	.1190	.1170
1.2	.1151	.1131	.1112	.1093	.1075	.1056	.1038	.1020	.1003	.0985
1.3	.0968	.0951	.0934	.0918	.0901	.0885	.0869	.0853	.0838	.0823
1.4	.0808	.0793	.0778	.0764	.0749	.0735	.0721	.0708	.0694	.0681
1.5	.0668	.0655	.0643	.0630	.0618	.0606	.0594	.0582	.0571	.0559
1.6	.0548	.0537	.0526	.0516	.0505	.0495	.0485	.0475	.0465	.0455
1.7	.0446	.0436	.0427	.0418	.0409	.0401	.0392	.0384	.0375	.0367
1.8	.0359	.0351	.0344	.0336	.0329	.0322	.0314	.0307	.0301	.0294
1.9	.0287	.0281	.0274	.0268	.0262	.0256	**.0250**	.0244	.0239	.0233
2.0	.0228	.0222	.0217	.0212	.0207	.0202	.0197	.0192	.0188	.0183
2.1	.0179	.0174	.0170	.0166	.0162	.0158	.0154	.0150	.0146	.0143
2.2	.0139	.0136	.0132	.0129	.0125	.0122	.0119	.0116	.0113	.0110
2.3	.0107	.0104	.0102	.0099	.0096	.0094	.0091	.0089	.0087	.0084
2.4	.0082	.0080	.0078	.0075	.0073	.0071	.0069	.0068	.0066	.0064
2.5	.0062	.0060	.0059	.0057	.0055	.0054	.0052	.0051	.0049	.0048
2.6	.0047	.0045	.0044	.0043	.0041	.0040	.0039	.0038	.0037	.0036
2.7	.0035	.0034	.0033	.0032	.0031	.0030	.0029	.0028	.0027	.0026
2.8	.0026	.0025	.0024	.0023	.0023	.0022	.0021	.0021	.0020	.0019
2.9	.0019	.0018	.0018	.0017	.0016	.0016	.0015	.0015	.0014	.0014
3.0	.0013	.0013	.0013	.0012	.0012	.0011	.0011	.0011	.0010	.0010

The answer is contained in a normal distribution probability table, such as that presented in Table 6-3. Examine Table 6-3 for a moment. Each entry in Table 6-3 reports the proportion of the normal distribution that lies *above* the absolute value of Z. Before we test the club president's mean using Table 6-3, let's look up the probability for a value of Z with which you are already conversant, $Z = +1.96$. Read down the left-hand column until the integer value and first decimal of Z is reached (1.9), and then read across until you find the cell associated with the second decimal place of Z (.06). The cell's entry is .0250. This number confirms what we already knew: 2.5 percent, or .025, of the curve lies above a Z score of $+1.96$. Notice that, because the normal distribution is symmetrical, Table 6-3 can be used to find probabilities associated with negative values of Z, as well. For example, since .025 of the curve lies above

$Z = +1.96$, then .025 must lie below $Z = -1.96$. Thus, in using Table 6-3, you can safely ignore the sign on Z and look up its absolute value. Remember, though, that for negative values of Z, the entries in Table 6-3 will tell you the proportion of the curve that lies *below* the value of Z.

Now apply Table 6-3 to the task of finding the proportion of the curve that lies above the absolute value of $Z = 2.8$. Again, read down the left-hand column until you find the integer and first digit of Z (2.8), then read across to find the proportion corresponding to the second decimal place of Z (.00). The entry in this cell is .0026. This number may be interpreted this way: the club president is (very) probably wrong. Why can we say this? Under the assumption that the president is correct, only .0026, or .26 percent, of all possible sample means will fall as far off the mark as the researchers' sample mean. If, as the president claims, the population mean really is 66, we would observe a sample mean of 59, by chance, only 26 times out of 10,000. Because such an event is unlikely, the student group would reject the president's hypothetical claim and infer that their sample statistic is closer to the true population mean.

Let's summarize this example, which demonstrates how inference using the normal distribution, often called *normal estimation* for short, is used for testing hypothetical claims about a population mean. This inferential process has a definite logic. The researcher begins by assuming that the hypothetical claim is correct. Thus, when the club president suggests that the mean Democratic rating in the student population is "at least 66," the student pollsters say, "All right, let's assume that the population mean is at least 66." The researcher then sees how well this assumption holds up, given the observed results obtained from the sample: "If the population mean really is 66, how often, by chance, will we obtain a sample mean of 59?" The answer, which is always a probability, determines the inferential decision: "If the population mean is at least 66, we would obtain a sample mean of 59 about 26 times out of 10,000 by chance. Therefore, we infer that the population mean is *not* 66 or higher." These steps—assuming the hypothetical claim to be correct, testing the claim using normal estimation, and then making an inference based on probabilities—define the logic of hypothesis testing using inferential statistics. In most research situations you will encounter, normal estimation will serve you well.

There is, however, one important way in which the student researcher example has been unrealistic—and this point must now be addressed. In figuring out the variation component of random sampling error, the standard deviation, it was assumed that the population standard deviation (σ) was a known quantity. This is not realistic. As a practical matter, a researcher rarely knows any of the population's parameters. If the population's parameters were known, there would be no need to take a sample to begin with! Fortunately, a lack of knowledge about the population's standard deviation is not as serious a problem as it may seem. A different distribution—one that, under many circumstances, is very similar to the normal distribution—can be applied to problems of inference when the population standard deviation is not a known quantity. This distribution is our next topic of discussion.

INFERENCE USING THE STUDENT'S *T*-DISTRIBUTION

As was just noted, in most realistic sampling situations the researcher has a random sample—and that's it. The researcher uses this sample to calculate a sample mean, just as the student pollsters calculated a mean Democratic rating from their sample. But to determine the standard error of the mean—the degree to which the sample mean varies, by chance, from the population mean—the researcher needs to know the population standard deviation. If that parameter is unavailable, as it usually is, then the researcher at least needs an estimate of the population standard deviation. A reasonable estimate is available. Why not simply calculate the standard deviation of the sample? One could then use the sample standard deviation as a stand-in for σ in calculating the standard error. So, easily enough, the standard error of the sample mean would become:

Sample standard deviation / Square root of the sample size.

Or, using the ordinary letter s to denote the sample standard deviation, the standard error is:

$$s / \sqrt{n}.$$

Recall that, in arriving at the population standard deviation, first we calculated the variance by adding up the squared deviations from the mean and dividing by the population size, N: population variance = (Summation of squared deviations)/N. The population standard deviation, σ, is the square root of the population variance. In calculating the sample variance, we also begin by adding up the squared deviations from the sample mean. In determining the sample variance, however, we divide by $n - 1$: sample variance = (Summation of squared deviations)/$(n - 1)$. The sample standard deviation, s, is equal to the square root of the sample variance.[6] Now, for fair-sized samples, substituting s for σ will work fine, and normal estimation can be used.[7] But when you are using smaller samples, or if you have divided up a large sample into smaller subsamples for separate analyses, the exact properties of the normal distribution may no longer be applied in making inferences. Fortunately, a similar distribution, the Student's t-distribution, can be applied. The **Student's t-distribution** is a probability distribution that can be used for making inferences about a population mean when the sample size is small.

The normal distribution always has the same shape. The Z scores of −1.96 and +1.96 always mark the boundary lines of the 95 percent confidence interval. The shape of the Student's t-distribution, by contrast, depends on the sample size. The boundaries of the 95 percent confidence interval are not fixed. Rather, they vary, depending on how large a sample is being used for making inferences. There is an undeniable logic here. When the population standard deviation is not known and the sample size is small, the t-distribution sets wider boundaries on random sampling error and permits less confidence in the accuracy of a sample statistic. When the sample size is large, the t-distribution adjusts these boundaries accordingly, narrowing the limits of random sampling error and allowing more confidence in the measurements made from the sample. Although the terminology used to describe the t-distribution is different from that used to describe the normal distribution, the procedures for drawing inferences about a population parameter are essentially the same.

Again we use the student researchers' sample mean to illustrate the similarities—and the differences—between the inferential properties of the Student's t-distribution and the normal curve. Let's keep most elements of the example the same as before. The pollsters' sample size is $n = 100$, and their sample mean Democratic rating, the value of $\bar{x}$, is 59. This time, however, we will make the example reflect a practical reality: the student group does not know the population standard deviation. They must rely, instead, on the standard deviation of their sample, which their computer calculates to be 27. So $s = 27$. In just the manner described above, the standard error of the sample mean now becomes:

$$s / \sqrt{n} = 27/\sqrt{100}$$
$$= 27/10$$
$$= 2.7$$

This is the same methodology used before. The campus group just substitutes the sample standard deviation for the population standard deviation, does the math, and arrives at the standard error of the sample mean, 2.7. And, just as with normal estimation, the pollsters know that their sample mean is equal to the population mean, within acceptable boundaries of random sampling error. What are those boundaries? What is the 95 percent confidence interval for the sample mean? The answer is contained in a Student's t-distribution table, such as the one shown in Table 6-4.

Before following through on the example, let's become familiar with Table 6-4. Because the specific shape of the Student's t-distribution depends on the sample size, Table 6-4 looks different from Table 6-3, which showed the area under the normal curve for different values of Z. In normal estimation, we do not have to worry about the size of the sample, so we calculate a value of Z and then find the area of the curve above that value. In estimation using Student's t, however, the sample size determines the shape of the distribution.

First consider the left-hand column of Table 6-4, labeled "Degrees of freedom." What are degrees of freedom? **Degrees of freedom** refers to a statistical property of a large family of distributions, including the Student's t-distribution. The number of degrees of freedom is equal to the sample size, n, minus the number of parameters being estimated by the sample. If we are using $n = 100$ observations to estimate one population parameter, μ, then we would have $n - 1$ degrees of freedom: $100 - 1 = 99$ degrees of freedom.

Now examine the columns of Table 6-4, under the heading "Area under the curve." The columns are labeled with different proportions: .10, .05, .025, and .01. The entries in each column are values of t. Each cell tells you the value of t above which that proportion of the curve lies. For example, the top-most cell in the .025 column says that, with one degree of freedom, .025, or 2.5 percent, of the distribution falls above a t-value of 12.706. That sets a very wide boundary for random sampling error. Imagine starting at the mean of the normal curve and having to progress over 12 units of Z above the mean before hitting the .025 boundary! The signature of the Student's t-distribution is that it adjusts the confidence interval, depending on the size of the sample. Indeed, notice what happens to the t-values in the .025 column as the sample size (and thus degrees of freedom) increases. As sample size increases, the value of t that marks the .025 boundary begins to decrease. More degrees of freedom mean less random sampling error and, thus, more confidence in the sample statistic. For illustrative purposes, the value of Z that is associated with the .025 benchmark, $Z = 1.96$, appears in the bottom cell of the .025 column of Table 6-4. Notice that the value of t for a large sample (degrees of freedom = 1,000) is $t = 1.962$. So, for large samples with many degrees of freedom, the Student's t-distribution closely approximates the normal distribution.

We can now return to the example and use Table 6-4 to find the 95 percent confidence interval of the student pollsters' sample mean. We first determine the number of degrees of freedom, which is tied to the sample size (degrees of freedom = $n - 1$.) Since the student organization's sample has $n = 100$, it has degrees of freedom equal to 99. There is no row in Table 6-4 that corresponds exactly to 99 degrees of freedom, so we will use the closest lower number, 90 degrees of freedom. Now read across to the column labeled ".025." This number, 1.987, tells us that .025, or 2.5 percent, of the curve falls above $t = 1.987$. The Student's t-distribution, like the normal distribution, is perfectly symmetrical. Thus, 2.5 percent of the curve must lie below $t = -1.987$. These t-values give us the information we need to define the 95 percent confidence interval of the sample mean:

$$\text{Lower confidence boundary} = \bar{x} - 1.987 \text{ standard errors}$$
$$= 59 - 1.987 \,(2.7)$$
$$= 53.64$$
$$\text{Upper confidence boundary} = \bar{x} + 1.987 \text{ standard errors}$$
$$= 59 + 1.987 \,(2.7)$$
$$= 64.36.$$

Table 6-4   The Student's t-Distribution

Degrees of freedom	Area under the curve			
	.10	.05	.025	.01
1	3.078	6.314	12.706	31.821
2	1.886	2.920	4.303	6.965
3	1.638	2.353	3.182	4.541
4	1.533	2.132	2.776	3.747
5	1.476	2.015	2.571	3.365
6	1.440	1.943	2.447	3.143
7	1.415	1.895	2.365	2.998
8	1.397	1.860	2.306	2.896
9	1.383	1.833	2.262	2.821
10	1.372	1.812	2.228	2.764
11	1.363	1.796	2.201	2.718
12	1.356	1.782	2.179	2.681
13	1.350	1.771	2.160	2.650
14	1.345	1.761	2.145	2.624
15	1.341	1.753	2.131	2.602
16	1.337	1.746	2.120	2.583
17	1.333	1.740	2.110	2.567
18	1.330	1.734	2.101	2.552
19	1.328	1.729	2.093	2.539
20	1.325	1.725	2.086	2.528
21	1.323	1.721	2.080	2.518
22	1.321	1.717	2.074	2.508
23	1.319	1.714	2.069	2.500
24	1.318	1.711	2.064	2.492
25	1.316	1.708	2.060	2.485
26	1.315	1.706	2.056	2.479
27	1.314	1.703	2.052	2.473
28	1.313	1.701	2.048	2.467
29	1.311	1.699	2.045	2.462
30	1.310	1.697	2.042	2.457
40	1.303	1.684	2.021	2.423
60	1.296	1.671	2.000	2.390
90	1.291	1.662	**1.987**	2.368
100	1.290	1.660	1.984	2.364
120	1.289	1.658	1.980	2.358
1,000	1.282	1.646	1.962	2.330
Normal (Z)	1.282	1.645	1.960	2.326

Thus the pollsters can be 95 percent confident that the true population mean lies between 53.64 on the low end and 64.36 on the high end. There is a 5 percent chance that the population mean falls outside these boundaries—lower than 53.64 or higher than 64.36.

There are two bits of comforting news about the Student's *t*-distribution. One of these features has already been pointed out: As sample size grows, the *t*-distribution increasingly resembles the normal curve. Again note the numbers along the bottom row of Table 6-4, which report the values of *Z* for each probability. You can see that, for samples with 100 degrees of freedom, the *t*-distribution looks just about "normalized." And for samples with 1,000 or more degrees of freedom, the two distributions become virtually identical. A second, related point has to do with the rule of thumb that you learned for normal estimation. Recall that the 95 percent confidence interval can be quickly determined by multiplying the standard error by 2, then subtracting this number from the sample mean to find the lower confidence boundary, and adding this number to the sample mean to find the upper confidence boundary. This is a good rule of thumb, because it works well in most situations. As you can see from Table 6-4, even for fairly small samples—those having 60 degrees of freedom—the rule of thumb will provide an adequate estimate of the 95 percent confidence interval.

WHAT ABOUT SAMPLE PROPORTIONS?

One attractive feature of inferential statistics is its methodological versatility. If certain preconditions are met, the rules of inference apply. We have reviewed the logic of random sampling, and we have seen that the rule of thumb for estimating the standard error of a sample mean generally applies, even when the population standard deviation is not known—provided, of course, that the sample is fair-sized. However, many of the variables of interest to political researchers are not measured at the interval level. Rather, they are measured at the nominal or ordinal level of measurement. Thus, we may take a sample and compute the percentage of respondents who "favor" increased military spending or the percentage who "oppose" gun control. In this case, we are not using an interval-level sample mean ($\bar{x}$) to estimate a population mean (μ). Instead we would be working with a **sample proportion**, the number of cases falling into one category of the variable divided by the number of cases in the sample. But again the rules of inference are robust. Within reasonable limits, the same procedures apply. You already know these procedures, so only a brief exposition is required.

Let's say that the campus organization asked the students in its sample ($n = 100$) whether or not they voted in the last student government election. Suppose that 72 students answered, "Yes, voted in the election," and the remaining 28 students answered, "No, did not vote in the election." What is the sample proportion of voters?

Sample proportion of voters   =   (Number answering "Yes") / Sample size
                             =   72 / 100
                             =   .72

Similarly, the sample proportion of nonvoters is

Sample proportion of nonvoters   =   (Number answering "No") / Sample size
                             =   28 / 100
                             =   .28.

So the sample proportion of voters is .72, and the proportion of nonvoters is .28. How closely does the sample proportion, .72, estimate the proportion of voters in the student population? What is the standard error of the observed sample statistic, .72?

The mean of a sample, as you know, is denoted by the symbol $\bar{x}$. The proportion of a sample falling into one category of a nominal or ordinal variable is denoted by the ordinary letter p. In this case, the proportion of students falling into the "Yes, voted in the election" category is .72, so $p = .72$. The proportion of a sample falling into all other categories of a nominal or ordinal variable is denoted by the letter q. This proportion, q, is equal to one minus p, or $q = 1 - p$. In this example, then, q would be equal to $1 - .72$, or .28, the proportion of students who responded "No, did not vote in the election." Let's assemble the numbers we have so far:

Sample proportion of voters (p)	=	.72
Sample proportion of nonvoters (q)	=	.28
Sample size (n)	=	100.

This information—the value of p, the value of q, and the sample size—permits us to estimate the standard error of the sample proportion, p. First recall that the general formula for random sampling error is the variation component divided by the sample size component:

Random sampling error = (Variation component) / (Sample size component).

In figuring out the random sampling error associated with a sample proportion, the sample size error component is the same as before, the square root of n. For sample proportions, however, the variation component of random sampling error is different. For a sample proportion, p, the variation component is equal to the square root of the product p times q. That is:

Variation component of random sampling error = Square root (pq), or $\sqrt{pq}$.

Therefore, the standard error of a sample proportion, p, is equal to:

$$= \sqrt{pq} \, / \, \sqrt{n}.$$

Now let's plug in the numbers from the example and see what we have:

$$pq = (.72)(.28) = .20$$
$$\sqrt{pq} = \sqrt{.20} = .45$$
$$\sqrt{n} = \sqrt{100} = 10$$
$$\sqrt{pq} \, / \, \sqrt{n} = .45/10 = .045.$$

Thus, the proportion of voters in the student population is equal to the sample proportion, .72, with a standard error of .045. So, the student pollsters know that the true population proportion is equal to .72, give or take .045. Indeed, just as with normal estimation, there is a 68 percent chance that the population parameter falls within one standard error of the sample proportion, between .72 minus .045 (which is equal to .675) and .72 plus .045 (which is equal to .765). The 95 percent confidence interval would be defined by:

$$\text{Lower confidence boundary} \; = \; p - 1.96 \text{ standard errors}$$
$$= \; .72 - 1.96 \, (.045)$$
$$\approx .63$$

$$\text{Upper confidence boundary} \; = \; p + 1.96 \text{ standard errors}$$
$$= \; .72 + 1.96 \, (.045)$$
$$\approx .81.$$

The student pollsters can be 95 percent confident that the true percentage of voters in the student population is between 63 percent and 81 percent.

A final statistical caveat. We have just demonstrated how normal estimation can be applied in determining the standard error of a sample proportion. Under most circumstances, this method works quite well. However, we know that normal estimation works best for sample proportions closer to .50, and it begins to lose its applicability as p approaches .00 or 1.00. How can one know if normal estimation may be used? Here is a general procedure. Multiply p by the sample size, and multiply q by the sample size. If both numbers are 10 or higher, then normal estimation will work fine. (Actually, if both numbers are 5 or more, normal estimation will still work.) The student researchers are on solid inferential ground, since $(100)(.72) = 72$ and $(100)(.28) = 28$. Trouble could begin to brew, however, if the pollsters were to subset their sample into smaller groups—subdividing, say, on the basis of gender or class rank—and were then to make inferences from these smaller subsamples.

As consumers of popular media, we are much more likely to encounter percentages or proportions than arithmetic means. Sometimes, of course, the sampling procedures used by media-based organizations are questionable, and their reported results should be consumed with a large grain of salt. Reputable pollsters always report some measure of random sampling error (normally following the form "margin of error ± . . ."), which typically defines the boundaries of the 95 percent confidence interval. You are now well equipped to interpret such percentages on your own. You are also armed with all the inferential statistics you will (probably) ever need in order to make sense of more advanced topics in political analysis.

SUMMARY

Just how accurately does a sample statistic estimate a population parameter? You now know that the answer is a resounding: "It depends on three factors." First, it depends on whether the sample was randomly selected from the population. By ensuring that each member of the population has an equal chance of being included in the sample, the researcher eliminates "bad" error, systematic error, from the sample statistic. A random sample permits the researcher to estimate the amount of "good" error, random error, contained in the sample statistic. Second, it depends on the size of the random sample. Larger samples yield better estimates than smaller samples. But you now understand why samples that seem small can nonetheless provide a solid basis for inference. Third, it depends on the amount of variation in the population. You are now familiar with a key measure of variation for interval-level variables, the standard deviation. And you know how the standard deviation works together with sample size in bracketing the confidence interval for a sample mean.

Many symbols and terms were discussed in this chapter. Table 6-5 provides a list, arranged in roughly the order in which these terms and symbols were introduced.

Table 6-5 Terms and Symbols and the Roles They Play in Inference

Term or symbol (pronunciation)	What it is or what it does	What role it plays in sampling and inference
μ ("mew")	Population mean	Usually μ is unknown and is estimated by $\bar{x}$.
N	Population size	
σ ("sigma")	Population standard deviation	Measures variation in a population characteristic. The variation component of random sampling error.
Z score	Converts raw deviations from μ into standard units	Defines the tick marks of the normal distribution; 68 percent of the distribution lies between $Z = -1$ and $Z = +1$; 95 percent of the distribution lies between $Z = -1.96$ and $Z = +1.96$.
$\bar{x}$ ("x bar")	Sample mean	Sample statistic that estimates μ.
n	Sample size	The sample size component of random sampling error is $\sqrt{n}$.
s	Sample standard deviation	Substitutes for σ as the variation component of random sampling error when σ is unknown.
Standard error of the sample mean	Measures how much $\bar{x}$ departs, by chance, from μ	Random sampling error. Equal to $\sigma/\sqrt{n}$, if σ is known. Equal to $s/\sqrt{n}$, if σ is unknown.
95 percent confidence interval	The interval in which 95 percent of all possible values of $\bar{x}$ will fall by chance	Defined by $\bar{x} \pm 1.96$ standard errors in normal estimation. Can usually be determined by rule of thumb: $\bar{x} \pm 2$ standard errors in all estimation.
p	Proportion of a sample falling into one value of a nominal or ordinal variable	Sample estimate of a population proportion.
q	Proportion of a sample falling into all other values of a nominal or ordinal variable	Equal to $1 - p$.
Standard error of a sample proportion	Measures how much p departs, by chance, from a population proportion	Defined by $\sqrt{pq}/\sqrt{n}$. Ordinarily can be applied in finding the 95 percent confidence interval of p, using normal estimation.

Let's review them. The population mean (μ) is the parameter of chief concern—the measure of central tendency that the researcher is most interested in estimating. Variation around the population mean is determined by the standard deviation (σ), the measure of dispersion that summarizes how much clustering or spread exists in the population. The relative position of any number in a list can be expressed as a Z score, the number of standard units that the number falls above or below the mean. The standard deviation and the standard error are standard units of deviation from the mean that can be used to convert raw values into Z scores. By knowing Z and applying the inferential properties of the normal distribution, we can decide how closely the sample mean, symbolized by $\bar{x}$, approximates the population mean.

We know that the sample mean will be equal to the population mean, plus any random sampling error that was introduced in drawing the sample. The size of this error, termed the standard error of the sample mean, is determined by σ and the sample size (n). Again applying the normal distribution, the researcher can estimate the 95 percent confidence interval for $\bar{x}$, the boundaries within which 95 percent of all possible sample means will fall by chance. Z scores are directly applied here. By multiplying the standard error by $Z = 1.96$—or rounding up to 2 by rule of thumb—the researcher can figure the probable boundaries of the true population mean. In practice, the population standard deviation is rarely known, so the researcher uses the sample standard deviation, denoted by s, as a stand-in for σ, and then applies the Student's t-distribution. As you know, much political research, especially survey research, involves nominal and ordinal variables. In this chapter we also discussed how normal estimation may be usefully applied in estimating the 95 confidence interval for a sample proportion.

KEY TERMS

census (p. 114)

central limit theorem (p. 127)

degrees of freedom (p. 136)

inferential statistics (p. 113)

95 percent confidence interval (p. 131)

normal distribution (p. 128)

population (p. 114)

population parameter (p. 114)

probability (p. 130)

random sample (p. 115)

random sampling error (p. 117)

random selection (p. 115)

range (p. 122)

response bias (p. 115)

sample (p. 114)

sample proportion (p. 138)

sample statistic (p. 114)

sampling frame (p. 115)

selection bias (p. 115)

standard deviation (p. 119)

standard error (p. 126)

standardization (p. 128)

Student's t-distribution (p. 135)

variance (p. 122)

Z score (p. 128)

EXERCISES

1. A polling firm wants to know which candidate is likely to win the presidential election. The major party candidates are Republican Dewey Cheatum and Democrat Andy Howe. The pollsters set up an Internet site and ask site visitors to indicate their preferred choice. The results: After months of polling and a sample of hundreds of thousands of individuals, Republican Cheatum held a huge edge. The only problem: On election day, Howe won handily.

   A. There was obviously a big difference between the preference of the population about which the polling firm wanted to draw inferences and the preference of the sample they obtained. What was the population the firm was interested in? What was the firm's sampling frame? Describe two reasons why the sampling technique was a poor one.

B. Internet polls like the one described here almost always find a preference for the Republican candidate over the Democratic candidate. Why do you suppose this is?

2. The leadership of a large interest group is trying to decide whether to raise its membership dues. The average income of the group's members is an important consideration. Although the group lacks current information on member incomes, they do have data from a previous census of the membership. According to this census, mean income is $60,000, with a standard deviation of $15,000.

A. This chapter discussed how to convert a raw value, such as income measured in dollars, into a Z score. A Z score is obtained by dividing the raw value's deviation from the mean by a standard unit. The standard deviation is a standard unit that can be used to convert deviations from the mean into Z scores. Provide the Z scores for each of the following incomes: $30,000, $37,500, $60,000, and $80,000.

B. A statistician hired by the group picks three members at random from the census data. The first has a Z score of +1.5. The second has a Z score of −.6. The third has a Z score of 0. What are the incomes of each of these three members?

C. The group decides that the census is too old, so they discard it. Since they lack the resources to conduct a new census, they ask the statistician to take a sample. Assuming that the statistician has access to the membership rolls, describe how the statistician would go about obtaining a random sample of $n = 400$.

3. The individuals in a population have been asked to register their opinions on health care reform. Opinions are gauged by a scale that ranges from 0 (the respondent favors a plan based on private medical insurance) to 10 (the respondent favors a government-based plan). The health care opinion scale scores for each population member: 0, 3, 4, 5, 6, 7, 10.

A. What is the value of μ, the mean health care opinion score for the population?

B. Write down three column headings on a sheet of paper: "Health care opinion score," "Deviation from the mean," and "Squared deviation from the mean." Fill in the columns.

C. Based on your calculations in part B, what is the population variance? What is σ, the population standard deviation?

D. Instead of having only seven members, imagine a population having tens of thousands of individuals. Assume that this more realistic population has the same population mean you calculated in part A and the same standard deviation you calculated in part C. (For this part of the exercise, round the standard deviation to the nearest whole number.) According to the central limit theorem, if one were to draw an infinite number of random samples of $n = 25$ from the population, 68 percent of the sample means would fall between what two health care opinion scores? 95 percent of the sample means would fall between what two health care opinion scores?

4. The sheriff is concerned about speeders on a certain stretch of county road, which has a posted speed limit of 45 miles per hour. The sheriff sets up a radar device and, over a long period of time, obtains data on the entire population of vehicles using the road. The mean vehicle speed: 52 miles per hour.

A. The sheriff cracks down on speeders. Following the crackdown, the sheriff takes a random sample ($n = 144$) of vehicle speeds on the roadway. *The sample data:* sample mean, 47 mph; sample standard deviation, 6 mph. Using the sample standard deviation as a substitute for the population standard deviation, what is the standard error of the sample mean? Using the $\pm$ 2 rule of thumb, what is the 95 percent confidence interval of the sample mean?

B. A skeptical county commissioner claims that the crackdown had no effect and that average speed on the roadway is still 52 mph. Is the skeptic on solid statistical ground? Explain how you know.

5. Sociologists have conducted much interesting research on gender stereotypes in American society. A curious aspect of stereotypes is that people tend to perceive differences between groups to be greater than they actually are. This suggests, for example, that when asked about the heights of men and women, survey respondents would tend to perceive men to be taller than women. Suppose you wanted to test this notion that individuals perceive a greater height difference between men and women than exists in the population. Let's say that, in the population, men, on average, are 4 inches taller than women. So the true population difference between men and women is 4 inches.

You obtain a random sample of 400 individuals. For each respondent you record his or her perceptions of the difference between male and female heights. In your sample, you find that the mean difference in perceived heights is 5 inches. So respondents perceive that men are 5 inches taller than women. The sample standard deviation is 4 inches. Figure out the standard error of the sample mean. Using the $\pm$ 2 rule of thumb, calculate the 95 percent confidence interval. If the true gender difference is 4 inches, can you infer from your sample that individuals perceive a greater difference than actually exists? Explain.

6. Each of the following proportions is based on survey responses. For each proportion, use the $\pm$ 2 rule of thumb to determine the 95 percent confidence interval.
   A. When asked if they are satisfied with their jobs, .49 said "very satisfied" ($n = 100$).
   B. When asked if people can be trusted, .67 said "yes" ($n = 225$).
   C. Of the individuals in a survey, .39 still live in the same city they lived in when they were sixteen years of age ($n = 81$).

7

Tests of Significance and Measures of Association

LEARNING OBJECTIVES

In this chapter you will learn:
- How the empirical relationship between two variables is affected by random sampling error
- How to use an informal test in making inferences about relationships
- How to use formal statistical tests in making inferences about relationships
- How measures of association gauge the strength of an empirical relationship
- Which measure of association to use in performing and interpreting political analysis

In Chapter 6 we discussed the fundamentals of inference and looked briefly at the role of inferential statistics in testing claims about a population mean. In this chapter we show how to use inferential statistics to test hypotheses. For example, consider this hypothesis: In a comparison of individuals, women will give the Democratic Party higher feeling thermometer ratings than will men. Working with a random sample, we set up the analysis by dividing subjects on gender, the independent variable, and then comparing values of the dependent variable, mean thermometer ratings of Democratic Party. Suppose we find that the mean rating among women is 61, while the mean for men is 56. This yields a mean difference of 5 points between female and male respondents. In any event, that is the difference we observe in our random sample. But just how closely does this difference between women and men in the sample reflect the true difference between women and men in the population from which the sample was drawn? Think about this question for a moment. Do you notice an analogy to the problems of sampling and inference discussed in Chapter 6?

In addressing inferential questions about the relationship between an independent variable and a dependent variable, the logic you learned in Chapter 6 can be applied directly. You already know about the standard error of a sample mean—the extent to which the sample mean departs, by chance, from the population mean. In this chapter you will learn that, just as one mean contains random sampling error, the difference between two sample means contains random error, as well. Just as the standard error is used to estimate the likely boundaries of an unseen population mean, so too is the standard error of the difference between two sample statistics used to estimate a difference in the population. And, just as the researcher can use the normal curve or the Student's *t*-distribution to test a hypothetical claim about a population mean, so too the researcher can use those techniques to test a hypothetical claim about a relationship between two variables in the population. In all these ways, this chapter builds on the skills you already have.

This chapter focuses and expands your skills in two ways. First, we concentrate more directly on tests of statistical significance. A **test of statistical significance** helps you decide whether an observed relationship between an independent variable and a dependent variable really exists in the population or whether it could have happened by chance when the sample was drawn. Second, we discuss how to gauge the strength of the relationship between an independent variable and a dependent variable by considering some of the more widely used measures of association. A **measure of association** tells the researcher how well the independent variable works in explaining the dependent variable.

STATISTICAL SIGNIFICANCE

Let's begin with the finding described above—the 5-point difference between the mean Democratic rating among women (a rating of 61) and men (a rating of 56). Because this difference corresponds to our hypothetical expectations, it would appear that the hypothesis has merit, that females do, in fact, rate the Democrats more highly than males do. But adopt a skeptic's stance. Suppose that in the population there is no relationship between gender and ratings of the Democratic Party. Assume that if we were to measure the mean rating for all women and all men in the population and then calculate the difference between these population means, we would end up with 0—no gender difference.

This skeptical assumption, called the **null hypothesis**, plays a vital role in hypothesis testing. The null hypothesis states that, in the population, there is no relationship between the independent and dependent variables. Furthermore, any relationship observed in a sample was produced by random sampling error. So, the null hypothesis offers a two-part defense of its name: (1) no relationship exists in the population, and (2) any apparent relationship found in a sample was produced by random processes when the sample was drawn. The null hypothesis, which is labeled H_0, reminds us that observed sample results could have been produced by random sampling error. The hypothesis that we have formulated, the hypothesis suggesting that there *is* a relationship between gender and thermometer ratings, is considered the *alternative hypothesis,* and is labeled H_A. If we are to have confidence in H_A, then it must be shown that H_0's version of events—it all happened by chance—is sufficiently implausible. Table 7-1 displays this inferential tension.

Table 7-1 Type I and Type II Error

Inferential decision, based on sample data	In the population, is H_0 true or is H_0 false?	
	H_0 is true	H_0 is false
Do not reject H_0	Correct inference	Type II error
Reject H_0	Type I error	Correct inference

Table 7-1's column headings reflect what is really going on in the unseen population. The null hypothesis could, in fact, be true or it could, in fact, be false. The rows communicate our inferential decision, reached by observing a relationship in a sample. Based on the sample, the researcher might decide not to reject the null hypothesis, or the researcher might decide to reject the null hypothesis. Obviously, every researcher wishes to make the correct inference about the data, rejecting the null hypothesis when it is false and not rejecting it when it is true. But there are two ways to get it wrong. **Type I error** occurs when the researcher concludes that there is a relationship in the population when, in fact, there is none. **Type II error** occurs when the researcher infers that there is no relationship in the population when, in fact, there is.

Plainly enough, if we had an unobstructed view of the population, we would always reach the correct inference. Clearly, any relationship that we find in a sample—a mean thermometer difference of 5 points, 15 points, 50 points, or whatever—could have been generated serendipitously when the cases were randomly picked. Because we operate in an environment of uncertainty, we need to be conservative in our inferential decisions and allow random chance every plausible opportunity to account for the observed results. So, we accord H_0 privileged status and stack the deck against committing Type I error. We do this by adopting a decision rule that minimizes the probability that it will occur.

We always begin by assuming that random sampling error explains our results. That is, we assume the null hypothesis is true, and we proceed by setting a fairly high threshold for rejecting it. The minimum standard is the **.05 level of significance**. If we decide to reject H_0, we want to commit Type I error fewer than 5 times out of 100. Spoken in the language of random error, the .05 threshold frames the inferential fate of H_0 and H_A: If H_0 is true, how often, by chance, will we obtain the relationship observed in the sample? If the answer is "more than 5 times out of 100," we do not reject the null hypothesis. That is, if there is a plausible likelihood—defined as more than 5 percent of the time—that random sampling error produced the relationship we see in the sample, then we do not reject H_0. If the answer is "5 times out of 100 or less," we reject the null hypothesis. That is, if it is unlikely—defined as 5 percent of the time or less—that random sampling error produced the relationship we see in the sample, then we reject H_0.

For the relationship between gender and Democratic ratings, the null hypothesis says that the mean difference in the population is equal to 0 and that the observed difference, 5 points, merely represents the handiwork of hit-and-miss random sampling. Would random processes produce the observed difference more than 5 times out of 100? Five times out of 100 or less? How do we test for the .05 threshold? The answer depends on a familiar factor: random error.

Comparing Two Sample Means

Even though we are now dealing with two sample means, one based on female respondents and one based on male respondents, the idea behind sampling error is the same as before. In Chapter 6 we saw that a single sample mean ($\bar{x}$) is equal to the population mean, give or take some random sampling error. The size of this error depends on two components, the size of the sample (n) and the magnitude of the population standard deviation (σ). Because the population standard deviation is usually unknown, the standard deviation of the sample, s, substitutes for σ. The standard error of the mean is equal to the standard deviation divided by the square root of n, or $s/\sqrt{n}$. The same logic applies, as well, to the difference between two sample means. The difference between two sample means is equal to the real but unobserved difference in the population, give or take some random sampling error. How large is this error?

Table 7-2 Mean Democratic Party Thermometer Ratings, by Gender

Gender	Sample mean	Standard error of sample mean	Squared standard error
Female	60.5	1.00	1.00
	(625)		
Male	55.9	.98	.96
	(553)		
Mean difference	4.6		
Sum of squared standard errors			1.96
Standard error of the mean difference			1.40

Source: University of Michigan, Center for Political Studies, American National Election Study, 2004: Pre- and Post-Election Survey [Computer file]. ICPSR04245–v1. Ann Arbor, Mich.: University of Michigan, Center for Political Studies [producer], 2004; Inter-university Consortium for Political and Social Research [distributor], 2004.

Note: Numbers of cases are in parentheses. The standard deviation of the female mean is equal to 24.9. The standard deviation of the male mean is equal to 23.1.

Consider Table 7-2, which displays mean Democratic thermometer ratings and related statistics for the 625 women and 553 men in the 2004 American National Election Study. On average, females rated the Democrats at 60.5, compared with 55.9 for males—a mean difference of 4.6 points. The standard errors for these means were derived in the now familiar way: by dividing the standard deviation of each sample by the square root of each sample's size. Thus, the female mean has a standard error of 1.00, and the male mean has a standard error of .98. So far we are on familiar terrain. Indeed, these two standard errors, 1.00 for the female mean and .98 for the male mean, provide prima facie evidence against the null hypothesis. What is the reasoning here? Apply the logic of the central limit theorem and use the ±2 rule of thumb. If we were to take an infinite number of random samples from a population of females, only 2.5 percent of those samples would yield means of 60.5 minus 2 standard errors: $60.5 - 2(1.00) = 58.5$. The remaining 97.5 percent of the sample means would yield mean ratings of greater than 58.5. Thus, we could reasonably say that this number, 58.5, represents a minimum mean Democratic rating for women in the population. Similarly, if we were to draw an infinite number of random samples from a population of males, only 2.5 percent of those samples would produce sample means of 55.9 plus two standard errors: $55.9 + 2(.98) \approx 57.9$. The remaining 97.5 percent of the sample means would yield ratings less than 57.9. This number, 57.9, represents a maximum mean Democratic rating for men in the population. If the minimum female mean is 58.5 and the maximum male mean is 57.9, then it is highly unlikely that the true mean difference is equal to 0, H_0's favorite number.

To see whether the statistical evidence supports the prima facie case against H_0, a more formal derivation of the standard error of the mean difference is required. To calculate the **standard error of the difference** between two sample means, follow these steps:

1. Square each mean's standard error. As shown in Table 7-2, the square of the female mean's standard error is equal to 1.00. The square of the male mean's standard error is equal to .96.

2. Sum the squared standard errors. In the example: $1.00 + .96 = 1.96$.
3. Take the square root of the sum obtained in step 2: $\sqrt{1.96} = 1.40$.

So the standard error of the difference between the female mean and the male mean is equal to 1.40.[1] Let's figure out how to use this information to test the null hypothesis using the .05 level of significance. We obtained a mean difference on the dependent variable of 4.6 points. The null hypothesis claims that, in the population, women and men do not differ on the thermometer scale, that the true mean difference is 0. If the null hypothesis is correct, how often will a random sample yield a mean difference of 4.6 by chance? If the answer is "more than 5 percent of the time," then we cannot reject H_0. If the answer is "5 percent of the time or less," then we can reject H_0.

There are two correct ways to apply the .05 standard—or, for that matter, any standard of statistical significance. In the **confidence interval approach,** the investigator uses the standard error to determine the smallest plausible mean difference in the population. If the smallest plausible difference is greater than 0, then the null hypothesis can be rejected. If smallest plausible difference is equal to or less than 0, then the null hypothesis cannot be rejected. In the **P-value approach,** the researcher determines the exact probability of obtaining the observed sample difference, under the assumption that the null hypothesis is correct. If the probability value, or P-value, is less than or equal to .05, then the null hypothesis can be rejected. If the P-value is greater than .05, then the null hypothesis cannot be rejected.

The ±2 rule of thumb is an informal but effective confidence interval approach. As you know, the rule's rationale is rooted in the rounding of one of the normal curve's iconic values, 1.96. In a normal distribution, 95 percent of the cases fall between 1.96 standard units below the mean and 1.96 standard units above the mean. Apply the rule to our example: there is a 95 percent chance the true difference between women and men in the population falls in the interval between the sample difference, 4.6, minus two standard errors and 4.6 plus two standard errors. The lower bound is equal to $4.6 - 2(1.4) = 1.8$. The upper bound is equal to $4.6 + 2(1.4) = 7.4$. Because the difference asserted by H_0, a population difference equal to 0, lies below the lower boundary of 1.8, the null hypothesis can be safely rejected. In practice, many researchers, upon obtaining a sample statistic and standard error, apply an *eyeball test of statistical significance.* According to this informal test, if the sample difference is at least twice as large as its standard error, then the result surpasses the .05 threshold and H_0 can be rejected. So, for example, the sample difference of 4.6 passes the eyeball test, given that 4.6 is more than twice 1.4. This is the correct logic, and it would not steer you in the wrong direction. However, for reasons we will discuss, this informal test exceeds the .05 standard and tilts the rules more strongly in H_0's favor.

When we perform the informal eyeball test using the ±2 rule of thumb, we are finding the upper *and* lower limits of random sampling error. There is only a .025 probability that the true population difference falls below the lower limit of 1.8 and a .025 chance that it falls above the upper limit of 7.4. The informal test applies what is termed a **two-tailed test of statistical significance**. In finding the upper and lower limits of random sampling error, it divides the .05 rejection region in half, reporting the value above which .025 of the curve falls and the value below which .025 of the curve falls. Note, however, that the upper boundary is not relevant to the alternative hypothesis, the hypothesis being tested against the null hypothesis. We do not care whether the population difference could be greater than 7.4. The null hypothesis has no vested interest in the upper boundary. H_0 is strictly a guardian of the statistical real estate bounded at 0. Accordingly, instead of finding upper and lower limits and dividing the .05 rejection region in half, we need find only a lower limit, the lowest plausible difference between the female mean and the male mean in the population. If this limit is greater than 0, we defeat H_0 and gain confidence in the idea that, in the population, the independent and dependent variables are related.

In most of the research situations you will encounter, the .05 threshold will be the standard and a **one-tailed test of statistical significance** will be the order of the day. In performing a one-tailed test of statistical significance, we do not divide .05 in two and find upper and lower confidence boundaries. Rather, we place the entire rejection region in null hypothesis territory, the region of the curve containing a population difference equal to 0. By moving the inferential decision to H_0's side of the curve, we can ask: "What is the probability that the population difference could be as low as 0? Is this probability greater than .05?" In normal estimation, which can reasonably be used with fair-sized samples, the absolute value of Z that marks the boundary between .95 of the curve and .05 in one tail is 1.645. Therefore, the lowest plausible difference is defined by the sample statistic minus 1.645 standard errors. If this value is greater than 0, then we can reject the null hypothesis. In our example, this lower limit would be 4.6 minus 1.645(1.4), which is equal to 4.6 minus 2.3, or 2.3. Thus, for women and men in the population, the lowest plausible difference in Democratic ratings is equal to 2.3. Because 2.3 is greater than 0, we can reject H_0. Notice that the .05 boundary, 2.3, is farther from H_0's property line than the .025 boundary, 1.8, which provided a more stringent test of the alternative hypothesis.

Sometimes when you calculate the difference between two sample statistics, you will obtain a positive number, as we did in the current example. Other times, the calculated difference will be a negative number. Students are sometimes confused about how to apply the one-tail, 1.645 rule when the difference between two sample statistics is negative. Remember that the normal curve is perfectly symmetrical. At $Z = -1.645$, .05 of the curve lies below Z and .95 of the curve lies above Z. At $Z = +1.645$, the symmetry is reversed, with .95 lying below Z and .05 lying above Z. Therefore, for the sole purpose of determining statistical significance, you can safely ignore the sign of the sample difference and deal, instead, with its absolute value. Follow these steps:

1. Multiply the standard error of the difference by 1.645.
2. Subtract this number from the absolute value of the sample difference.
3. If the result is greater than 0, then reject H_0. If the result is not greater than 0, do not reject H_0.

Confidence interval approaches to statistical significance work fine and involve a bare minimum of calculation, but they are blunt. A more precise way of doing it is to find out how many standard errors lie between the sample statistic and the population parameter hypothesized by H_0, and then figure out the likelihood of obtaining this difference if H_0 is true. This is how all computer programs do it, so we need to understand the principle involved.

To determine the exact probability of obtaining a given sample difference if the true population difference is 0, we need to know three things: (1) the difference associated with the alternative hypothesis, H_A; (2) the difference claimed by the null hypothesis, H_0; and (3) the standard error of the difference. This information allows us to calculate a **test statistic,** which tells us exactly how many standard errors separate the sample difference from zero, the difference claimed by H_0. The general formula for any test statistic is:

Test statistic $= (H_A - H_0)$ / Standard error of the difference.

H_A is the observed sample difference, H_0 is the assumed difference in the population (which equals 0), and the standard error is the same as before. Now, the particular test statistic that the researcher uses will depend on the type of estimation being performed. If we use normal estimation, then the test statistic is expressed in units of Z:

$$Z = (H_A - H_0) \text{ / Standard error of the difference.}$$

If we use the Student's t-distribution, then the test statistic is expressed in units of t:

$$t = (H_A - H_0) \text{ / Standard error of the difference.}$$

Which test statistic is appropriate? As a statistical matter, normal estimation may be used for samples of $n = 100$ or more.[2] As a practical matter, most data analysis software supplies statistics for the Student's t-distribution. Let's use the example to demonstrate both sorts of estimation, beginning with Z. Under normal estimation:

$$\begin{aligned} Z &= (H_A - H_0) \text{ / Standard error of the difference} \\ &= (4.6 - 0) / 1.4 \\ &= 3.3. \end{aligned}$$

So, the observed difference, 4.6, and the difference asserted by the null hypothesis, 0, are 3.3 standard errors apart. Yet again, things look bad for the null hypothesis. But how bad? If the population difference really is 0, how often, by chance, will we observe a sample difference of 4.6? Most computer programs answer this question by reporting a P-value. A P-value, or probability value, is the precise magnitude of the region of the curve above the absolute value of Z. If the P-value is greater than .05, we cannot reject H_0. If it is less than or equal to .05, we can reject H_0. Stata, a popular data analysis program, returned a P-value equal to .0005 for $Z = 3.3$. If the null hypothesis is correct, then random sampling error would create the observed relationship between the independent and dependent variables 5 times out of 10,000 by chance. From the eyeball test we already knew that the P-value was less than .025. The computer gives us much greater precision. The null hypothesis represents a highly remote random event. Reject H_0.[3]

For large samples, the normal distribution and the t-distribution are virtually identical. Even so, computer programs usually report the P-values that are associated with a Student's t-distribution test statistic. This test statistic, sometimes called a **t-ratio**, is calculated just like Z:

$$\begin{aligned} t &= (H_A - H_0) \text{ / standard error of the difference} \\ &= (4.6 - 0) / 1.4 \\ &= 3.3. \end{aligned}$$

Unlike the normal curve, however, the P-value for $t = 3.3$ will depend on degrees of freedom. As we saw in Chapter 6, the number of degrees of freedom for a single sample mean is equal to the sample size, n, minus 1. In comparing two sample means, the number of degrees of freedom is equal to the size of the first sample plus the size of the second sample, minus 2. According to Table 7-2, there are 625 individuals in the female sample and 553 individuals

Table 7-3   Opinions on Adoptions by Homosexuals, by Gender

Favor adoptions by homosexuals?	Gender		Total
	Female	*Male*	Total
Yes	54.1%	44.8%	49.7%
	(285)	(217)	(502)
No	45.9%	55.2%	50.3%
	(242)	(267)	(509)
Total	100.0%	100.0%	100.0%
	(527)	(484)	(1,011)

Source: 2004 American National Election Study.

Note: Question: "Do you think gay or lesbian couples, in other words, homosexual couples, should be legally permitted to adopt children?"

in the male sample. So the test statistic, 3.3, has $625 + 553 - 2$, or 1,176 degrees of freedom. As if to confirm Student *t*'s large-sample kinship with *Z*, Stata returned a *P*-value of .0005 for a *t*-ratio of 3.3 with 1,176 degrees of freedom. Reject H_0.

Comparing Two Sample Proportions

The same principles that apply to the comparison of two sample means also apply to the comparison of two sample proportions. This is not too surprising, given that sample proportions, in a statistical sense, *are* sample means.[4] To illustrate the similarities, we consider another gender gap hypothesis: In a comparison of individuals, women are more likely than men to favor laws that permit homosexual couples to adopt children. Table 7-3 shows a cross-tabulation analysis of the dependent variable, whether or not respondents favor allowing adoptions by homosexuals, by the independent variable, gender. Are the data consistent with the hypothesis? It would seem so. Reading across the columns at the "Yes" value of the dependent variable, 54.1 percent of the females are in favor, compared with 44.8 percent of the males, a difference of 9.3 percentage points. Expressed in proportions, the difference would be .541 minus .448, or .093. The always redundant null hypothesis, of course, claims that this sample difference of .093 resulted from random sampling error, that the true difference in the population is .000. Thus, the inferential question becomes: Assuming that the null hypothesis is correct, how often will we observe a sample difference of .093? Again the answer depends on the size of the standard error.

Chapter 6 introduced conventional labeling for sample proportions. The ordinary letter *p* represents the proportion of cases falling into one value of a variable, and the letter *q* represents the proportion of cases falling into all other values of a variable. The proportion *q*, which is called the complement of *p*, is equal to $1 - p$. So that we don't confuse the female sample statistics with the male sample statistics, let's use different subscripts: p_1, q_1, and n_1 refer to the female sample; p_2, q_2, and n_2 refer to the male sample. For the female sample shown in Table 7-3: $p_1 = .541$, $q_1 = .459$, and $n_1 = 527$. For males: $p_2 = .448$, $q_2 = .552$, and $n_2 = 484$. To find the standard error of the difference between proportions, plug the six numbers into the formula that follows:

Table 7-4 Proportions Favoring Adoptions by Homosexuals, by Gender

Gender	Sample proportion (p)	Complement of Sample proportion (q)	Squared standard error (pq/n)
Female	.541 (527)	.459	.00047
Male	.448 (484)	.552	.00051
Difference in proportions	.093		
Sum of squared standard errors			.00098
Standard error of the difference			.031

Source: 2004 American National Election Study.
Note: Numbers of cases are in parentheses.

$$\text{Standard error of the difference in proportions} = \sqrt{p_1 q_1/n_1 + p_2 q_2/n}.$$

The formula implies the following steps:

1. Multiply each proportion (p) by its complement (q) and divide the result by the sample size. For the female sample: (.541*.459)/527 = .00047. For the male sample: (.448*.552)/484 = .00051.
2. Sum the two numbers from step 1. Summing the female number and the male number: .00047 + .00051 = .00098.
3. Take the square root of the number from step 2. In the example: $\sqrt{.00098}$ = .031.

Table 7-4 presents in tabular format the relevant calculations for determining the standard error of the difference between the proportions of females and males who favor allowing adoptions by homosexuals. The female proportion (p_1), .541, and the male proportion (p_2), .448, appear in the left-hand column, with the difference, .093, appearing at the bottom. The complements (q_1 and q_2) are in the next column. Notice that the right-most column is labeled "Squared standard error (pq/n)." By multiplying p times q, and then dividing by the sample size, we are in fact deriving the squared standard error of each proportion.[5] Thus, the derivation of the standard error of the difference in proportions is directly analogous to the derivation of the standard error of the difference in means: the square root of the summation of squared standard errors. In the current example, we found a difference in the sample proportions of .093, with a standard error equal to .031.

Does the observed difference, .093, pass muster with the informal eyeball test? Yes, it does, given that .093 is at least twice its standard error, .031. How about the 1.645 test? To find the lowest plausible difference between the proportions of women and of men who favor adoptions by homosexuals, we first would multiply 1.645 times the standard error: 1.645(.031) = .051. We would then subtract this amount from the difference in sample proportions: .093 − .051 = .042. Since .042 is greater than 0, we know that the sample difference would occur fewer than 5 times out of 100 by chance. What is the precise probability, the *P*-value, associated with the difference in sample proportions? Using normal estimation:

$$Z = (H_A - H_0) \text{ / standard error of the difference}$$
$$= (.093 - 0) / .031$$
$$= 3.0.$$

The computer returned a *P*-value of .0013 for $Z = 3.0$. If the null hypothesis is correct, then chance processes would produce a sample difference of .093 a bit more frequently than 1 time in 1,000. Reject H_0.

The Chi-square Test of Significance

You are now versed in the most common statistical tests—the comparison of two means and the comparison of two proportions. A different but complementary test looks at how the cases are dispersed across the values of the dependent variable. The **chi-square test of significance** determines whether the observed dispersal of cases departs significantly from what we would expect to find if the null hypothesis were correct. Developed by British statistician Karl Pearson in 1900, chi-square (pronounced "ki-square," like the "ki" in "kite," and denoted by χ^2) is the oldest statistical test still in use.[6] It is simple and versatile, two traits that probably account for its longevity.

Chi-square is perhaps the most commonly used statistical test in cross-tabulation analysis. It is especially useful for this purpose because, in determining statistical significance, it takes all the tabular data into consideration. It does not evaluate a specific difference between two percentages or proportions. This makes it easier to analyze relationships between nominal or ordinal variables that may have several categories or values. To illustrate chi-square, we will test this hypothesis: In a comparison of individuals, women will be more likely than men to favor diplomacy instead of military force in solving international problems.

Table 7-5 sets up the cross-tabulation for testing the hypothesis. The values of the independent variable (gender) are on the columns, and the values of the dependent variable (opinions about diplomacy and military force) are on the rows. However, in acquainting yourself with Table 7-5, ignore—for the moment—the main body of the table, the distributions of cases down the "Female" and "Male" columns. Focus exclusively on the right-most "Total" column, which shows the distribution for all 1,041 individuals in the sample. The null hypothesis hones in on this column. In its chi-square guise, the null hypothesis states that the distribution of cases down each column of a table should be the same as the total distribution for all cases in the sample. According to this reasoning, for example, if 40.4 percent of all respondents fall into the "Diplomacy" category, then 40.4 percent of the 530 females will be expected to hold this view, as will 40.4 percent of the 511 males. The null hypothesis applies the same relentless logic to every value of a dependent variable. Since, overall, 25.6 percent of the sample falls into the "Middle" category of the dependent variable, then 25.6 percent of the females and 25.6 percent of the males can be expected to take this position. And, since 33.9 percent said "Force," then 33.9 percent of both genders can be expected to give this response. As usual, of course, the null hypothesis hedges its bet: Any observed departures from this expected pattern are accounted for by random sampling error.

Reflect for a moment on this particular model of the null hypothesis. H_0 offers the view that, in the population, gender and opinions about how to solve international problems have

Table 7-5 Opinions on Diplomacy versus Military Force, by Gender

Favor diplomacy or force?	Gender		Total
	Female	*Male*	
Diplomacy	46.2%	34.4%	40.4%
	(245)	(176)	(421)
Middle	23.6	27.8	25.6
	(125)	(142)	(267)
Force	30.2	37.8	33.9
	(160)	(193)	(353)
Total	100.0%	100.0%	99.9%[a]
	(530)	(511)	(1,041)

Source: 2004 American National Election Study.

Note: Question: "Some people believe the United States should solve international problems by using diplomacy and other forms of international pressure and use military force only if absolutely necessary. Suppose we put such people at "1" on this scale. Others believe diplomacy and pressure often fail and the US must be ready to use military force. Suppose we put them at number 7. And of course others fall in positions in-between, at points 2, 3, 4, 5, and 6. Where would you place yourself on this scale, or haven't you thought much about this?" "Diplomacy" is defined as scale positions 1–3; "Middle" is position 4; "Force" is defined as scale positions 5–7.

[a] Does not total to 100% due to rounding.

nothing to do with each other, that the values of one are not related to the values of the other. So any way you break down a random sample—into smaller groups of 530 or 511 or whatever—you can expect the same pattern, the same distribution of cases, that you see in the overall sample. The alternative hypothesis (H_A) says that, in the population, the two variables have a lot to do with each other, that the values of one variable (opinions about diplomacy and force) *depend on* the values of the other variable (gender). The face-off between H_0 and H_A, then, centers on the question of how well H_0's expectations are met by the observed data.

Chi-square is based on the difference between two numbers: (1) the observed frequency, which is the actual number of cases falling into each cell of a cross-tabulation; and (2) the expected frequency, the hypothetical number of cases that should fall into each cell, if H_0 is correct. For any given cell, an observed frequency is labeled f_o. For example, 245 individuals fall into the "Female–Diplomacy" cell of Table 7-5. So f_o for this cell is 245. An expected frequency for any given cell, the number of cases that the null hypothesis claims should fall into that cell, is labeled f_e. How would one figure out this hypothetical number? Looking at the distribution of all cases across the values of the dependent variable, the null hypothesis would see that 40.4 percent (or .404) of the sample fall into the "Diplomacy" category. According to H_0's logic, then, 40.4 percent of the 530 females should be in the "Diplomacy" category. Thus, .404 times 530 individuals, or about 214 people, should be in this cell, if the null hypothesis is correct. Now, the building block of the chi-square test statistic is the difference between an observed frequency and an expected frequency: f_o minus f_e. For the "Female–Diplomacy" cell, this difference is 245 minus 214, or 31. So, in this cell anyway, the null hypothesis missed the mark by 31 cases. There are 31 more women in the "Diplomacy" category than we expected to find, based on the null hypothesis.

Table 7-6 Chi-square for Opinions on Diplomacy versus Military Force, by Gender

Favor diplomacy or force?		Gender	
		Female	*Male*
Diplomacy	Observed frequency (f_o)	245	176
	Expected frequency (f_e)	214.3	206.7
	$f_o - f_e$	30.7	−30.7
	$(f_o - f_e)^2$	942.5	942.5
	$(f_o - f_e)^2 / f_e$	**4.4**	**4.6**
Middle	Observed frequency (f_o)	125	142
	Expected frequency (f_e)	135.9	131.1
	$f_o - f_e$	−10.9	10.9
	$(f_o - f_e)^2$	118.8	118.8
	$(f_o - f_e)^2 / f_e$	**.9**	**.9**
Force	Observed frequency (f_o)	160	193
	Expected frequency (f_e)	179.7	173.3
	$f_o - f_e$	−19.7	19.7
	$(f_o - f_e)^2$	388.1	388.1
	$(f_o - f_e)^2 / f_e$	**2.2**	**2.2**

Source: 2004 American National Election Study.

The difference between f_o and f_e is the starting place for the chi-square test statistic. In calculating chi-square, however, several additional steps are required. And the chi-square test statistic looks much different from the other formulas we have discussed:

$$\chi^2 = \Sigma (f_o - f_e)^2 / f_e.$$

Before we illustrate how to apply this formula, consider the central role played by the difference between f_o and f_e. Notice that, if the null's expectations, on the whole, are more or less accurate (that is, if $f_o - f_e$ is close to 0), then χ^2 is a small number. As the real-world situation departs from H_0's model world, however, χ^2 grows in size, and we can begin to entertain the idea of rejecting the null hypothesis. How do we obtain the chi-square test statistic?

According to this formula, one obtains a chi-square test statistic for an entire cross-tabulation by following five steps. (For ease of exposition, all numbers used for illustration are rounded. More precise calculations appear in Table 7-6.)

1. Find the expected frequency for each cell. As shown above, this can be accomplished by applying the proportion falling into a given category of the dependent variable to each column total. *Example:* Since .339 of the total sample fall into the "Force" category of the dependent variable, the expected frequency (f_e) for the "Female-Force" cell would be .339 times 530, or about 180.[7]
2. For each cell, subtract the expected frequency from the observed frequency. *Example:* f_o for the "Female-Force" cell is equal to 160 and f_e (from step 1) is equal to 180. Subtracting 180 from 160 yields −20.

3. Square this number for each cell. *Example:* Squaring −20 yields 400.
4. Divide this number by the expected frequency (f_e). *Example:* 400 divided by 180 is equal to 2.2.
5. To arrive at the chi-square test statistic, add up all the cell-by-cell calculations.

Table 7-6 shows the cell-by-cell building blocks of the chi-square statistic for the gender–diplomacy opinions example. The first number in each cell, f_o, is the actual number of cases from the sample data, and the next number, f_e, is the number of cases expected by H_0. The remaining values show each step in the calculation: The difference between f_o and f_e, the square of the difference, and the squared difference divided by f_e. Notice the pattern in the "$f_o - f_e$" cells. The null hypothesis underestimates the number of females in the "Diplomacy" category (and overestimates the number of males), and H_0 overestimates the number of females in the "Force" category (and underestimates the number of males). The overall pattern of cases may accord more comfortably with H_A than H_0. Does Table 7-6 defeat the null hypothesis?

The chi-square test statistic is calculated by summing the cell-by-cell calculations. For Table 7-6: 4.4 + 4.6 + .9 + .9 + 2.2 + 2.2 = 15.2. The null hypothesis says that, after squaring each departure and dividing by f_e, and then aggregating all the numbers, we should get a result that is close to 0, give or take random sampling error. But in Table 7-6 the calculated value of chi-square is 15.2. It might be another poor showing for H_0. What are the limits of random sampling error for a calculated value of chi-square?

A χ^2 statistic has its own distribution, but its inferential interpretation is roughly analogous to a Z score or t-statistic: the larger it gets, the lower the probability that the results were obtained by chance. However, the precise value of χ^2 that marks the boundary of the .05 threshold is not fixed. Rather, it depends, as does the Student's t-distribution, on degrees of freedom. For chi-square, the number of degrees of freedom is determined by the number of rows and columns in a cross-tabulation. Specifically, the number of degrees of freedom is equal to the number of rows in the cross-tabulation, minus one, multiplied by the number of columns, minus one:

Number of degrees of freedom = (number of rows − 1)(number of columns − 1).

So Table 7-6 has $(3 − 1)(2 − 1) = 2$ degrees of freedom.

Now consider Table 7-7, a table of critical values of χ^2. A **critical value** marks the upper plausible boundary of random error and so defines H_0's limit. A specific critical value depends on degrees of freedom (the left-hand column of Table 7-7) and level of significance chosen by the researcher. In our example we have 2 degrees of freedom, and we will apply the .05 standard. Find 2 in the degrees of freedom column and read across to the .05 column. The critical value, 5.991, tells us this: If the null hypothesis is correct that, in the population, gender and diplomacy opinions are completely independent of each other, we would obtain a χ^2 of less than 5.991 more frequently than 5 times out of 100 by chance. So the null hypothesis reigns over all the territory between 0 (chi-square cannot assume negative values) and the critical value of chi-square, 5.991. For any chi-square test statistic in this region, between 0 and the critical value, the null hypothesis cannot be rejected. But of course our χ^2 test statistic, 15.2, exceeds critical value of 5.991. Again, the computer can tell us precisely how often we would get a value this huge if H_0 were true. Stata returned this P-value for 15.2 (and 2 degrees of freedom): .0005. If the null hypothesis is correct, then random sampling

Table 7-7 Critical Values of χ^2

Degrees of freedom	Area to the right of critical value			
	.10	.05	.025	.01
1	2.706	3.841	5.024	6.635
2	4.605	5.991	7.378	9.210
3	6.251	7.815	9.348	11.345
4	7.779	9.488	11.143	13.277
5	9.236	11.070	12.833	15.086
6	10.645	12.592	14.449	16.812
7	12.017	14.067	16.013	18.475
8	13.362	15.507	17.535	20.090
9	14.684	16.919	19.023	21.666
10	15.987	18.307	20.483	23.209
11	17.275	19.675	21.920	24.725
12	18.549	21.026	23.337	26.217
13	19.812	22.362	24.736	27.688
14	21.064	23.685	26.119	29.141
15	22.307	24.996	27.488	30.578
16	23.542	26.296	28.845	32.000
17	24.769	27.587	30.191	33.409
18	25.989	28.869	31.526	34.805
19	27.204	30.144	32.852	36.191
20	28.412	31.410	34.170	37.566
21	29.615	32.671	35.479	38.932
22	30.813	33.924	36.781	40.289
23	32.007	35.172	38.076	41.638
24	33.196	36.415	39.364	42.980
25	34.382	37.652	40.646	44.314
26	35.563	38.885	41.923	45.642
27	36.741	40.113	43.195	46.963
28	37.916	41.337	44.461	48.278
29	39.087	42.557	45.722	49.588
30	40.256	43.773	46.979	50.892
40	51.805	55.758	59.342	63.691
50	63.167	67.505	71.420	76.154
60	74.397	79.082	83.298	88.379
70	85.527	90.531	95.023	100.425
80	96.578	101.879	106.629	112.329
90	107.565	113.145	118.136	124.116
100	118.498	124.342	129.561	135.807

error would create the observed relationship between the independent and dependent variable 5 times out of 10,000 by chance. Reject H_0.

Chi-square is remarkably adaptable. It can be used to test for independence between any categorical variables, nominal or ordinal, in any combination, for tables of any size. Chi-square can also be used to test whether the distribution of a single variable departs from chance—or from an expected distribution devised by the researcher. George Bohrnstedt and David Knoke cite the (somewhat whimsical) example of the distribution of birth months.[8] Are births more likely in some months than others? In the absence of any alternative reasoning, one would expect an even distribution of cases across the 12 months of the year. In a random sample of individuals, one-twelfth should report being born in each of the 12 months, January through December. So, for a random sample of, say, 120 individuals, an expected frequency of 10 should fall into each category. Chi-square, in precisely the manner we have discussed, would compare these expected frequencies with the observed data and report a calculated value of χ^2, which could be compared with the plausible limit of random error. (When examining the distribution of a single variable, the number of degrees of freedom is simply the number of categories minus one. Because there are 12 months, there are 11 degrees of freedom in this example.) Bohrnstedt and Knoke's analysis reveals no statistically significant departure from the expected pattern.[9]

As durable as it is, chi-square also has shortcomings. When expected frequencies are small (generally 5 or fewer), it does not work well. This problem can often be addressed by collapsing the categories of the offending variable, thereby repopulating the cells with higher numbers of cases. Statisticians have developed other solutions to this problem.[10] More generally, chi-square is quite sensitive to sample size. In fact, χ^2 is directly proportional to n. If we were to double the observed frequencies in Table 7-6, for example, the χ^2 statistic also would double, producing a P-value even more remote than the one we obtained originally. Thus, sizable samples often produce statistically significant results even though, from a substantive viewpoint, the relationship between the independent and dependent variables may appear rather anemic. Remember that chi-square, like any test of significance, can reveal only the likelihood that the observed results occurred by chance. It does not interpret the data for us, and it does not give us a direct reading of the strength of the relationship. Interpretation requires thought and imagination, attributes not supplied by tests of significance. However, other statistics can help the researcher in the strength-of-relationship department.

MEASURES OF ASSOCIATION

Determining the statistical significance of interesting relationships is an essential first step in testing hypotheses and exploring explanations. Measures of association are an additional resource for the investigator. A measure of association communicates the strength of the relationship between an independent variable and a dependent variable. Although statistical significance is usually the main thing you want to know about a relationship, measures of association always add depth of interpretation—and they can be of central importance in testing hypotheses. Suppose that you are investigating the hypothesized relationship between gender and partisanship: women are more likely than men to be Democrats. Although you expect this relationship to hold among all age groups, you theorize that the relationship will be stronger for younger people than for older people. The first stop along the investigatory line, then, is to find out whether the relationship is statistically significant for each age cohort. The second stop is to find out whether the relationship is stronger for the younger cohort.

Statisticians have developed a large number of measures of association. Are some preferred to others? A preferred measure of association has two main features. First, it uses a **proportional reduction in error** (**PRE**) approach for gauging the strength of a relationship. A PRE measure is a prediction-based metric that varies in magnitude between 0 and 1. The precise value of the measure tells you how much better you can predict the dependent variable

Table 7-8 Gun-Control Attitudes and Gender

Gun ban?	Male	Female	Total
	Without knowledge of the independent variable		
Oppose	?	?	807
Favor	?	?	707
Total	?	?	1,514
	With knowledge of the independent variable		
Oppose	449	358	807
Favor	226	481	707
Total	675	839	1,514

Source: Steven J. Rosenstone, Donald R. Kinder, Warren E. Miller, and the American National Election Studies, American National Election Study, 1996: Pre- and Post-election Survey, 4th version (Ann Arbor: University of Michigan, Center for Political Studies [producer], 1999; Inter-University Consortium for Political and Social Research [distributor], 2000).

Note: Question: "Do you favor or oppose a ban on the sale of all handguns, except those that are issued to law enforcement officers?"

by knowing the independent variable than by not knowing the independent variable. If knowledge of the independent variable does not provide any help in predicting the dependent variable, then a PRE statistic will assume a value of 0. If knowledge of the independent variable permits perfect prediction of the dependent variable, then a PRE statistic will assume a magnitude of 1. Second, asymmetric measures are preferred to symmetric measures. A **symmetric measure of association** takes on the same value, regardless of whether the independent variable is used to predict the dependent variable or the dependent variable is used to predict the independent variable. An **asymmetric measure of association,** by contrast, models the independent variable as the causal variable and the dependent variable as the effect. Because asymmetric measures are better suited to the enterprise of testing causal hypotheses, they are preferred to symmetric measures, which are agnostic on the question of cause and effect. In the sections that follow, we describe two asymmetric PRE measures. **Lambda** is designed to measure the strength of a relationship between two categorical variables, at least one of which is nominal-level.[11] **Somers' d_{yx}** is appropriate for gauging the strength of ordinal-level relationships.[12]

Lambda

Suppose you were presented with the upper half of Table 7-8 and asked to summarize the relationship between gender and gun-control opinions. Obviously, this cross-tabulation provides no information about the independent variable. You can, however, identify the dependent variable: the distribution of oppose/favor responses for all 1,514 respondents. You can further report its mode, "oppose," the response given by 807 individuals. Assume that, based only on this knowledge, you randomly drew respondents one at a time and tried to guess their gun-control opinions. What is your better guess, "oppose" or "favor"? Because there are more respondents who oppose gun control than favor it, your better guess is the modal response, "oppose." In the long run, by guessing "oppose" for each randomly chosen case you are guaranteed 807 correct guesses. But you will record a missed guess for every case that is not in the modal category: the 707 respondents who support gun control. In the lambda logic, this number, 707, measures prediction error without knowledge of the independent variable. *Prediction error without knowledge of the independent variable* is the number of errors made when using the overall distribution of the dependent variable as a predictive instrument.

Now suppose that the relationship between the dependent variable and the independent variable is revealed to you, as in the lower half of Table 7-8, which shows a cross-tabulation of gender, the independent variable, and attitudes toward gun control. Notice that, among males, 449 oppose gun control (the modal response for men), and 226 favor gun restrictions. For females we see a different distribution: 481 in favor (the modal response for women) and 358 oppose. Suppose that, based on knowledge of how the dependent variable is distributed within categories of the independent variable—more males oppose than support, more females support than oppose—you randomly drew respondents one at a time and tried to guess his or her gun-control opinion. What is your better guess, "oppose" or "favor"? For each randomly picked male, the better bet is to guess "oppose." This stratagem will guarantee you 449 correct hits. You will have some missed guesses, too: the 226 male respondents who support gun control. For each randomly chosen female, your better guess is "support." Based on what you know about the distribution of opinion among women, this approach will give you 481 hits, along with 358 misses. All told, how many errors will there be? Adding the 226 errors for males and the 358 misses for females: 226 + 358 = 584. In the lambda logic, this number, 584, measures prediction error with knowledge of the independent variable. *Prediction error with knowledge of the independent variable* is the number of errors made when using the distribution of the dependent variable within each category of the independent variable as a predictive instrument. Now consider the formula for lambda:

$$\text{Lambda} = \frac{\text{Prediction error without knowledge of the independent variable} - \text{Prediction error with knowledge of the independent variable}}{\text{Prediction error without knowledge of the independent variable}}$$

The numerator, prediction error without knowledge of the independent variable minus prediction error with knowledge of the independent variable, is the reduction-in-error part of PRE. If the independent variable does not provide much improvement, if you make about as many prediction errors with the independent variable as without it, then the numerator approaches 0. But if many fewer prediction errors are made by knowing the independent variable, then the size of the numerator stays close to its first term, prediction error without knowledge of the independent variable. In the gun-control example, there were 707 errors without knowledge of the independent variable and 584 missed guesses when the independent variable was taken into account, improving the prediction by 707 minus 584, or 123. The denominator, prediction error without knowledge of the independent variable, translates error reduction into a ratio, providing the proportional part of PRE. For the gun-control example:

$$\text{Lambda} = (707 - 584) / 707 = 123 / 707 \approx .17$$

Of the 707 errors that were made using only the dependent variable as a guide, the independent variable helped us pick up 123, or 123/707, which is about .17. So the independent variable provided a 17 percent reduction in prediction error.

Is this relationship weak? Or is it moderate, or strong? In the analysis of social science data, especially individual-level survey data, large PRE magnitudes (of, say, .5 or above) are uncommon. Lesser values (of about .3 or lower) are more frequent. Although one independent variable may do part of the work in predicting a dependent variable, the prediction can almost always be improved by taking into account additional independent variables. Keeping these qualifications in mind, the following is a useful guideline for labeling the strength of a PRE statistic:

Weak	=	Less than or equal to .1
Moderate	=	Greater than .1 but less than or equal to .2
Moderately strong	=	Greater than .2 but less than of equal to .3
Strong	=	Greater than .3

According to this guideline, the lambda we obtained from Table 7-8 (lambda = .17), falls in the moderate range.

A Problem with Lambda

Lambda is based on a simple PRE model that has considerable intuitive appeal. Lambda gauges strength by comparing the predictive power of the overall mode of the dependent variable with the predictive power of the mode of the dependent variable within categories of the independent variable. However, this approach sometimes creates a problem. When the within-category modes are the same as the overall mode, then lambda will return a value of 0, even for tables that clearly show a relationship between the independent and dependent variables. Let's review the problem and settle on a workable solution.

Table 7-9 tests the relationship between a nominal independent variable, gender, and campaign interest, which has been collapsed into two categories. To make things interesting, and to illustrate a separate point about using measures of association, Table 7-9 displays the relationship between gender and campaign interest, controlling for race. By applying your table-reading skills, you could well describe the relationship between the independent and dependent variables for each category of the control. Among whites, gender has some effect, since men are somewhat more likely than women to say that they are "very interested" in the campaign. This difference, about 6 percentage points, is interesting if not particularly earth-shattering. Among blacks, however, the gender difference is huge. A nearly 20-percentage-point difference exists between the percentages of black males (44.3%) and black females (25.8%) who expressed a high level of campaign interest. Clearly, too, interaction is at work, since the effect of the independent variable, gender, depends on the value of the control, race. The calculated values of chi-square are statistically significant for both whites and blacks.[13]

But notice that lambda remains mute on these relationships. For both whites and blacks, this PRE measure says that gender does not help at all in predicting campaign interest. Why is this? Because, guessing the modal value of the dependent variable, "some/not very," produces the same number of errors with knowledge of the independent variable as without knowledge of the independent variable. Among all whites, for example, the modal response is "some/not very," and the modal response among white males and white females also is "some/not very." This is the problem with lambda. When the dependent variable has a low amount of variation—one category is populated much more heavily with cases than are the other categories—lambda becomes an insensitive measure of strength.

In such situations, another measure of association, **Cramer's V**, is recommended. Cramer's V, which is based on chi-square, takes a value of between 0, no relationship, and 1, a perfect relationship.[14] Because Cramer's V is not a PRE measure, the researcher may not

Table 7-9   Campaign Interest, by Gender, Controlling for Race

Interested in campaign?	Race					
	White			Black		
	Gender		Total	Gender		Total
	Male	*Female*		*Male*	*Female*	
Very	30.0%	24.1%	26.8%	44.3%	25.8%	32.9%
	(198)	(191)	(389)	(35)	(33)	(68)
Some/not very[a]	70.0%	75.9%	73.2%	55.7%	74.2%	67.1%
	(462)	(603)	(1,065)	(44)	(95)	(139)
Total	100.0%	100.0%	100.0%	100.0%	100.0%	100.0%
	(660)	(794)	(1,454)	(79)	(128)	(207)
Lambda			.000			.000
Cramer's V			.067			.192

Source: 1996 American National Election Study.

Note: Numbers of cases are in parentheses. Question: "Some people don't pay much attention to political campaigns. How about you? Would you say that you have been very much interested, somewhat interested, or not much interested in the political campaigns so far this year?"

[a] Respondents who are "somewhat" or "not very" interested have been collapsed into a single category.

interpret its value as a gauge of predictive accuracy. However, when it comes to interpreting controlled comparisons, such as that presented in Table 7-9, Cramer's V is entirely appropriate and quite useful. Just as we learned from our substantive interpretation, Cramer's V reflects a modest relationship between gender and campaign interest for whites (V = .067) but a much stronger relationship among blacks (V = .192).

Somers' d_{yx}

Somers' d_{yx} and lambda bear a strong PRE family resemblance. As you know, the values of a nominal variable only communicate a difference between two cases. The nominal measure of choice, lambda, tells you if this difference matters in helping to predict differences on the dependent variable. The values of an ordinal variable communicate the direction of a difference between two cases. If two cases differ, then one case has more or less of the measured characteristic than the other case. Ordinal measures of association, such as Somers' d_{yx}, tell you whether the direction of the difference between cases on the independent variable matters in helping to predict the direction of their difference on the dependent variable.

Let's say that we are investigating the hypothesis that people who attend religious services less frequently will be more supportive of abortion rights than those who attend more frequently. We measure the independent variable, religious attendance, with three ordinal categories: frequent, occasional, and infrequent. The dependent variable, abortion-rights support, is measured in two ordinal categories: low or high. Now imagine a pair of observations— one person attends frequently and one attends infrequently. Our hypothesis would predict that the two observations also will have different values on the dependent variable: the respondent who attends frequently will have "low" abortion-rights support and the respondent who

Table 7-10 Religious Attendance and Abortion Rights Support

Abortion rights support	Religious attendance			Total
	Frequent	*Occasional*	*Infrequent*	
Low	7	5	4	16
High	3	5	6	14
Total	10	10	10	30

Note: Hypothetical data.

infrequently attends will fall into the "high" support category. To the extent that a cross-tabulation is populated with pairs of observations like this one—the observation with lower religious attendance has higher abortion-rights support than the observation with higher religious attendance—Somers' d_{yx} will have a magnitude that approaches 1. But think of the other pairings we might find. For some paired observations, the respondent with higher religious attendance might have higher support of abortion than the respondent with lower attendance. Or two respondents may differ in how often they attend religious services but have the same abortion-rights opinion. To the extent that a cross-tabulation is populated with pairs of observations like these—pairs that detract from the predictive power of the independent variable—Somers' d_{yx} will assume a value closer to 0.

Table 7-10, which uses fabricated data to address the religious attendance–abortion opinion hypothesis, will help to illustrate the logic and computation of Somers' d_{yx}. (For simplicity, the table shows raw frequencies only.) Notice that the table's 30 observations tend to be distributed from upper-left to lower right. Seven of the 10 frequent attendees are in the upper-left cell; occasional attendees are evenly split; and 6 of the 10 infrequent attendees are in the lower-right cell. Somers' d_{yx} uses the following template for determining the direction of a cross-tabulation relationship. If the cases are distributed from the upper left to the lower right, then the relationship is measured as positive. If the cases are distributed from the lower left to the upper right, then the relationship is measured as negative. As you can see, the direction of the relationship is determined by how the analysis is set up. Table 7-10 is arranged so that frequency of religious attendance declines from left to right and abortion-rights support increases from top to bottom. Based on this arrangement, our hypothesis suggests that Somers' d_{yx} will report a positive relationship. Of course, it would make just as much sense to set up the table so that the values of religious attendance increase from left to right. If this were the case, our hypothesis would be looking for a negative sign on the measure of association. Because of this sort of arbitrariness, it is important not to become too concerned about whether Somers' d_{yx} has a plus sign or a minus sign. Concentrate on the magnitude of the measure. The formula for Somers' d_{yx} is as follows:

$$\text{Somers' } d_{yx} = (C - D) / (C + D + T_y),$$

Where:

C = Number of concordant pairs
D = Number of discordant pairs
T_y = Number of pairs that are tied on the dependent variable.

In the language of ordinal measures of association, a **concordant pair** is a pair of observations that is consistent with a positive relationship. A **discordant pair** is a pair of observations that is consistent with a negative relationship. A **tied pair** has different values on the independent variable but the same value on the dependent variable. According to the formula, if concordant pairs greatly outnumber discordant pairs, and there are few ties, Somers' d_{yx} will approach +1. By the same token, if discordant pairs are much more prevalent than concordant pairs, then Somers' d_{yx} will approach –1. But as the pattern becomes less consistent—lots of offsetting concordant and discordant pairs that weaken the numerator, too many tied pairs that inflate the denominator—the measure of association slumps toward zero.

Let's calculate Somers' d_{yx} for the relationship in Table 7-10. Consider the upper-left cell, the frequent-low cell that contains 7 observations. Every individual in this cell has a concordant relationship with every individual who has lower religious attendance, those who attend occasionally or infrequently, and who express higher abortion-rights support. So the 7 observations in the frequent-low cell form concordant pairs with the 5 observations in the occasional-high cell and with the 6 cases in the infrequent-high cell. The number of concordant pairings with the frequent-low cell: $(7 * 5) + (7 * 6) = 35 + 42 = 77$. Now move to the 5 people in the occasional-low cell. Every case in this cell bears a positive relationship with every individual who has lower religious attendance and a higher level of abortion-rights support: the 6 individuals in the infrequent-high cell. Calculating the number of these concordant pairs: $5 * 6 = 30$. Now move to the infrequent-low cell. Are there any observations with which the cases in this cell can be positively paired? No. We have exhausted the number of concordant pairs in the table. Let's aggregate the numbers we have: $77 + 30 = 107$. So Table 7-10 has 107 concordant pairs of observations, pairs that fit a positive relationship.

According to the Somers' d_{yx} logic, the number of concordant observations must be reduced by the number of discordant observations. Indeed, if a large number of people fit a negative pattern—frequent attendees who support abortion rights and less frequent attendees who oppose abortion rights—then the numerator needs to record this offsetting tendency and report a value closer to 0. How many discordant pairs are there in Table 7-10? Start with the 3 observations in the frequent-high cell, religiously observant individuals who nonetheless support abortion rights. Every observation in this cell bears a discordant relationship with every individual who attends less frequently and professes lower abortion-rights support: the 5 cases in the occasional-low cell and the 4 cases in the infrequent-low cell. The products and sums: $(3* 5) + (3 * 4) = 15 + 12 = 27$. Move to the occasional-high cell. The 5 occupants of this cell are negatively paired with the 4 cases in infrequent-low cell: $5 * 4 = 20$. That exhausts the number of discordant pairs in the table. Totaled we have: $27 + 20 = 47$ discordant pairs, pairs that fit a negative relationship between religious attendance and support for abortion rights.

Somers' d_{yx}, like all ordinal measures of association, arrives at the numerator by subtracting the number of discordant pairs from the number of concordant pairs. For our illustrative example: $107 – 47 = 60$. Consistent with our hypothesis, there are more concordant pairs than discordant pairs in Table 7-10. How do we convert this number, 60, to a proportion? How many pairs—and what pairs—belong in the denominator? The various ordinal measures offer different answers to this question.[15] Somers' d_{yx} is based on the intuitively appealing idea that any two observations having different values on the causal variable provide information about the effect of the independent variable on the dependent variable. Perhaps the effect is positive, perhaps the effect is negative, or perhaps the effect is zero. Observations that are tied on the dependent variable, zero-effect observations, represent weakness in the relationship. In Table 7-10, for example, there are quite a few individuals for whom the independent variable is changing—from frequent, to occasional, to infrequent— but the dependent variable is not. These cases need to be taken into account.

Begin with the 7 observations in the frequent-low cell. These observations are tied on the dependent variable with the 5 cases in the occasional-low cell and the 4 cases in the infrequent-low cell: 7 * 5 = 35 and 7 * 4 = 28. The occasional-low and the infrequent-low cells are tied, too: 5 * 4 = 20. Move to the next value of the dependent variable, high abortion-rights support, and find these ties: 3 * 5 = 15, 3 * 6 = 18, and 5 * 6 = 30. Totaled, Table 7-10 contains a large number of cases that are tied on the dependent variable: 35 + 28 + 20 + 15 + 18 + 30 = 146. We now have all the numbers we need to calculate Somers' d_{yx}:

$$\text{Somers' } d_{yx} = (107 - 47) / (107 + 47 + 146) = 60 / 300 = .20.$$

Somers' d_{yx} communicates how much better we can predict the dependent variable by using the independent variable as a predictive instrument than by not using it as a predictive instrument. Compared with how well we can predict abortion-rights support by not knowing frequency of religious attendance, we can improve our prediction by .20, or 20 percent, by knowing how often individuals attend religious services. In terms of strength, this relationship falls in the moderate range.

SUMMARY

As we have seen in previous chapters of this book, political researchers are interested in describing political variables and constructing explanations for political phenomena. They propose hypotheses and analyze relationships to find out whether their explanations are consistent with what is happening in the political world. Yet we have also seen that political researchers often do not have complete information about each and every unit of analysis they seek to understand. Because they often rely on sample data, researchers learn to be cautious: "I have performed the analysis and found this relationship. If I assert that my findings describe what is really going on 'out there' in the population, will I be making a mistake?"

With random samples, of course, we can never know the answer to this question with absolute certainty. But, thanks to tests of statistical significance, we can determine the probability of making a mistake. As discussed in this chapter, a test of significance tells the researcher the probability of committing an inferential error—asserting the existence of a relationship in the population if, in reality, no relationship exists. The approaches we have discussed—the informal eyeball test, the 1.645 test, the chi-square test—are based on the same idea. If the sample results can be accounted for by random sampling error, then we must say that there is no relationship in the population. We accept the null hypothesis. If, by contrast, the sample results cannot plausibly be attributed to random sampling error, then we can say that a relationship probably does exist in the population. We would take an acceptable chance and reject the null hypothesis. A computer-reported P-value, the exact probability of obtaining the observed relationship if the null hypothesis is correct, affords the researcher a high degree of precision in making this inferential decision.

We have also discussed measures of association, statistics that help the researcher gauge the strength of the relationship between an independent variable and a dependent variable. Over the years, social scientists and statisticians have devised a wide array of these measures.

Preferred measures provide a proportional reduction in error interpretation. Asymmetric measures are preferred to symmetric measures. A PRE measure tells you how much better you can predict the values of the dependent variable by knowing the independent variable than by not knowing the independent variable. Asymmetric measures model the independent variable as the causal variable and the dependent variable as the effect. For the analysis of nominal relationships, lambda is the measure of choice. For those situations in which lambda cannot detect a relationship, Cramer's V is recommended. For the analysis of ordinal relationships, Somers' d_{yx} is the preferred PRE measure of association. In this chapter we considered the PRE logic as it applies to nominal and ordinal relationships. Yet the same logic applies to relationships between interval-level variables. The analysis of interval-level relationships is our next topic of discussion.

KEY TERMS

.05 level of significance (p. 147)
asymmetric measure of association (p. 160)
chi-square test of significance (p. 154)
concordant pair (p. 165)
confidence interval approach (p. 149)
Cramer's V (p. 162)
critical value (p. 157)
discordant pair (p. 165)
lambda (p. 160)
measure of association (p. 146)
null hypothesis (p. 146)
one-tailed test of significance (p. 150)

proportional reduction in error (p. 159)
P-value approach (p. 149)
Somers' d_{yx} (p. 160)
standard error of the difference (p. 148)
symmetric measure of association (p. 160)
test of statistical significance (p. 146)
test statistic (p. 150)
tied pair (p. 165)
t-ratio (p. 151)
two-tailed test of statistical significance (p. 149)
Type I error (p. 147)
Type II error (p. 147)

EXERCISES

1. How could current election laws be changed so that more citizens go to the polls on Election Day? One suggestion is to hold elections on nonworkdays, such as holidays or weekends, instead of workdays, as is the current practice in the United States. Interestingly, however, the mean turnout in democracies holding elections on workdays is actually *higher* than the mean turnout in democracies holding elections on nonworkdays. The following table reports the mean turnouts and standard errors for 18 workday democracies and 40 nonworkday democracies:[16]

Workday or nonworkday?	Mean turnout	Standard error
Workday	71.8	2.75
Nonworkday	68.5	2.24

A. What is the difference between the mean turnout in workday countries and the mean turnout in nonworkday countries? State the null hypothesis for the workday/nonworkday comparison.

B. What is the standard error of the difference between the workday mean and the nonworkday mean? Calculate the *t*-statistic.

C. Does the mean difference pass the eyeball test of significance? What is your inferential decision: reject the null hypothesis or do not reject the null hypothesis? Explain.

D. Upon seeing that workday democracies have a higher mean turnout than do nonworkday democracies, an election reformer suggests that, to boost turnout, countries currently holding elections on weekends or holidays should switch to workdays. Is this reformer's suggestion supported by the data? Explain.

2. Scholars of political development have investigated the so-called oil curse, the idea that oil-rich countries tend not to develop democratic systems of governance, that "oil and democracy do not mix."[17] Framing this idea in the form of a hypothesis: In a comparison of countries, those having economies less dependent on oil wealth will be more democratic than will countries having economies more dependent on oil wealth. Consider the following data, which present information for 20 countries with non–oil-dependent economies and 15 countries with oil dependent economies. The dependent variable is a 7-point level-of-democracy scale, with higher scores denoting more democracy. Here are the mean values and standard errors of the democracy scale for non–oil-dependent and oil-dependent countries:

Oil-dependent economy?	Mean democracy scale	Standard error
No	4.6	.48
Yes	2.6	.28

A. State the null hypothesis for the non–oil-dependent/oil-dependent comparison.
B. Calculate the difference between the means of the democracy scale for non–oil-dependent and oil-dependent countries. Calculate the standard error of the difference. Calculate the *t*-statistic.
C. Does the mean difference pass the 1.645 test of significance? Explain how you know.
D. Based on these results, suppose a researcher decides to reject the null hypothesis. Is this decision supported by the statistical evidence? Explain.

3. Below are two conventional wisdoms, statements generally believed to be accurate descriptions of the world. Accompanying each statement are data (from the General Social Surveys) that allow you to test each conventional wisdom. For each statement:
A. State the null hypothesis.
B. Calculate the difference between the proportions or means.
C. Calculate the standard error of the difference in proportions or means.
D. Using the eyeball test, (i) state whether you would accept the null hypothesis or reject the null hypothesis and (ii) explain your reasoning.
    CONVENTIONAL WISDOM ONE: *Catholics have larger families than Protestants.* The 2006 General Social Survey asked respondents how many brothers and sisters they have. Catholic respondents reported an average of 4.11 siblings (standard error = .11), whereas Protestants averaged 3.81 siblings (standard error = .09).
    CONVENTIONAL WISDOM TWO: *Southerners are more likely than nonsoutherners to own a gun.* According to the 2006 General Social Survey, .35 of southern respondents owned a gun (*n* = 755), compared with .33 of nonsoutherners (*n* = 1,216).

4. In 2001, the U.S. Senate voted on the question of whether to allow oil and gas drilling in the Gulf of Mexico. The relationship between party affiliation and vote is shown in the following table:

Allow drilling?	Party		Total
	Democrat	*Republican*	
No	33	0	33
Yes	18	49	67
Total	51	49	100

A. Calculate chi-square for this table. Show your work. Draw a table just like the one above, leaving room in each cell to record these numbers: observed frequency (f_o), expected frequency (f_e), $f_o - f_e$, $(f_o - f_e)^2$, and $(f_o - f_e)^2/f_e$.

B. Use chi-square to test the null hypothesis that, in the population from which the sample was drawn, there is no relationship between party and vote. Using Table 7-7, find the appropriate critical value (use the .05 level of significance). (i) Write down the critical value. (ii) Should you reject the null hypothesis or not reject the null hypothesis? (iii) Explain your reasoning.

C. Calculate lambda for this table. (i) Show your work. (ii) Write a sentence explaining exactly what the value of lambda means. (iii) State whether the relationship is weak, moderate, moderately strong, or strong.

5. The following table of raw frequencies can be used to test this hypothesis: In a comparison of individuals, people with lower levels of education will express stronger support for the death penalty than will people with higher levels of education.[18]

Support for death penalty	Education		
	High school or less	*Some college*	*College or higher*
Not strong	47	43	56
Strong	49	50	35

A. Based on the way the table is arranged, would the hypothesis expect a positive sign on Somers' d_{yx} or a negative sign on Somers' d_{yx}? Explain how you know.

B. Calculate Somers' d_{yx} for this table. Show your work. On a sheet of paper, label three columns: Concordant pairs (C), Discordant pairs (D), and Tied pairs (T_y). Work your way through the table, recording and computing each concordant pair, discordant pair, and tied pair.

C. Examine the value of Somers' d_{yx} that you calculated in B. Exactly what does this value tell you about your ability to predict death-penalty opinions using education as a predictive instrument?

D. State whether the relationship is weak, moderate, moderately strong, or strong.

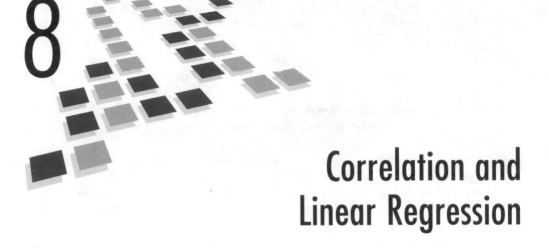

8

Correlation and Linear Regression

LEARNING OBJECTIVES

In this chapter you will learn:
- How to use correlation analysis to describe the relationship between two interval-level variables
- How to use regression analysis to estimate the effect of an independent variable on a dependent variable
- How to perform and interpret dummy variable regression
- How to use multiple regression to make controlled comparisons
- How to analyze interaction relationships using multiple regression

By this point you have covered a fair amount of methodological ground. In Chapter 3 you learned two essential methods for analyzing the relationship between an independent variable and a dependent variable: cross-tabulation analysis and mean comparison analysis. Chapters 4 and 5 covered the logic and practice of controlled comparison—how to set up and interpret the relationship between an independent variable and a dependent variable, controlling for a third variable. In Chapters 6 and 7 you learned about the role of inferential statistics in evaluating the statistical significance of a relationship, and you became familiar with measures of association. You can now frame a testable hypothesis, set up the appropriate analysis, interpret your findings, and figure out the probability that your observed results occurred by chance.

In many ways, correlation and linear regression are similar to the methods you have already learned. **Correlation analysis** produces a measure of association, known as Pearson's correlation coefficient or Pearson's r, that gauges the direction and strength of a relationship between two interval-level variables. **Regression analysis** produces a statistic, the regression coefficient, that estimates the size of the effect of the independent variable on the dependent variable.

Suppose, for example, we were working with survey data and wanted to investigate the relationship between individuals' ages and the number of hours they spend watching television each day. Is the relationship positive, with older individuals watching more hours of TV than younger people? Or is the relationship negative, with older people watching less TV than younger people? How strong is the relationship between age and the number of hours devoted to TV? Correlation analysis would help us address these questions.

Regression analysis is similar to mean comparison analysis. In performing mean comparison analysis, you learned to divide subjects on the independent variable—females and males, for example—and compare values on the dependent variable—such as mean Democratic Party thermometer ratings. Furthermore, you learned how to test the null hypothesis, the assumption that the observed sample difference between females and males was produced by random sampling error. Similarly, regression analysis communicates the mean difference on the dependent variable, such as thermometer ratings, for subjects who differ on the independent variable, females compared with males. And, just as in the comparison of two sample means, regression analysis provides information that permits the researcher to determine the probability that the observed difference was produced by chance.

However, regression is different from the previously discussed methods in two ways. First, regression analysis is very precise. It produces a statistic, the regression coefficient, that reveals the exact nature of the relationship between an independent variable and a dependent variable. A regression coefficient reports the amount of change in the dependent variable that is associated with a one-unit change in the independent variable. Strictly speaking, regression may be used only when the dependent variable is measured at the interval level.[1] The independent variable, however, can come in any form: nominal, ordinal, or interval. In this chapter we show how to interpret regression analysis in which the independent variable is interval level. We also discuss a technique called dummy variable regression, which uses nominal or ordinal variables as independent variables.

A second distinguishing feature of regression is the ease with which it can be extended to the analysis of controlled comparisons. In setting up a cross-tabulation analysis, controlling for a third variable, we performed mechanical control—separating units of analysis on the control variable, then reexamining the relationships between the independent and dependent variables. Regression also analyzes the relationship between a dependent variable and a single independent variable. This is simple or bivariate regression. However, regression easily accommodates additional control variables that the researcher may want to include. In multivariate or multiple regression mode, regression uses statistical control to adjust for the possible effects of the additional variables. Regression is remarkably flexible in this regard. Properly applied, regression can be used to detect and evaluate spuriousness, and it allows the researcher to model additive relationships and interaction effects.

CORRELATION

Figure 8-1 shows the relationship between two variables: the percentage of a state's population having at least a high school diploma (the independent variable) and the percentage of the eligible population that voted in the 2006 elections (the dependent variable). This display is called a **scatterplot.** In a scatterplot, the independent variable is measured along the horizontal axis and the dependent variable is measured along the vertical axis. Figure 8-1, then, locates each of the fifty states in two-dimensional space. Consider the overall pattern of this relationship. Would you say that the relationship is strong, moderate, or weak? What, if any, is the direction of the relationship—positive or negative? Even without the help of a numeric measure of strength and direction, you can probably arrive at reasonable answers to these questions.

Figure 8-1 Scatterplot: Education and Turnout (in percentages)

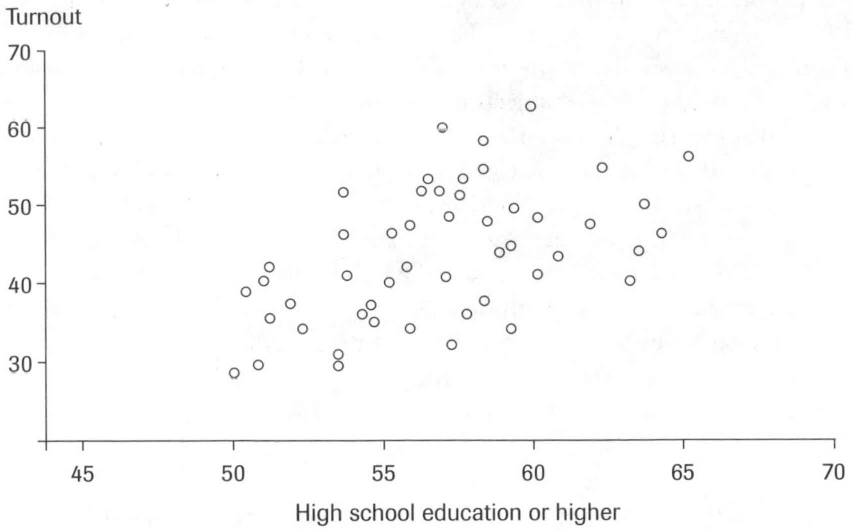

Source: Percentage high school or higher calculated from U. S. Census Bureau, www.census.gov/compendia/ smadb/TableA-22.pdf. Turnout is percentage of voting eligible population (VEP) in the 2006 elections, calculated by Michael McDonald, Department of Public and International Affairs, George Mason University, Fairfax, Virginia, and made available through his Web site: http://elections. gmu.edu/voter_turnout.htm.

Direction is easily ascertained. Clearly, as you move from lower to higher values of the independent variable—comparing states with lower levels of education to states with higher education levels—the values of the dependent variable tend to adjust themselves accordingly, clustering a bit higher on the turnout axis. So the direction is positive. How strong is the relationship? In assessing strength, you need to consider the consistency of the pattern. If the relationship is strong, then just about every time you compare a state with lower education levels against a state with higher levels, the higher education state would also have higher turnout. An increase in the independent variable would be associated with an increase in the dependent variable most of the time. If the relationship is weak, you would encounter many cases that do not fit the positive pattern, many higher education states with turnouts that are about the same as, or perhaps less than, lower education states. An increase in the independent variable would not consistently occasion an increase in the dependent variable. Now suppose you had to rate the relationship in Figure 8-1 on a scale from 0 to 1, where a rating close to 0 denotes a weak relationship, a rating of around .5 denotes a moderate relationship, and a rating close to 1 denotes a strong relationship. What rating would you give? A rating close to 0 would not seem correct, because the pattern has some predictability. Yet the other pole, a rating of 1, does not seem right either, because there are quite a few states that are in the "wrong" place on the turnout variable, given their levels of education. A rating of around .5, somewhere in the moderate range, would seem a reasonable gauge of the strength of the relationship.

The **Pearson's correlation coefficient**, which is symbolized by a lowercase r, uses just this approach in determining the direction and strength of an interval-level relationship. Pearson's r always has a value that falls between -1, signifying a perfectly negative association between the variables, and $+1$, a perfectly positive association between them. If no relationship exists between the variables, Pearson's r takes on the value of 0. The correlation coefficient is a pure number—that is, it is impervious to the units in which the variables are measured.[2] Correlate age measured in years with TV watching measured in hours and Pearson's r will report a value of between -1 and $+1$. Correlate states' education and turnout, both measured in percentages, and again r will weigh in with a number between -1 and $+1$. This unit-free feature is one of Pearson r's main attractions. Consider the formula for the sample correlation coefficient:

$$\frac{\sum \left(\frac{x_i - \bar{x}}{s_x} \right) \left(\frac{y_i - \bar{y}}{s_y} \right)}{n-1}$$

where:

x_i = individual observations of x,
$\bar{x}$ = the sample mean of x,
s_x = the sample standard deviation of x,
y_i = individual observations of y,
$\bar{y}$ = the sample mean of y, and
s_y = the sample standard deviation of y.

Notice what is going on in the numerator: standardization. Standardization was discussed in Chapter 6. To standardize a number, first find its deviation from the mean. Numbers above the mean will have positive deviations, and numbers below the mean will have negative deviations. A number that is equal to the mean will have a deviation equal to zero. According to the formula for Pearson's r, deviations from the mean are calculated for each individual value of the independent variable (x_i) and dependent variable (y_i). Standardization is achieved by dividing each deviation from the mean by a standard unit of measurement. The standard deviation is the correlation coefficient's unit of standardization. Because we are dealing with two variables, the standard deviation of the independent variable is symbolized by s_x, and the standard deviation of the dependent variable is symbolized by s_y. When we divide each deviation from the mean by the standard deviation, we convert each individual value into a Z score. Z scores also were covered in Chapter 6. Each observation's Z score tells you the number of standard units it lies above (a positive value of Z) or below (a negative value of Z) the mean of the distribution. Z is equal to 0 for any observation having a raw score equal to the mean.

In the data displayed in Figure 8-1, the education variable has a mean equal to 56.9 and a standard deviation of 3.9. Turnout has a mean of 43.3 and a standard deviation of 7.8. For example, Montana's values on the independent variable (58.4 percent with a high school education or higher) and dependent variable (56.7 percent turnout) convert to Z scores of .4 and 1.7—slightly above the education mean and 1.7 standard deviations above the turnout mean. Similarly, Florida's raw values on the independent variable (55.2) and dependent variable (39.9) convert to Z scores of $-.4$ on education and $-.4$ on turnout. According to the formula for the correlation coefficient, we multiply each observation's Z scores together. For Montana we would have, .4 * 1.7 = .68. For Florida, $-.4$ * $-.4$ = .16. To arrive at r, we would sum all these products and take the average—more precisely, we divide the sum of the products by the sample size minus 1.

To see how the correlation coefficient summarizes direction and strength, examine Figure 8-2, which displays a scatterplot of the education and turnout variables, converted to Z scores. Reference lines are drawn at the means of the independent and dependent variables, that is, at a Z score of 0 on each axis. The reference lines create four quadrants. All the states in the lower-left quadrant fall below the means of both variables—negative Z scores on education and negative Z scores on turnout. Because the product of two negative numbers equals a positive number, cases in this quadrant help create a positive correlation coefficient. For example, the product of Florida's Z scores, $-.4$ and $-.4$, is a positive number, .16. All states in the upper-right quadrant have above-the-mean values on both variables and so also help produce a positive value of r. Multiplying Montana's Z scores, .4 and 1.7, yields a positive number, .68. The upper-left and lower-right quadrants are the negative-correlation quadrants. When we multiply the Z scores of cases that have below average education levels but above average turnout (upper-left quadrant) or above average education but below average turnout (lower-right quadrant), we obtain negative numbers. Since these negative numbers contribute to the summation of products, states in these quadrants detract from the positive correlation and weaken the relationship. Even within the positive-correlation quadrants you can see inconsistencies—for example, two states may have different Z scores on education but similar Z scores on turnout. Inconsistencies keep r in the positive range but weaken its magnitude. As the cases coalesce into a consistent pattern and the negative-quadrant occupants shrink in number, Pearson's r moves toward 1.0. Of course, for negative values of r these symmetries are reversed. Systematic groupings in the upper-left and lower-right quadrants, and few cases in the positive correlation quadrants, result in correlation coefficients that approach -1.0. So what is Pearson's r for Figure 8-2? After performing the obligatory multiplication and addition for all fifty states, we end up with the sum, 24.9. Dividing by the sample size minus one: $24.9 / 49 = .51$. Thus, for this relationship $r = +.51$.

It should be pointed out that Pearson's r is a symmetrical measure of association between two variables. This means that the correlation between the independent variable and the dependent variable is the same as the correlation between the dependent variable and the independent variable. So, if the axes in Figure 8-1 or Figure 8-2 were reversed—with education appearing on the vertical axis and turnout appearing on the horizontal axis—Pearson's r would still be $+.51$. This makes correlation analysis quite useful during the early stages of research, when the researcher is getting an idea of the overall relationship between two variables. However, Pearson's r is neutral on the question of which variable is the causal variable and which is the effect. Therefore, one cannot attribute causation based on a correlation coefficient. As always, the researcher's explanation—the process the researcher has described to explain the connection between the variables—is the source of causal reasoning. Furthermore, Pearson's r is not a PRE (proportional reduction in error) measure of association. It is neatly bounded by -1 and $+1$ and thus communicates strength and direction by a common metric. But the specific magnitude of r does not tell us how much better we can predict the dependent variable by knowing the independent variable than by not knowing the independent variable. Happily, another measure can do that job for us, but first we need to understand regression analysis, to which we now turn.

BIVARIATE REGRESSION

By way of introducing bivariate regression, consider a hypothetical example. Let's say that an instructor wishes to analyze the relationship between the scores that her students received on

Figure 8-2 Scatterplot: Education and Turnout (Z scores)

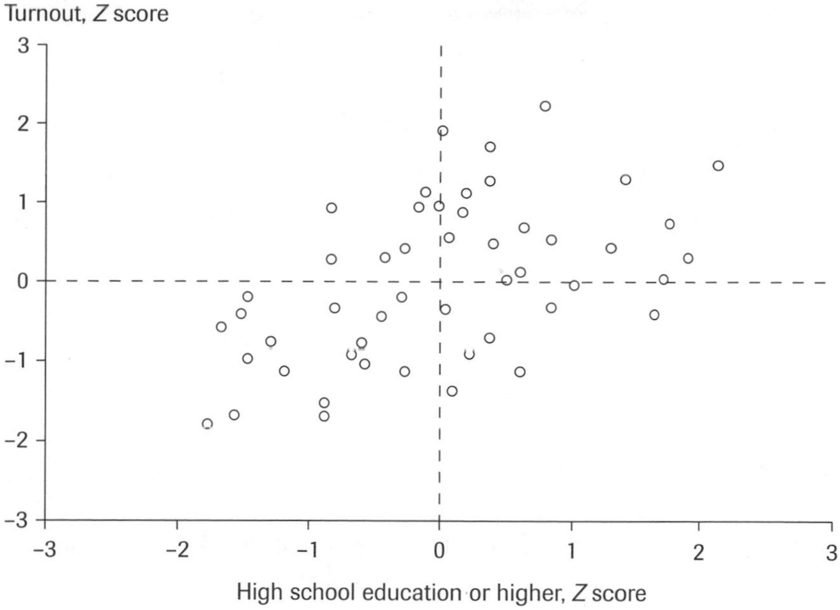

Turnout, Z score

High school education or higher, Z score

an exam (the dependent variable, y) and the number of hours they spent studying for the test (the independent variable, x). The raw data for each of her eight students are presented in Table 8-1. Examine these numbers closely. Can you identify a pattern to the relationship? You can see that the relationship is positive: Students who studied more received better scores than did students who studied less. And you can be sure that the correlation between the variables is indeed strong. But regression analysis permits us to put a finer point on the relationship. In this case, each additional hour spent studying results in exactly a 6-point increase in exam score. The student who studied 1 hour did 6 points better than the student who didn't study at all. And the student who studied 2 hours received a score that was 6 points better than the student who studied 1 hour, and so on. More than this, the xy pattern can be summarized by a line. Visualize this for a moment. What linear equation would summarize the relationship between hours spent studying and exam score?

To visualize or draw a line, you must know two things: the y-intercept and the slope of the line. The y-intercept is the point at which the line crosses the y axis—the value of y when x is 0. The slope of a line is "rise over run," the amount of change in y for each unit change in x. Memories of high school algebra may include the following formula for a line: $y = mx + b$. In this formulation, the two quantities of interest, the slope and the intercept, are symbolized by the letters m and the letter b, respectively. In representing the general formula for a **regression line**, statisticians prefer a different rendition of the same idea:

$$y = a + b(x)$$

In this formulation, the y-intercept or constant, usually symbolized by the letter a, is the first element on the right-hand side. Based on the information in Table 8-1, the y-intercept is 55, the score received by the student who did not study, for whom $x = 0$. The term b represents

Table 8-1 Hours Spent Studying (x) and Test Score (y)

Hours (x)	Score (y)
0	55
1	61
2	67
3	73
4	79
5	85
6	91
7	97

Note: Hypothetical data.

the slope of the line. The slope of the regression line, called the **regression coefficient**, is the workhorse of regression analysis. When you first considered Table 8-1, you probably figured out the regression coefficient right away. With each one-unit change in the independent variable, there is a six-unit change in the dependent variable. In the example, the regression coefficient (b) is equal to 6. Thus, the regression line for Table 8-1 is:

$$\text{Test score} = 55 + 6(\text{Number of hours}).$$

Notice a few aspects of this approach. First, the regression equation provides a general summary of the relationship between the independent and dependent variables. For any given student in Table 8-1, we can plug in the number of hours spent studying, do the arithmetic, and arrive at his or her exam score. Second, the formula would seem to hold some predictive power, the ability to estimate scores for students who do not appear in the data. For example, if information about the studying efforts of a new student were introduced—she or he spent, say, 3.5 hours studying—we would have a predictive tool for estimating that student's score. Our estimate: $55 + 6(3.5) = 76$. Using an established regression formula to predict values of a dependent variable for new values of an independent variable is a common application of regression analysis.

In any event, empirical relationships are never as well behaved as the data in Table 8-1, so we will modify the example to make it somewhat more realistic. Assume that the instructor is working with a sample of 16 students drawn at random from the student population. Table 8-2 groups these cases on the independent variable. According to Table 8-2, neither of the first two students spent any time studying (for each, $x = 0$). One received a 53, and the other did somewhat better, a 57. The next two students share the same value on the independent variable, 1 hour, but their scores, too, were different, a 59 and a 63, and so on for the other pairs of cases. Now suppose you had to calculate, for each pair of cases, a one-number summary of their value on the dependent variable. How would you proceed? You would probably do what regression does—calculate the mean value of the dependent variable for each value of the independent variable. For the two nonstudiers you would average their scores: $(53 + 57)/2 = 55$; for the two 1-hour cases, $(59 + 63)/2 = 61$, and so on. Notice that this averaging process does not reproduce the actual data. Instead, it produces *estimates* of the

Table 8-2   Hours Spent Studying (x), Test Score (y), and Estimated Score ($\hat{y}$)

Hours (x)	Score (y)	Estimated score ($\hat{y}$) for a given value of x
0	53	55
0	57	
1	59	61
1	63	
2	65	67
2	69	
3	71	73
3	75	
4	77	79
4	81	
5	83	85
5	87	
6	89	91
6	93	
7	95	97
7	99	

Note: Hypothetical data.

actual test scores for each number of hours spent studying. Because these estimates of y do not represent real values of y, they are given a separate label, $\hat{y}$ ("y-hat"), the letter y with a "hat" on top. Now If, based on Table 8-2, the instructor had to describe the relationship between the independent and dependent variables, she might say, "Based on my sample, each additional hour spent studying produced, *on average*, a 6-point increase in exam score." So the regression coefficient, b, communicates the average change in y for each unit change in x. A linear regression equation takes this general form:

$$\hat{y} = \hat{a} + \hat{b}(x) + e$$

$\hat{y}$ ("y-hat") is the estimated mean value of the dependent variable, $\hat{a}$ ("a-hat") is the average value of the dependent variable when the independent variable is 0, and $\hat{b}$ ("b-hat") is the average change in the dependent variable for each unit change in the independent variable. The term, e, symbolizes error—the inability of the linear formula to exactly predict the value of y for each case. The error term plays a central role in the underlying statistical assumptions of regression analysis.[3] For Table 8-2 we would have:

Estimated score = 55 + 6(Number of hours)

Regression analysis is built on the estimation of averages. Suppose that there were no students in the sample who had not studied at all. No matter. Regression will use the available information to calculate a y-intercept, an estimate of the average score that students would have received, had they not studied. By the same token, if no empirical examples existed for

x = 5 hours, regression nonetheless will yield an estimate, 55 + 6(5) = 85, an estimated average score. A regression line travels through the two-dimensional space defined by x and y, estimating mean values along the way.[4]

Regression relentlessly summarizes linear relationships. Feed it some sample values for x and y, and it will return estimates for a and b. But because these coefficients are means—means calculated from sample data—they will contain random sampling error, as do any sample means. Focus on the workhorse, the regression coefficient, $\hat{b}$. Based on Table 8-2, the instructor can infer that, in the population from which the sample was drawn, the average change in score for each additional hour spent studying is +6 points. But obviously this estimate contains some error, because the actual student scores fall a bit above or below the average, for any given value of x. Just like any sample mean, the size of the error in a regression coefficient is measured by a familiar statistic, its standard error. We know that the real value of b in the population (labeled with the Greek letter β ["beta"]) is equal to the sample estimate, $\hat{b}$, within the bounds of standard error:

$$\beta = \hat{b} \pm \text{(standard error of } \hat{b})$$

There is nothing mysterious or special about regression analysis in this regard. All the statistical rules you have learned—the informal ± 2 rule of thumb, the more formal 1.645 test, the calculation of P-values, the inferential set-up for testing the null hypothesis—apply to regression analysis. In evaluating the difference between two sample means, we tested the null hypothesis that the difference in the population is equal to 0. In its regression guise, the null hypothesis says much the same thing—that the true value of β in the population is equal to zero, that $\beta = 0$. Put another way, H_0 claims that a one-unit change in the independent variable produces zero units of change in the dependent variable, that the true regression line is flat and horizontal. Just as in the comparison of two sample means, we test the null hypothesis by calculating a t-statistic, or t-ratio:

$$t = (\hat{b} - \beta) / \text{(standard error of } \hat{b}), \text{ with degrees of freedom (d.f.)} = n - 2.$$

Informally, if the t-ratio is equal to or greater than 2, then we can reject the null hypothesis. And of course a precise P-value for t can be obtained in the familiar way.[5] For each 1-hour increase in studying time, we estimated a 6-point increase in exam score ($\hat{b}$ = +6). Calculated by computer, the standard error of $\hat{b}$ is .233. So the t-statistic is 6/.233 = 25.75, which has a P-value that rounds to .000. If the true value of β in the population really is 0, then the probability of obtaining a sample estimate of $\hat{b}$ = 6 is highly remote.

This hypothetical example has demonstrated some basic points. But let's return to a real-world relationship, the education-turnout example introduced in Figure 8-1, and discuss some further properties of regression analysis. Figure 8-3 again displays the scatterplot of states, only this time the estimated regression line has been superimposed within plot. Where did this line originate?

In the hypothetical example of exam scores, we figured out the regression line by averaging the scores for each value of the independent variable. The estimation of a regression line for actual data uses the same principle. Linear regression finds a line that provides the best fit to the data points. Using each case's value on the independent variable, x, it finds $\hat{y}$, an estimated value of y. It then calculates the difference between this estimate and the case's actual value of y. This difference is called **prediction error**. Each individual case's actual value of y is

Figure 8-3 Regression: Education and Turnout (in percentages)

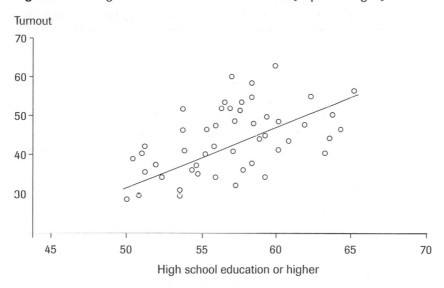

symbolized by y_i, "individual i's value of y." Prediction error is represented by the expression $y_i - \hat{y}$, the actual value of the dependent variable minus the estimated value of the dependent variable. Regression would use the values of the independent variable, percent high school or higher, to determine an estimated value on the dependent variable, percent turnout. Prediction error, for each state, would be the difference between the state's actual level of turnout, y_i, and the estimated turnout, $\hat{y}$. For example, based on its value on the education variable, Montana's predicted turnout, its value on $\hat{y}$, is equal to 44.8 percent. Montana's actual turnout rate is substantially higher, 56.7 percent. For Montana, then, prediction error is equal to $56.7 - 44.8 = +11.9$. Using education as a predictor, Florida has a predicted turnout of 41.6. Because Florida's actual turnout is equal to 39.9, prediction error for Florida is equal to $39.9 - 41.6 = -1.7$.

As the cases of Montana and Florida illustrate, prediction error for any given state may be positive—its actual turnout is higher than its estimated turnout—or it may be negative—its actual turnout is lower than its estimated turnout. In fact, if one were to add up all the positive and negative prediction errors across all states, they would sum to zero. When it finds the best-fitting line, therefore, regression does not work with the simple difference between y and $\hat{y}$. Rather, it works with the square of the difference, $(y_i - \hat{y})^2$. For each state, regression would square the difference between the state's actual level of turnout and its estimated level of turnout. In the regression logic, the best-fitting line is the one that minimizes the sum of these squared prediction errors across all cases. That is, regression finds the line that minimizes the quantity $\sum(y_i - \hat{y})^2$. This criterion of best fit—the line that minimizes the sum of the squared differences between real values of y and estimated values of y—is frequently used to distinguish ordinary least squares (OLS) regression from other regression-based techniques. The line represented in Figure 8-3, then, is an OLS regression line. True to form, regression reported the equation that provides the best fit for the relationship between the independent and dependent variables:

Estimated turnout = −14.53 + 1.02(High school education or higher)

How would you interpret each of the estimates for *a* and *b*? First consider the estimate for *a*, the level of turnout when *x* is 0. What can a turnout level of −14.53 possibly mean? In this case, not much. In the actual data, of course, no state has a value of 0 on the independent variable (0 percent with high school education or more). Nonetheless, regression produced an estimate for *â*, anchoring the line at a −14.53 turnout rate.[6] For some applications of regression, the value of the *y*-intercept, the estimate *â*, has no meaningful interpretation. (Sometimes *â* is essential, however. This is discussed below.) What about *b̂*, the estimated effect of education on turnout?

There are two rules for interpreting a regression coefficient. First, be clear about the units in which the independent and dependent variables are measured. In this example, the dependent variable, *y*, is measured by percentages—the percentage of each state's eligible population that voted. The independent variable, *x*, also is expressed in percentages—the percentage of each state's population that has at least a high school education. Second, remember that the regression coefficient, *b̂*, is expressed in units of the dependent variable, not the independent variable. Therefore, the coefficient, 1.02, tells us that turnout (*y*) increases, on average, by 1.02 percentage points for each 1 percentage-point increase in education (*x*).

Indeed, a common mistake made by beginners is to interpret *b̂* in terms of the independent variable. It might seem reasonable to say something like, "If the percentage of high school graduates increases by 1.02 percent, then turnout increases by 1 percent." This would be incorrect. Remember that *all* the coefficients in a regression equation are measured in units of the dependent variable. The intercept is the value of the dependent variable when *x* is zero. The slope is the estimated change in the dependent variable for a one-unit change in the independent variable.

The data in Figure 8-3 were calculated on a population, all fifty states, not a random sample of states. Strictly speaking, of course, when one is dealing with a population, questions of statistical inference do not enter the picture. However, we will continue to assume, for the sake of illustration, that we have a sample here, and that we obtained a sample estimate of 1.02 for the true population value of β. The null hypothesis would claim that β is really 0, and that the sample estimate we obtained, 1.02, is within the bounds of sampling error. As before, we enlist the regression coefficient's standard error, which the computer calculated to be .25, and arrive at a *t*-ratio:

$$t = (\hat{b} - \beta) \, / \, (\text{standard error of } \hat{b}), \text{ with d.f.} = n - 2$$
$$= 1.02 \, / \, .25$$
$$= 4.08, \text{ with d.f.} = 50 - 2 = 48.$$

Let's use these results to exercise our inferential skills. The informal ± 2 rule of thumb advises us to reject the null hypothesis. According to the statistics, 95 percent of all possible random samples will yield regression coefficients in the range between $1.02 - 2(.25)$ at the low end and $1.02 + 2(.25)$ at the high end, between .52 and 1.52. H_0's lucky number, 0, is not within this range. Use the 1.645 rule to find the lowest plausible value of β in the population: $1.02 - 1.645(.25) = .61$. Does this number fall above H_0's zero mark? Yes, it does. The computer returned a *P*-value of .0001, which punctuates our inference: reject the null hypothesis.

R-SQUARE

Regression analysis gives a precise estimate of the effect of an independent variable on a dependent variable. It looks at a relationship and reports the relationship's exact nature. So if someone were to inquire, "What, exactly, is the effect of education on turnout in the states?," the regression coefficient provides an answer: "Turnout increases by 1.02 percentage points for each 1 percent increase in the percentage of the states' population with at least a high school education. Plus, the regression coefficient has a *P*-value of .0001." By itself, however, the regression coefficient does not measure the completeness of the relationship, the degree to which the dependent variable is explained by the independent variable. A skeptic, on viewing Figure 8-3, might point this out. Certainly only a handful of states fall exactly on the regression line. Most have lower levels of turnout—or higher levels of turnout—than the regression line would predict. Overall, just how good a job does the independent variable do in explaining the dependent variable? How much better can we predict a state's turnout by knowing its education level than by not knowing its education level?

These are not questions about the precise nature of the relationship. They are questions about the size of the contribution x makes to the explanation of y. You can probably think of several variables that may explain why states are below or above the line. Perhaps below-the-line cases tend to be southern states, which historically have lower turnouts, and above-the-line cases are nonsouthern states. Or maybe states with lower-than-predicted turnout rates have more stringent voter registration requirements than do states with higher-than-predicted turnouts. Hard to tell. In any event, states' education levels, although clearly related to turnout, provide an incomplete explanation of it.

In regression analysis, the completeness of a relationship is measured by the statistic R^2 ("*R*-square"). **R-square** is a PRE measure, and so it poses the question of strength the same way as lambda or Somers' d_{yx}: "How much better can we predict the dependent variable by knowing the independent variable than by not knowing the independent variable?" Consider the state turnout data. If you had to guess a state's turnout without knowing its education level, what is your best guess? As we saw in the Chapter 7 illustration of lambda, the best guess for a nominal-level variable is the variable's measure of central tendency, its mode. In the case of an interval-level dependent variable, such as percent turnout, the best guess also is provided by the variable's measure of central tendency, its mean. This guess, the mean value of y, is represented by the symbol $\bar{y}$ ("y-bar"), y with a bar across the top.

Figure 8-4 again shows the scatterplot of states and the regression line. This time, however, a flat line is drawn at 43.3 percent, mean turnout for all fifty states. So $\bar{y} = 43.3$ percent. If we had no knowledge of the independent variable, we would guess a turnout rate of 43.3 for each state taken one at a time. Because more states fall closer to the mean than to any other single value of the dependent variable, this guess would serve us fairly well for many states. But, of course, this guessing strategy would also produce errors. Consider the case of one state, Maine, which appears as a solid dot in the scatterplot. Maine has a turnout rate of 53.5 percent. For Maine, our guess of 43.3 would underestimate turnout by an amount equal to actual turnout minus mean turnout: $53.5 - 43.3 = 10.2$ percentage points. Maine's turnout rate is 10.2 percentage points higher than predicted, based only on the mean turnout for all states. Formally, this error can be labeled as $y_i - \bar{y}$, the actual value of y minus the overall mean of y. This value, calculated for every case, is the starting point for *R*-square. Specifically, *R*-square finds $(y_i - \bar{y})$ for each case, squares it, and then sums these squared values across all observations. The result is the total sum of squares of all deviations from the mean value of y: $\sum (y_i - \bar{y})^2$. We encountered the total sum of squares before, in the Chapter 6 discussion of the variance and standard deviation. As a building block of *R*-square, the calculation of total sum of squares is identical. *R*-square determines the deviation from the mean of the distribution for each observation, squares this deviation, and then sums the squared deviations across all

Figure 8-4 Scatterplot with Mean of y and Regression Estimate $\hat{y}$: Education and Turnout

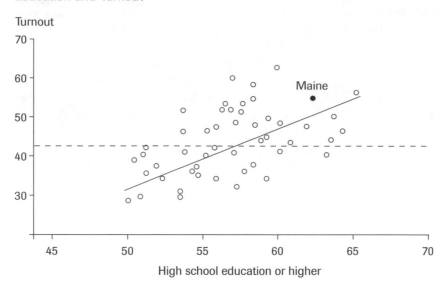

the cases. The **total sum of squares** is an overall summary of the variation in the dependent variable. It also represents all our errors in guessing the value of the dependent variable for each case, using the mean of the dependent variable as a predictive instrument.

Now reconsider the regression line in Figure 8-4 and see how much it improves the prediction of the dependent variable. The regression line is the estimated level of turnout calculated with knowledge of the independent variable, education. For each state, taken one at a time, we would not guess $\bar{y}$, the overall mean of the dependent variable. Rather, we would guess $\hat{y}$, the estimated value of y for a given value of the independent variable. Maine, for example, has a value of 62.4 on the independent variable, since 62.4 percent of its population has at least a high school education. What would be our estimate of its turnout level? Plugging 62.4 into the regression equation, for Maine we obtain:

$$-14.53 + 1.02(62.4) = 48.9.$$

By guessing the mean, 43.3, we missed Maine's actual turnout by 10.2 points. How much closer is our new estimate, 48.9? Formally, the amount of improvement due to the regression equation is equal to $\hat{y} - \bar{y}$, the predicted value of y minus the mean of y. For Maine, the difference between the estimate that uses the regression equation, 48.9, and the prediction based on the mean, 43.3, is equal to 48.9 − 43.3 = 5.6. Thus, we started out with error equal to 10.2 points. Of that 10.2 points, the regression model allows us to pick up 5.6 points. But the difference between the regression prediction, 48.9, and Maine's actual turnout rate, 53.5, remains unexplained. This is prediction error. As discussed earlier, prediction error is equal to $y_i - \hat{y}$. For Maine, the difference between actual turnout, 53.5, and predicted turnout, 48.9, is equal to 53.5 − 48.9 = 4.6.

Notice that the two numbers we just derived—5.6 (the improvement due to the regression) and 4.6 (prediction error)—sum to 10.2, the total distance between Maine's turnout rate and the mean for all states. Indeed, for each state in the data set, we could divide its total distance from the mean, $(y_i - \bar{y})$, into two parts: the amount explained by the regression, $\hat{y} - \bar{y}$, and prediction error, $y_i - \hat{y}$. More generally, in regression analysis the total sum of squares has two components:

$$\text{Total sum of squares} = \text{Regression sum of squares} + \text{Error sum of squares}$$
$$\Sigma (y_i - \bar{y})^2 = \Sigma (\hat{y} - \bar{y})^2 + \Sigma (y_i - \hat{y})^2$$

The total sum of squares, as we saw, measures all the variation in the dependent variable and summarizes the amount of error produced by using the overall mean of the dependent variable to guess each case. The **regression sum of squares** is the component of the total sum of squares that we pick up by knowing the independent variable. The **error sum of squares** is prediction error, the component of the total sum of squares that is not explained by the regression equation.[7] Obviously, if the regression sum of squares is a large chunk of the total sum of squares, then the independent variable is doing a lot of work in explaining the dependent variable. As the contribution of regression sum of squares declines, and the contribution of the error sum of squares increases, knowledge of the independent variable provides less help in explaining the dependent variable. R-square is simply the ratio of the regression sum of squares to total sum of squares:

$$R^2 = \text{Regression sum of squares} / \text{Total sum of squares}.$$

R-square measures the goodness of fit between the regression line and the actual data. If x completely explains y, if the regression sum of squares equals the total sum of squares, then R-square is 1. If the regression sum of squares makes no contribution—if we would do just as well in accounting for the dependent variable without knowledge of the independent variable as with knowledge of the independent variable—then R-square is 0. R-square is a PRE measure and is always bracketed by 0 and 1. Its value may be interpreted as the proportion of the variation in the dependent variable that is explained by the independent variable. The R-square for the state data is equal to .26. Therefore, 26 percent of the variation among states in their turnout rates is accounted for by their levels of education. The leftover variation among states, 74 percent, may be explained by other variables, but it is not accounted for by differences in education.[8]

R-square, also called the *coefficient of determination,* bears a family resemblance to r, Pearson's correlation coefficient. In fact, $R^2 = r^2$. So the value of R^2, equal to .26 for the state data, is the square of r, reported earlier to be +.51 for the same data. Pearson's r is well understood in political analysis, and it is a good measure to use when you want to examine the overall relationships between variables. R-square adds depth to the repertoire of regression statistics. Because it is tied to the PRE standard, R-square communicates how well the regression performs—how completely the model accounts for the variation in the dependent variable.

Adjusted R-square

Most research articles report a conservative version of R-square, **adjusted R-square**, which is often close to (but always less than) regular R-square. Why does R-square need to be adjusted? Like any sample statistic, a sample R-square estimates the true value of R-square in the population. The sample R-square is equal to the population R-square, within the boundaries of random sampling error. However, R-square's errors can only assume positive values—squaring any negative error, after all, produces a positive number—inflating the estimated value of R-squared. This inflation requires downward adjustment. Consider the formula for adjusted R-square:

$$\text{Adjusted } R\text{-square} = 1 - (1 - R^2)\left(\frac{n-1}{n-k-1}\right)$$

In the formula, n is the sample size and k is the number of independent variables in the regression model. The quantity inside the big brackets is the adjustment factor. Notice two aspects of the adjustment factor. First, it will always be greater than 1. Because all regression models have at least one independent variable, the factor will never be smaller than $(n-1)$ / $(n-2)$. (Below we discuss regression analysis with more than one independent variable.) Second, as sample size goes down, the adjustment factor goes up. Assuming one independent variable, samples of $n = 500$ have an adjustment equal to $(500 - 1)$ / $(500 - 2) = 1.002$. For samples of $n = 50$, such as the states data, the adjustment is equal to $(50 - 1)$ / $(50 - 2) = 1.021$. To apply the adjustment, we multiply the adjustment factor by $(1 - R^2)$ and then subtract the result from 1. For the education–turnout relationship: $1 - (1 - .26)(1.021) = 1 - (.74)(1.021) = 1 - .76 = .24$. So, after adjustment, we can conclude that level of education explains 24 percent of the variation in turnout across states.[9]

DUMMY VARIABLE REGRESSION

As noted earlier, one attraction of regression analysis is its adaptability to a variety of research problems. In one very common situation, the researcher has a nominal or ordinal independent variable, not an interval-level independent variable. Regression will adapt. Consider the following regression equation, which sets up an analysis of the relationship between turnout (again appearing as the dependent variable) and an independent variable labeled "South." The independent variable, which is categorical, identifies states as southern or nonsouthern.

$$\hat{y} = \hat{a} + \hat{b} \text{ (South)}$$

All the elements of the equation are by now familiar. $\hat{y}$ is the estimate of y, turnout. The intercept, $\hat{a}$, is the value of $\hat{y}$ when the independent variable is equal to 0. And $\hat{b}$, as before, is the average change in the dependent variable for each unit change in the independent variable. Now suppose that the independent variable, South, is coded so that nonsouthern states have a value of 0. So South = 0 for all nonsouthern states in the data set. Suppose further that South is coded 1 for all southern states, so that South = 1 for each southern state in the data set. South is a dummy variable. A **dummy variable** is a variable for which all cases falling into a specific category assume the value of 1, and all cases not falling into that category assume a value of 0. Apply the regression logic. Since $\hat{a}$ is the value of $\hat{y}$ when the independent variable is 0, $\hat{a}$ in the above equation will estimate the average level of turnout for nonsouthern states. Why? Nonsouthern states are coded 0 on South, so their estimated turnout is equal to

$$\hat{a} + \hat{b}\,(0) = \hat{a}.$$

Furthermore, the estimated turnout for southern states will be equal to $\hat{a} + \hat{b}$. Southern states are coded 1 on South. Our estimate for their turnout is:

$$\hat{a} + \hat{b}\,(1) = \hat{a} + \hat{b}.$$

Table 8-3   Voter Turnout in Southern and Nonsouthern States

$\hat{y}$	=	$\hat{a}$	+	$\hat{b}$	(x)
Estimated turnout	=	Estimated turnout when South is 0	+	Mean change in turnout	(South)
Estimated turnout	=	46.34	+	−9.44	(South)
Standard error of $\hat{b}$				1.98	
t-statistic				−4.78	
P-value				.000	
Adjusted R-square = .31					

Sources: Turnout is the percentage of voting eligible population in the 2006 elections, calculated by Michael McDonald, Department of Public and International Affairs, George Mason University, Fairfax, Va., and made available through his Web site: http://elections.gmu.edu/voter_turnout.htm. The independent variable is based on the definition of census regions: www.census.gov/geo/www/us_regdiv.pdf.

Since regression is in the business of estimating means, it can be applied, by using dummy variables, to problems that compare the means on a dependent variable for subjects that differ on a nominal or ordinal variable. Table 8-3 reports the estimates obtained for the comparison of turnouts in southern and nonsouthern states.

Let's interpret these numbers. The estimate for the y-intercept, in this case, is clearly meaningful. It communicates the average turnout for nonsouthern states, a turnout rate of 46.34. The coefficient for $\hat{b}$, −9.44, tells us to adjust our estimate for southern states relative to nonsouthern states by minus 9.4 percentage points. So southern states, on average, fall about 9 percentage points below nonsouthern states on the turnout scale. Remember that the regression coefficient, by itself, does not estimate turnout for southern states. Rather, it reflects the average difference between states that have a value of 0 on South and states that have a value of 1 on South. When South switches from 0 to 1, average turnout drops by 9.4 points. We can, of course, use this information to arrive at an estimate of the average turnout for southern states: $46.3 − 9.4 = 36.9$.

What would the null hypothesis have to say about these results? The null hypothesis would assert that, in the population, no difference exists in the turnout rates of southern and nonsouthern states, that $\beta = 0$. Is the estimated average difference of −9.4 large enough to reject the null hypothesis? The t-ratio says so. Dividing the regression coefficient by its standard error, −9.40/1.98, yields a t-statistic of −4.78, well beyond the acceptable realm of random error. Finally, adjusted R-square, which is equal to .31, gives us an overall reading of how well we can explain differences in turnout by knowing whether states are southern or nonsouthern. The independent variable accounts for 31 percent of the variation among states in turnout rates.[10]

The dummy logic extends to nominal or ordinal variables that have more than two values. In the state data, for example, region is measured by four categories: Northeast, Midwest, West, and South. In the dummy regression just discussed, northeastern, midwestern, and western states were lumped together and coded 0 on the dummy variable, South. Southern states were given a value of 1 on the dummy. But suppose that you wanted to compare the turnout rates of all four regions. Again regression adapts. Consider the following regression equation:

$$\text{Estimated turnout} = \hat{a} + \hat{b}_1 \text{ (Northeast)} + \hat{b}_2 \text{ (West)} + \hat{b}_3 \text{ (South)}.$$

Northeast is a dummy variable that takes on the value of 1 for northeastern states and a value of 0 for all other states. West, too, is a dummy, coded 1 for western states and 0 otherwise. Similarly, South is 1 for southern states and 0 for all nonsouthern states. What about midwestern states? Have we mistakenly excluded them from the equation? Not at all. Keep the "0 or 1" dummy idea in mind. A midwestern state would have a value of 0 on Northeast, a value of 0 on West, and a value of 0 on South. The estimated turnout for midwestern states, then, is captured by the y-intercept, $\hat{a}$. As always, the intercept gives you the average value of the dependent variable when the independent variable is equal to 0. In this example, there are three dummies, each of which identifies a different category of region. When all these categories are 0, the intercept records the dependent variable for all the cases not included in any of the other categories. In the language of dummy variable regression, the intercept estimates the value of the dependent variable for the *base category* or *excluded category* of the independent variable.

There are a few rules to keep in mind when setting up a regression with dummy variables. First, in doing dummy regression with an independent variable that has k categories, the number of dummies should be equal to $k - 1$. Since region has four categories, we include three dummies in the equation. We applied this rule in the earlier example, too. The independent variable had two categories, non-South and South, so the regression had one dummy, South. A second rule is that the categories must be mutually exclusive and jointly exhaustive. Mutually exclusive means that any case that is coded 1 in one category must be coded 0 in all other categories. Jointly exhaustive means that all the categories account for all the cases, so that no cases are left unclassified. Cases in the base or excluded category meet these conditions, since they are uniquely classified by their absence from all the other categories.

Now consider the notation and meaning of the regression coefficients. To keep their identities separate, each coefficient has its own subscript: $\hat{b}_1$, $\hat{b}_2$, and $\hat{b}_3$. What does each coefficient mean? The coefficient for $\hat{b}_1$ will tell us the average difference in turnout between northeastern states and the average turnout for the base category, Midwest. Similarly, $\hat{b}_2$ adjusts the value of the intercept for western states, and $\hat{b}_3$ gives us the average difference between southern states and the base category. Let's look at the results of the analysis, presented in Table 8-4, and clarify these points. According to the regression estimates, the turnout rate for states in the base category, Midwest, is 48.7, since $\hat{a} = 48.7$. Do the other three regions have higher or lower turnouts than this? The coefficient for Northeast is -2.7, telling us that, on average, the turnout rate for these states is 2.7 percentage points lower than the base, or $48.7 - 2.7 = 46.0$. The average for West, too, is a bit lower than the base rate of 48.7. Judging from the estimate form $\hat{b}_2$, -4.4, the mean turnout for western states is about 4 points lower than midwestern turnout. A much larger adjustment is called for in the case of South. Southern states, on average, score 11.8 units below the intercept, since $\hat{b}_3 = -11.8$. That places their mean turnout at $48.7 - 11.8 = 36.9$.

The interpretation of the t-statistics and P-values is exactly the same as before. You can see that the coefficient for $\hat{b}_1$ does not pass muster with the null hypothesis. The t-ratio for $\hat{b}_1$, $-.95$, and its accompanying P-value, .35, suggest that the average difference between Midwest and Northeast can be accounted for by sampling error. The same might be said for $\hat{b}_2$, with $t = -1.69$ and $P = .10$.[11] The coefficient for South, though, is significantly different from zero. Its t-ratio of -4.79 and P-value of .00 suggest that the estimate of $\hat{b}_3$ is too large to have occurred by chance.

Table 8-4 Estimating Voter Turnout in Four Regions

Estimated turnout	=	$\hat{a}$	+	$\hat{b}_1$ (Northeast)	+	$\hat{b}_2$ (West)	+	$\hat{b}_3$ (South)
		48.73	+	−2.69	+	−4.36	+	−11.82
Standard error of $\hat{b}$				2.85		2.58		2.47
t-statistic				−.95		−1.69		−4.79
P-value				.35		.10		.00
Adjusted *R*-square = .32								

Sources: Turnout is the percentage of voting eligible population in the 2006 elections, calculated by Michael McDonald, Department of Public and International Affairs, George Mason University, Fairfax, Va., and made available through his Web site: http://elections.gmu.edu/voter_turnout.htm. Dummy variables for region are based on the definition of census regions: www.census.gov/geo/www/us_regdiv.pdf.

MULTIPLE REGRESSION

By now you are getting a feel for regression analysis, and you may be gaining an appreciation for its malleability. Ordinary least squares regression is one of the most powerful techniques available to the political researcher. Indeed, once you become comfortable with this method, you can begin using it to model and estimate complex relationships. In **multiple regression**, for example, we are able to isolate the effect of one independent variable on the dependent variable, while controlling for the effects of other independent variables.

Consider some of the variables we have used in this chapter to explain differences in turnout. In discussing simple bivariate regression, we analyzed the relationship between turnout and state education level. In looking at dummy regression, we analyzed the relationship between turnout and the two-category dummy, South. In each case we found the independent variable to be related to turnout. But there is a potential problem. What if these two independent variables, education and South, are themselves related? For example, what if southern states have lower levels of education than nonsouthern states? If the two independent variables are related in this way, then when we compared southern states with nonsouthern states we were also comparing less-educated states (which have lower turnouts) with more-educated states (which have higher turnouts). Indeed, an inspection of the data reveals that southern states average 55.2 percent on the education variable, compared with 57.7 percent for nonsouthern states.

The logic of controlled comparison tells us how to sort this out. If we had categorical variables and were doing cross-tabulation analysis, we would separate the states into two categories, less-educated states and more-educated states. We would then reexamine the relationship between South/non-South and turnout separately for each category of the control. We would isolate the partial effect of region on turnout, controlling for education. As you know, this procedure would also permit us to examine the controlled effect of education on turnout—the partial effect of education level on turnout, controlling for region. Multiple regression does the same thing. Multiple regression produces a partial regression coefficient for each independent variable. A **partial regression coefficient** estimates the mean change in the dependent variable for each unit change in the independent variable, *controlling for the other independent variables in the model.* Therefore, a multiple regression analysis of turnout that uses education level and region as independent variables will report two partial regression coefficients—one that estimates the effect of education on turnout, controlling for

Table 8-5 Regression Estimates for Education Level and South, for Dependent Variable, Voter Turnout

Estimated turnout	=	$\hat{a}$	+	$\hat{b}_1$ (Education)	+	$\hat{b}_2$ (South)
		3.70	+	.74	+	−7.57
Standard error of $\hat{b}$				.23		1.90
t-statistic				3.24		−4.00
P-value				.00		.00
Adjusted R-square = .42						

Sources: Turnout is the percentage of voting eligible population in the 2006 elections, calculated by Michael McDonald, Department of Public and International Affairs, George Mason University, Fairfax, Va., and made available through his Web site: http://elections.gmu.edu/voter_turnout.htm. Percentage high school or higher is calculated from U. S. Census Bureau data, www.census.gov/compendia/smadb/TableA-22.pdf. The dummy variable for southern region is based on the definition of census regions: www.census.gov/geo/www/us_regdiv.pdf.

region, and one that estimates the effect of region on turnout, controlling for education. A multiple regression model takes the general form:

$$\hat{y} = \hat{a} + \hat{b}_1 (x_1) + \hat{b}_2 (x_2) \ldots + b_k(x_k) \ldots + e.$$

A multiple regression model with two independent variables will estimate two partial regression coefficients. The coefficient, $\hat{b}_1$, estimates the average change in the dependent variable for each unit change in x_1, controlling for x_2. The coefficient, $\hat{b}_2$, estimates the average change in the dependent variable for each unit change in x_2, controlling for x_1. Thus, $\hat{b}_1$ and $\hat{b}_2$ tell us the partial effects of each independent variable on the dependent variable. The intercept, $\hat{a}$, estimates the mean of the dependent variable when all the independent variables are equal to zero. To illustrate the basic properties of multiple regression—and to point out some of its limitations and pitfalls—we return to the state turnout data.

Here is the regression model we want to estimate:

Estimated turnout = $\hat{a} + \hat{b}_1$ (High school education or higher) + $\hat{b}_2$ (South).

The coefficient $\hat{b}_1$ will estimate the partial effect of education level, controlling for South. The coefficient $\hat{b}_2$ will estimate the partial effect of South, controlling for education level. Table 8-5 reports the results. According to Table 8-5, the regression coefficient for the effect of education on turnout is .74. In the context of multiple regression, what does this mean? Regression statistically divided the cases into regional categories, southern states and nonsouthern states, and then found the best estimate of the partial effect of education on turnout. So the estimate for $\hat{b}_1$, .74, means that, if South is held constant, a 1 percentage-point increase in education is associated with an increase of .74 of a point in turnout. Similarly, the coefficient for South, −7.57, says that, after we control for educational differences between southern and nonsouthern states, southern states averaged 7.57 points lower on the turnout scale.

Is the effect of education contaminated by the possible relationship between region and turnout? No, the regression has controlled for region. Might southern states have lower turnouts because they have proportionately fewer high school graduates than nonsouthern states? No, again. Multiple regression has taken this into account. When we used South as the sole predictor, we found that southern states averaged 9.4 points lower than nonsouthern states. After accounting for educational differences between states, however, the mean difference in turnout narrows to about 7.6 points. Thus, part of the regional difference in turnout is caused by educational differences between regions—southern states have lower levels of education than nonsouthern states. When we used education as the sole predictor, we found that turnout increased by 1.02 points for each 1 percentage-point increase in the education variable. After accounting for regional differences, however, this effect declines to .74, about three-quarters of a percentage point. Thus, part of the educational difference in turnout is caused by regional differences—lower-education states are more likely to be in the South than are higher-education states. Multiple regression has worked its methodological magic, isolating the partial effects of the independent variables.

The interpretation of the remaining numbers in Table 8-5—the standard errors, the t-ratios, and the P-values—follows the same protocol as for bivariate regression. The t-statistics and P-values suggest that each variable has an independent effect on turnout. In fact, with t-ratios of well over 2, both pass the eyeball test of significance. In multiple regression, adjusted R-square takes on a more expansive PRE meaning. Its value communicates how completely *all* the independent variables explain the dependent variable. An adjusted R-square of .42 says that by knowing two things about states—the percentage of the population with a high school education or more and whether they are located in the South—we can account for 42 percent of their differences in turnout.

Interaction Effects in Multiple Regression

The multiple regression technique is linear and additive. It estimates the partial effect of one independent variable by controlling for the effects of all other independent variables in the model. In doing this, regression assumes the effect of each independent variable is the same at all values of the other independent variables. In the example just discussed, for instance, regression estimated the partial effect of education level on turnout, controlling for South. The estimates produced by the technique were based on the assumption that the effect of education on turnout is the same for nonsouthern and southern states. According to the results, in nonsouthern and southern states alike, a one-unit increase in education is associated with a .74-unit increase in turnout. Multiple regression is ready-made for detecting spurious relationships and for modeling additive relationships. If the researcher thinks that the relationship between South and turnout might be spurious—that the apparent regional difference in turnout is the spurious result of educational differences between nonsouthern and southern states—then multiple regression will ferret out this spurious effect. If additive relationships are at work, with both variables contributing to our understanding of turnout, regression will estimate the partial effects of each one.

In Chapters 4 and 5 we discussed interaction relationships. In an interaction relationship, you will recall, the effect of the independent variable on the dependent variable is not the same for all values of the control variable. In multiple regression analysis, an interaction relationship is called an **interaction effect**. An interaction effect occurs when the effect of an independent variable cannot be fairly summarized by a single partial effect. Instead, the effect varies, depending on the value of another independent variable in the model. In the regression we just estimated, for example, multiple regression's linear-additive coefficients would tell us to estimate turnout by adding the partial effect of education level to the partial effect of South. But what if education had a bigger effect in southern states than in nonsouthern states? If this were

Table 8-6 Mean Prochoice Scores by Partisanship, Controlling for Political Knowledge

| Partisanship | Political knowledge | | Total |
	Low	High	
Democrat	4.3	5.7	4.8
	(184)	(111)	(295)
Independent	3.7	4.6	4.1
	(205)	(142)	(347)
Republican	2.9	2.8	2.8
	(144)	(156)	(300)
Total	3.7	4.2	3.9
	(533)	(409)	(942)

Source: 2004 American National Election Study.

the case, then the model we have described and estimated would be an incorrect depiction of the relationship. Much of the time multiple regression's linear-additive assumptions work quite well. However, if interaction is going on in the data, or the researcher has described an explanation or process that implies interaction, then a different model needs to be identified.

By way of introducing interaction effects in regression analysis, consider an interesting theory of public opinion, which we will label the *polarization perspective*. Based on the influential work of John R. Zaller, the polarization perspective says that people who are the most attuned to politics are also the most deeply divided by it.[12] For example, when we compare the abortion opinions of most Democrats and Republicans, we may find a weak if discernible pattern: Democrats will be more supportive of abortion rights than will Republicans. However, among the politically aware—people who pay close attention to politics and have high levels of political knowledge—partisans will be arrayed into warring ideological camps, with Democrats much more prochoice than Republicans. Morris P. Fiorina, Samuel J. Abrams, and Jeremy C. Pope think that the partisan polarization among this small yet highly visible group fuels the myth that a "culture war" is occurring in American politics. They argue that partisan polarization among most citizens is but a faint echo of the battle being waged among the political *cognoscenti*.[13]

From a methodological standpoint, the polarization perspective implies interaction. A comparison of Democrats and Republicans on an important issue should reveal a larger partisan difference for individuals with high political knowledge than for the less knowledgeable. Before we model these relationships using regression, let's turn to a familiar friend, mean comparison analysis, and get a better idea of what is going on in the data. Table 8-6 uses data from the American National Election Study to compare the abortion opinions of Democrats, Independents, and Republicans—at low and high levels of political knowledge. The abortion scale measures attitudes by a 10-point metric, from 0 (most anti-abortion rights) to 9 (most pro-abortion rights). Political knowledge is based on the number of political facts that respondents correctly identified.[14] Examine Table 8-6 for a few minutes. Is interaction taking place?

Indeed interaction's signature is evident in the data. Evaluate the party–opinion relationship for low-knowledge respondents. Democrats average 4.3, Independents 3.7, and Republicans 2.9.

At low levels of political knowledge, the mean declines by 1.4 points across the values of the independent variable. Now consider the relationship among those with higher political knowledge. Democrats average 5.7, Independents 4.6, and Republicans 2.8—nearly a 3-point drop across the values of partisanship. Also notice that the effect of the control variable differs across the values of the independent variable. Whereas high-knowledge Democrats are more supportive of abortion rights than low-knowledge Democrats, the control variable has no effect among Republicans.

Clearly, we are not dealing with linear-additive effects here. What regression equation would accurately depict these relationships? Consider the following model:

$$\hat{y} = \hat{a} + \hat{b}_1(\text{partisanship}) + \hat{b}_2(\text{high knowledge}) + \hat{b}_3(\text{partisanship} * \text{high knowledge}), \text{ where}$$

$\hat{y}$ is estimated prochoice scale score;

partisanship is coded 0 for Democrats, 1 for Independents, and 2 for Republicans; and

high knowledge is a dummy variable, coded 1 for respondents with high political knowledge and 0 for respondents with low political knowledge.

Switch political knowledge to 0 and see what the estimates say. When the knowledge dummy is equal to 0, then b_2 and b_3 drop out—$b_2(0) = 0$ and $b_3(0) = 0$—leaving $a + b_1$ (partisanship). Therefore, b_1 will estimate the effect of partisanship for low-knowledge respondents. Although the effect appeared modest, the mean comparison analysis revealed that low-knowledge Independents scored lower than Democrats, and low-knowledge Republicans scored lower still. In the way that partisanship is coded (higher codes are more Republican), b_1 should be negative, adjusting the intercept downward as partisanship moves from Democrat to Independent to Republican. Now switch high knowledge to 1. When the knowledge dummy is equal to 1, b_2 goes active. What does this quantity estimate? The coefficient, b_2, tells us the partial effect of knowledge on the dependent variable. In examining the "Total" row of Table 8-6, you may have noticed that high-knowledge respondents are more prochoice than low-knowledge respondents. We do not know whether this effect is statistically significant—the regression statistics will provide more information—but we can expect a positive sign on b_2.

The coefficients, b_1 and b_2, build the linear-additive base of the model. For low-knowledge Democrats, for example, the base estimate of the dependent variable is equal to $a + b_1(0) + b_2(0)$. For high-knowledge Democrats: $a + b_1(0) + b_2(1)$. For low-knowledge Republicans: $a + b_1(2) + b_2(0)$. For high-knowledge Republicans: $a + b_1(2) + b_2(1)$. Note that these linear-additive components tell us by how much to adjust the intercept, a, the mean value of the dependent variable when all independent variables are equal to 0. We can use the coefficients to estimate abortion scale scores for any combination of partisanship and political knowledge. If we were not modeling interaction, the regression analysis would begin and end with these base estimates. According to Table 8-6, however, the base effect of partisanship, b_1, by itself is not potent enough to describe the effect among the politically knowledgeable. For low-knowledge respondents, Republicans score 1.4 points lower than Democrats. For high-knowledge respondents, Republicans score 2.9 points lower—the 1.4-point base effect *plus* another 1.5-point effect.

Turn your attention to the interaction variable, "partisanship * high knowledge." An **interaction variable** is the multiplicative product of two (or more) independent variables. Note that for high-knowledge respondents the interaction variable takes on a value that is

Table 8-7 Modeling Interaction: Partisanship, Political Knowledge, and Abortion Opinions

Variable	Coefficient	Standard error	t-statistic	Significance
Constant	4.33			
Partisanship	−.70	.15	−4.67	.000
High political knowledge	1.50	.29	5.17	.000
Party identification *				
High political knowledge	−.76	.22	−3.45	.001
Adjusted R-square = .11				

Source: 2004 American National Election Study.

equal to partisanship. For Democrats this is equal to 0 (0 * 1 = 0), for Independents equal to 1 (1 * 1 = 1), and for Republicans equal to 2 (2 * 1 = 2). The coefficient, b_3, estimates the partial effect of the interaction variable. Specifically, b_3 tells us by how much to adjust the additive effects of partisanship and knowledge, $b_1 + b_2$, for each one-unit change in partisanship among high-knowledge individuals. If b_3 is close to 0, then the interaction effect is weak and we may just as well use the base effects to estimate abortion scores for high-knowledge people. As b_3 departs from 0, then the interaction effect becomes stronger and we will need to adjust the additive effects up or down, depending on b_3's sign. Table 8-6 shows that the effect of partisanship has the same tendency at low and high knowledge—as partisanship increases, mean scores decline—but that the effect is stronger for the more knowledgeable. Thus we would expect a negative sign on b_3.

Table 8-7 reports the results of the regression analysis. Table 8-7's layout, which departs from the tabular style of this chapter's previous examples, is closer to the format you will encounter in computer output and research articles. The variables appear on the left. The estimates and statistics for each variable appear to the right. With the possible exception of the label "Significance"—another name for "*P*-value"—all these elements are familiar. Let's interpret them. Start by switching everything to 0. The constant, 4.33, is the estimated abortion score for low-knowledge Democrats:

$$4.33 - .70(0) + 1.50(0) - .76(0) = 4.33.$$

The coefficient on partisanship, −.70, is negative and statistically significant. For each one-unit increase in party identification, abortion scores decline .70 of a point, on average. Our estimate for low-knowledge Independents: 4.33 − .70(1) = 3.63. For low-knowledge Republicans: 4.33 − .70(2) = 2.93. The coefficient on the high knowledge dummy, +1.50, is positive and statistically significant. Let's use the coefficients to obtain base model estimates for high-knowledge respondents. Using the base effects of party and knowledge, we would have 5.83 for Democrats (4.33 + 1.50), 5.13 for Independents (3.63 + 1.50), and 4.43 for Republicans (2.93 + 1.50).

Now let's bring the interaction effect into play. Check the significant coefficient on the interaction variable, −.76. This coefficient tells us that we must revise our estimates for high-knowledge people downward by .76 of a point for each one-unit increase in partisanship. Because Democrats are coded 0 on party, their high-knowledge estimate remains unchanged at 5.83, a full 1.50 points higher than low-knowledge Democrats. For Independents, coded 1 on partisanship, we deduct .76 from the initial additive estimate: 3.63 − .76 = 2.87. For Republicans, coded 2 on party, the deduction is twice the value of the coefficient: 4.43 − .76(2) = 2.91. Among Republicans, the negative interaction effect completely neutralizes the base effect of high knowledge, bringing high-knowledge (2.91) and low-knowledge Republicans (2.93) into agreement on this issue. [15]

Multicollinearity

In several examples in this chapter, multiple regression was used to estimate the partial effects of two independent variables on a dependent variable. However, the technique is not limited to two independent variables. In fact, the researcher may include any number of independent variables in the equation. In explaining state turnout, for example, we could hypothesize that yet another state-level variable—say, the degree of interparty competition in the state—has an effect on turnout. We could easily enter this variable into the equation and estimate its partial effects. By the same token, in accounting for differences in prochoice opinions, we could come up with other plausible variables, such as religiosity or gender, and further expand the model. Multiple regression would oblige by returning regression coefficients, t-statistics, R-square, and all the rest. If the researcher becomes too enamored of the power and flexibility of multiple regression, however, a serious statistical problem may be overlooked.

This problem can best be appreciated by thinking of controlled comparisons in cross-tabulations. Consider a realistic example. Suppose that you were using survey data, and you wanted to figure out the relationship between an independent variable, race (white/black), and a dependent variable, turnout (voted/did not vote). The bivariate analysis would be easy enough: Compare the percentage of whites who voted with the percentage of blacks who voted. But suppose further that you wanted to control for partisanship (Democrat/Republican). Logically, this too is easily accomplished. You would divide the subjects on the basis of partisanship—Democrats in one group, Republicans in the other—and then reexamine the race–turnout relationship separately for Democrats and Republicans. At that point you would encounter a classic problem. When you looked at the group of Republicans and sought to compare whites with blacks, you would find very few blacks—too few, in fact, to make a reasonable comparison. The problem, of course, is that the two independent variables, race and partisanship, are very strongly related. Blacks are overwhelmingly Democratic. So when you divide the sample into Democrats and Republicans, most of the blacks remain in one category of the control, Democrat. You might have a few black Republicans, but the percentages you calculated would be highly suspect because they would be computed on so few cases.

When this problem occurs in multiple regression, it is called **multicollinearity**. Multicollinearity occurs when the independent variables are so strongly related to each other that it becomes difficult to estimate the partial effect of each independent variable on the dependent variable. In its attempt to statistically control for one independent variable so that it can estimate the partial effect of the other independent variable, regression runs into the same problem encountered in the cross-tabulation example: too few cases. It is okay for the independent variables to be related. After all, the beauty of multiple regression is its ability to partial out the shared variance of the independent variables and arrive at estimates of the regression coefficients. The problem lies in the degree of relationship between the independent variables.

How can you tell if multicollinearity is a problem? If the magnitude of the correlation coefficient between the independent variables is less than .8, then multiple regression will work fine. If the correlation is .8 or higher, then multiple regression will not return good estimates. Regression models with interaction effects are especially susceptible to multicollinearity since, by design, interaction variables are combinations of other independent variables. In the model we just estimated, for example, the interaction variable (partisanship * high knowledge) correlates at .55 with partisanship and .73 with high knowledge. Happily, methodologists have worked out ways to ameliorate this problem.[16] Another multicollinearity clue is provided by comparing the value of adjusted R-square when one independent variable is included in the regression to the value of adjusted R-square when both variables are included. This comparison tells you how much better you can account for the dependent variable by knowing both independent variables. If the two independent variables are strongly related, then there will not be much improvement in R-square. In the state turnout examples, using education alone returned an adjusted R-square of .24, and using the South dummy alone returned .31. Both variables together gave us .42, a fair improvement and a statistical indication that multicollinearity was not a serious issue in the multiple regression results.

SUMMARY

In this chapter we introduced two powerful techniques of political analysis, correlation and regression. Correlation and regression together provide the answers to four questions about a relationship: (1) How strong is the relationship? (2) What is the direction of the relationship? (3) What is the exact nature of the relationship? (4) Could the observed relationship have occurred by chance? As we have seen, correlation speaks to questions of strength and direction. If the researcher has interval-level measurements for two variables, Pearson's r will summarize direction with a positive or negative sign on r, and it will give a reading of strength, from -1 to $+1$. Researchers typically use correlation analysis during the early stages of the research process, to explore the overall relationships between variables of interest. Regression analysis is more specialized than correlation. Regression, too, reveals the direction of the relationship between an independent variable and a dependent variable. A positive or negative sign on a regression coefficient indicates which way the relationship runs, with a positive slope indicating a positive relationship and a negative slope indicating an inverse relationship. Regression is often used to examine the causal connection between the variables. The regression coefficient, as we have seen, reveals the exact nature of this connection: the mean change in the dependent variable for each unit change in the independent variable. Thanks to the statistics provided by regression, the researcher also can test the null hypothesis that the true regression coefficient is equal to zero. Regression analysis provides a measure of strength as well. R-square is a PRE measure that tells the researcher how completely the independent variable explains the dependent variable.

KEY TERMS

adjusted R-square (p. 183)
correlation analysis (p. 170)
dummy variable (p. 184)
error sum of squares (p. 183)
interaction effect (p. 189)
interaction variable (p. 191)
multicollinearity (p. 193)
multiple regression (p. 187)
partial regression coefficient (p. 187)

Pearson's correlation coefficient (p. 172)
prediction error (p. 178)
regression analysis (p. 170)
regression coefficient (p. 176)
regression line (p. 175)
regression sum of squares (p. 183)
R-square (p. 181)
scatterplot (p. 171)
total sum of squares (p. 182)

EXERCISES

1. A researcher is investigating the relationship between economic development (x) and level of religiosity (y) in 10 countries. (The researcher has interval-level measurements for both variables.) The researcher theorizes that citizens of countries at the lower end of the development

scale will profess higher levels of religiosity than will citizens of countries at the higher end of the development scale. As development increases, religiosity decreases. Draw and label four sets of axes, like the one below:

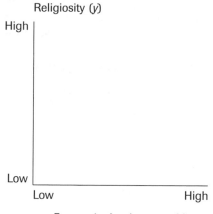

Religiosity (y)

A. Is the researcher hypothesizing a positive correlation between x and y, a negative correlation between x and y, or no correlation between x and y? Explain.

B. In the first set of axes you have drawn, and using a dot to represent each of the 10 countries, depict a correlation of –1 between economic development and level of religiosity.

C. In the second set of axes, depict a correlation of +1 between economic development and level of religiosity.

D. In the third set of axes, depict a correlation of 0 between economic development and level of religiosity.

E. Suppose the researcher finds a correlation of –.7 between the independent and dependent variables. In the fourth set of axes, show what a correlation of –.7 would look like. (Just make it look plausible. It doesn't have to be precise.)

F. Based on a correlation of –.7, the researcher concludes, "Level of economic development explains 70 percent of the variation in level of religiosity." Is this correct? Explain. (*Hint:* Review the difference between Pearson's r and R-square.)

2. Below are two columns of numbers: values of x and y for five observations. The variable x has a mean of 3.0 and a standard deviation equal to 1.6. The variable y has a mean of 13.0 and a standard deviation of 3.9. In this exercise you will hand-calculate the Pearson's correlation coefficient, R-square, and adjusted R-square for relationship between x and y.

x	y
1	17
2	14
3	16
4	10
5	8

A. Copy the columns onto a sheet of paper. Label three additional columns: "Z score of x," "Z score of y," and "Z score of x * Z score of y."

B. Calculate and write in the appropriate values in each column.

C. Show your calculation of Pearson's r for the relationship.

D. Calculate R-square for this relationship.

E. Showing your work, calculate adjusted R-square for this relationship.

3. *Environmental equity* is an interesting, and somewhat controversial, area of research. Some observers have argued that when state and local governments decide where to locate undesirable facilities, such as hazardous waste treatment plants, they choose areas more heavily populated with minorities. So, in this view, the racial composition of an area (x) should predict the area's proximity (y) to environmental and health hazards. Below are fabricated data for 10 census blocks. The independent variable is percent black; the dependent variable is the distance (in miles) between each block and the nearest hazardous waste treatment facility. The hypothesis: In a comparison of census blocks, those with higher percentages of blacks will be closer to hazards than will blocks having lower percentages of blacks.

Census block:	Percent black (x)	Distance in miles (y)
Block 1	0	27
Block 2	0	23
Block 3	10	22
Block 4	10	18
Block 5	20	17
Block 6	20	13
Block 7	30	12
Block 8	30	8
Block 9	40	6
Block 10	40	4

A. What is the regression equation for this relationship? Interpret the regression coefficient. What is the effect of x on y?
B. Interpret the y-intercept. What does the intercept tell you, exactly?
C. Based on this equation, what is the predicted value of y for census blocks that are 15 percent black? Census blocks that are 25 percent black?
D. Adjusted R-square for these data is .93. Interpret this value.

4. Are Catholics more likely to oppose abortion rights than are non-Catholics? To find out, a researcher first constructs an abortion scale from responses to the General Social Survey (GSS). Scores range from 0 (abortion should be permitted under all circumstances) to 7 (abortion should not be permitted under any circumstances). Unlike the abortion scale analyzed in this chapter, respondents with lower scores on the GSS scale are more pro-abortion rights, and respondents with higher scores are anti-abortion rights. The researcher uses the GSS scale as the dependent variable, y. The researcher then creates a dummy variable. Catholics are coded 1 on this dummy, and non-Catholics are coded 0. This is the independent variable, x, which the researcher names "Catholic." Here are the regression results:

$$y = 2.6 + .56(\text{Catholic})$$
$$\text{Standard error of } b = .15$$
$$\text{Adjusted } R^2 = .01$$

A. Based on these findings, the researcher concludes: "While non-Catholics averaged 2.6 on the abortion scale, Catholics averaged only .56 on the scale. Therefore, Catholics are more pro-abortion rights than are non-Catholics." Is this inference correct? Why or why not?
B. Another conclusion reached by the researcher: "The independent variable does not have a statistically significant effect on the dependent variable." Is this inference correct? Why or why not?
C. Yet another of the researcher's conclusions: "The independent variable explains very little of the variation in the dependent variable." Is this inference correct? Why or why not?

5. Another researcher, after viewing the puny value of adjusted R^2 in Exercise 4, suggests that another variable—the frequency with which individuals attend religious services—may contribute to the explanation of abortion beliefs. This researcher defines a dummy variable, which is coded 1 for individuals who report high levels of religious attendance and coded 0 for people who have low levels of attendance. The regression to be estimated: $y = a + b_1$(Catholic) + b_2(high attendance), where "Catholic" is the Catholic/non-Catholic dummy (Catholics are coded 1, non-Catholics coded 0) and "high attendance" is the high attendance/low attendance dummy. Here are the results (the standard errors for the regression coefficients are in parentheses):

$$y = 2.14 + .50 \text{ (Catholic)} + 1.95 \text{ (high attendance)}$$
$$(.14) \qquad\qquad (.14)$$
$$\text{Adjusted } R^2 = .12$$

A. What is the *partial* effect of Catholicism on the abortion scale, controlling for attendance at religious services? Is it reasonable to infer that, in the population, Catholics are more opposed to abortion than are non-Catholics? Explain.

B. What is the partial effect of attendance on the abortion scale, controlling for differences between Catholics and non-Catholics? Is it reasonable to infer that, in the population, people who attend services more frequently are more opposed to abortion than are people who attend less frequently? Explain.

C. Based on this regression, what is the mean abortion score for non-Catholic low-attenders? For Catholic high-attenders?

D. The adjusted R^2 value is .12. This means that _____ percent of the variation in abortion scores is explained by both variables in the model. It also means that _____ percent is explained by variables *not* in the model.

E. Name one other variable that may account for differences in the dependent variable. Briefly describe why you think this variable may contribute to the explanation of abortion attitudes.[17]

6. Suppose you want to model a set of interaction relationships between Catholicism, religious attendance, and abortion beliefs. You think that the positive effect of religious attendance on anti-abortion attitudes is significantly stronger for Catholics than non-Catholics. To construct the interaction model, you will build on the base effects of the model shown in Exercise 5: $y = a + b_1$(Catholic) + b_2(high attendance), where "Catholic" is a Catholic/non-Catholic dummy (Catholics are coded 1, non-Catholics coded 0) and "high attendance" is a high attendance/low attendance dummy (frequent attenders are coded 1, infrequent attenders are coded 0). Before you specify the model, you will need to compute an interaction variable.

A. The interaction variable is computed by multiplying _____ times _____. Which of the following groups of respondents will have a value of 0 on the interaction variable: Catholic low-attenders, non-Catholic low-attenders, Catholic high-attenders, non-Catholic high-attenders? Write down all answers that apply.

B. Which of the following groups of respondents will have a value of 1 on the interaction variable: Catholic low-attenders, non-Catholic low-attenders, Catholic high-attenders, non-Catholic high-attenders? Write down all answers that apply.

C. Write out the interaction model to be estimated.

D. Focus on the coefficient that estimates the interaction effect. If your idea is correct—that the positive effect of religious attendance on anti-abortion attitudes is significantly stronger for Catholics than non-Catholics—then would you expect the sign on the coefficient to be negative, positive, or close to 0? Explain your answer.

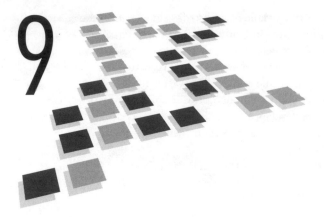

9

Logistic
Regression

LEARNING OBJECTIVES

In this chapter you will learn:
- How to use logistic regression to describe the relationship between an interval-level independent variable and a dichotomous dependent variable
- How logistic regression is similar to—and different from—ordinary least squares regression
- How maximum likelihood estimation works
- How to use logistic regression with multiple independent variables

Political analysis is not unlike using a toolbox. The researcher looks at the substantive problem at hand, selects the methodological tool most appropriate for analyzing the relationship, and then proceeds with the analysis. Selection of the correct tool is determined largely by the levels of measurement of the variables of interest. If both the independent and dependent variables are measured by nominal or ordinal categories—a common situation, particularly in survey research—the researcher would most likely select cross-tabulation analysis. If both the independent and dependent variables are interval level, then ordinary least squares (OLS) regression would be applied. Finally, if you wanted to analyze the relationship between an interval-level dependent variable and a categorical independent variable, then you might use mean comparison analysis or, alternatively, you could specify and test a linear regression model using dummy variables. These techniques, all of which have been discussed in earlier chapters, add up to a well-stocked toolbox. Even so, one set of tools is missing.

Logistic regression is part of a family of techniques designed to analyze the relationship between an interval-level independent variable and a categorical dependent variable, a dependent variable measured by nominal or ordinal values. The dependent variable might have any number of categories—from several, to a few, to two. In this chapter we discuss how to use and interpret logistic regression in the simplest of these situations, when the dependent variable takes on only two values. For example, suppose we are using survey data to investigate the relationship between education and voter turnout. We think that a positive relationship exists here: As education increases, so does the likelihood of voting. The independent variable, education, is measured in precise 1-year increments, from 0 (no formal education) to 20 (20 years of schooling). The dependent variable, however, takes on only two values—respondents either voted or they did not vote. In this situation, we have a binary dependent variable. A **binary variable** is a dichotomous variable, one that can assume only two values. Binary variables are identical to dummy variables, discussed in Chapter 8. Thus, voted/did not vote, smoker/nonsmoker, approve/do not approve, and married/unmarried are all examples of dummy variables or binary variables.[1]

In some ways, logistic regression is similar to OLS regression. Like OLS, logistic regression gauges the effect of the independent variable by estimating an intercept and a slope, both familiar fixtures of linear regression. Plus logistic regression provides a standard error for the slope, which allows the researcher to test hypotheses about the effect of the independent variable on the dependent variable. And like OLS, logistic regression is remarkably flexible, permitting the use of multiple independent variables, including dummy independent variables.

In one fundamental way, however, logistic regression is a different breed of cat. When we perform OLS regression, we can reasonably assume a linear relationship between an independent variable (x) and a dependent variable (y). For example, for the relationship between years of schooling (x) and income in dollars (y), we can use a linear model to estimate the average dollar-change in income for each 1-year increase in education. OLS would give us an idea of how closely the relationship between x and y fits this linear pattern. However, when we have a binary dependent variable, we must assume that it bears a nonlinear relationship to x. So as education (x) increases from 8 years to 9 years to 10 years, we most plausibly assume that the likelihood of voting (y) is low and increases slightly for each of these 1-year increments. But as education increases from 11 years to 12 years to 13 years, we would expect voter turnout to show large increases for each 1-year increment in this range of x. In the higher values of education—say, beyond 13 years—we would assume that turnout is already high and that each additional year of schooling would have a weaker effect on voting. A logistic regression analysis would give us an idea of how closely the relationship between x and y fits this nonlinear pattern.

This chapter is divided into four sections. In the first section we use both hypothetical and real-world data to illustrate the logic behind logistic regression. Here you will be introduced to some unfamiliar terms—such as odds and logged odds—that define the workings of the technique, and you will learn what to look for in your own analyses and how to describe and interpret your findings. In the second section we take a closer look at maximum likelihood estimation, the method logistic regression uses to estimate the effect of the independent variable (or variables) on the dependent variable. Here you will see how logistic regression is similar to other techniques and statistics we discussed previously, particularly chi-square. In the third section we demonstrate how the logistic regression model, much like multiple linear regression, can be extended to accommodate several independent variables. Finally, we will consider some additional ways to present and interpret logistic regression results. By the end of this chapter you will have added another powerful technique to your toolbox of political research methods.

THE LOGISTIC REGRESSION APPROACH

Let's begin with a hypothetical example. Suppose we are investigating whether education (x) affects voter turnout (y) among a random sample of respondents ($n = 500$). For purposes of

Table 9-1 Education and the Probability of Voting

Did respondent vote?	Education					
	0. Low	*1. Middle-low*	*2. Middle*	*3. Middle-high*	*4. High*	*Total*
1. Yes, voted	6	20	50	80	94	250
0. No, did not vote	94	80	50	20	6	250
Total (*n*)	100	100	100	100	100	500
Probability of voting	.06	.20	.50	.80	.94	.50

Note: Hypothetical data.

illustration, let's assume that the independent variable, education, is an interval-level variable that varies from 0 (low) to 4 (high), and that voter turnout is a binary dependent variable, coded 1 if the individual voted and 0 if he or she did not vote. Table 9-1 shows the results from a cross-tabulation analysis of the hypothetical sample data. Although column percentages have not been supplied in Table 9-1, they are easy to figure out because each value of education contains exactly 100 cases. For example, of the 100 people in the low-education category, 6 voted—a percentage equal to 6 or a proportion equal to .06. Twenty percent (.20) of the 100 middle-low education individuals voted, 50 percent (.50) of the middle group voted, and so on. The bottom row of Table 9-1 presents the proportion of voters for each value of education, but it uses the label "Probability of voting." Why use *probability* instead of *proportion*? The two terms are synonymous. Think of it this way: If you were to randomly select one individual from the group of 100 low-education respondents, what is the probability that this randomly selected person voted? Because random selection guarantees that each case has an equal chance of being picked, there are 6 chances in 100—a probability of .06—of selecting a voter from this group. Similarly, you could say that there is a random probability of voting equal to .06 for any individual in the low-education category, a probability of .20 for any respondent in the middle-low group, and so on. It is important to shift your thinking from proportions to probabilities, because logistic regression is aimed at determining how well an independent variable (or set of independent variables) predicts the probability of an occurrence, such as the probability of voting.

Consider the Table 9-1 probabilities and make some substantive observations. Clearly a positive relationship exists between education and the probability of voting: As education (*x*) goes up, so does the probability of voting (*y*). Now examine this pattern more closely and apply the logic of linear regression. Does a one-unit increase in the independent variable produce a consistent increase in the probability of voting? Starting with the interval between low and middle-low, the probability goes from .06 to .20—an increase of .14. So by increasing the independent variable by 1 in this interval, we see a .14 increase in the probability of voting. Between middle-low and middle, however, this effect increases substantially, from .20 to .50—a jump of .30. The next increment, from middle to middle-high, produces another .30 increase in the probability of voting, from .50 to .80. But this effect levels off again between the two highest values of the independent variable. Moving from middle-high to high education occasions a more modest increase of .14 in the probability of voting. Thus the linear logic does not work very well. A unit change in education produces a change in the probability of voting of either .14 or .30, depending on the range of the independent variable examined. Put another way, the probability of voting (*y*) has a nonlinear relationship to education (*x*).

Rest assured that there are very good statistical reasons the researcher should not use OLS regression to estimate the effect of an interval-level independent variable on a binary dependent variable.[2] Perhaps as important, there are compelling substantive reasons you would not expect a linear model to fit a relationship such as the one depicted in Table 9-1. Think about this for a moment. Suppose you made $10,000 a year and were faced with the decision of whether to make a major purchase, such as buying a home. There is a good chance that you would decide not to make the purchase. Now suppose that your income rose to $20,000, a $10,000 increase. To be sure, this rise in income might affect your reasoning, but most likely it would not push your decision over the purchasing threshold, from a decision not to buy to a decision to buy. Similarly, if your initial income were $95,000, you would probably decide to buy the house, and an incremental $10,000 change, to $105,000, would have a weak effect on this decision—you were very likely to make the purchase in the first place. But suppose that you made $45,000. At this income level, you might look at your decision a bit differently: "If I made more money, I could afford a house." Thus that $10,000 pay raise would push you over the threshold. Going from $45,000 to $55,000 greatly enhances the probability that you would make the move from "do not buy" to "buy." So at low and high initial levels of income, an incremental change in the causal variable has a weaker effect on your dichotomous decision (do not buy/buy) than does the same incremental change in the middle range of income.

Although fabricated, the probabilities in Table 9-1 show a plausible pattern. Less-educated individuals are unlikely to vote, and you would not expect a small increment in education to make a huge difference in the probability of voting. The same idea applies to people in the upper education range. Individuals in the middle-high to high category are already quite likely to vote. It would be unreasonable to suggest that, for highly educated people, a one-unit increase in the independent variable would have a big effect on the likelihood of voting. It is in the middle intervals of the independent variable—from middle-low to middle-high—where you might predict that education would have its strongest effect on the dependent variable. As people in this range gain more of the resource (education) theoretically linked to voting, a marginal change in the independent variable is most likely to switch their dichotomous decision from "do not vote" to "vote." Logistic regression allows us to specify a model that takes into account this nonlinear relationship between education and the probability of voting.

As we have seen, the first step in understanding logistic regression is to think in terms of the probability of an outcome. The next step is to get into the habit of thinking in terms of the odds of an outcome. This transition really is not too difficult, because odds are an alternative way of expressing probabilities. Whereas probabilities are based on the number of occurrences of one outcome (such as voting) divided by the total number of outcomes (voting plus nonvoting), **odds** are based on the number of occurrences of one outcome (voting) divided by the number of occurrences of the other outcome (nonvoting). According to Table 9-1, for example, among the 100 people in the middle-high education group, there were 80 voters—a probability of voting equal to 80/100 or .80. What are the odds of voting for this group? Using the raw numbers of voters and nonvoters, the odds would be 80 to 20, or, to use a more conventional way of verbalizing odds, 4 to 1, four voters to every nonvoter. In describing odds, we ordinarily drop the ". . . to 1" part of the verbalization and say that the odds of voting are 4. So for the middle-high education group, the probability of voting is .80 and the odds of voting are 4. In figuring odds, you can use the raw numbers of cases, as we have just done, or you can use probabilities to compute odds. The formula for converting probabilities to odds is as follows:

$$\text{Odds} = \text{Probability} / (1 - \text{Probability}).$$

Table 9-2 Probability of Voting, Odds of Voting, and Logged Odds of Voting at Five Levels of Education (hypothetical data)

Education (x)	Probability of voting (y)	Odds of voting (y)	Logged odds of voting (y)
0. Low	.06	.06/.94 = .06	−2.8
1. Middle-low	.20	.20/.80 = .25	−1.4
2. Middle	.50	.50/.50 = 1	0
3. Middle-high	.80	.80/.20 = 4	+1.4
4. High	.94	.94/.06 = 16	+2.8

Note: Hypothetical data.

Apply this conversion to the example just discussed. For middle-high education respondents, the odds would be .80 divided by (1 minus .80), which is equal to .80 / .20, or 4. The "Odds of voting" column in Table 9-2 shows this conversion for the five education groups.

Consider the numbers in the "Odds of voting" column and note some further properties of odds. Note that probabilities of less than .50 produce odds of less than 1 and probabilities of greater than .50 convert to odds of greater than 1. The probabilities for low and middle-low education respondents (.06 and .20, respectively) convert to odds of .06 and .25, and the probabilities among the highest education groups translate into odds of 4 and 16. If an event is as likely to occur as not to occur, as among the middle education people, then the probability is .50 and the odds are equal to 1 (.50 / .50 = 1).

Now examine the "Odds of voting" column in Table 9-2 more closely. Can you discern a systematic pattern in these numbers, as you proceed down the column from low education to high education? Indeed, you may have noticed that the odds of voting for the middle-low education group is (very nearly) four times the odds of voting for the low education category, since 4 times .06 is about equal to .25. And the odds of voting for the middle education group is four times the odds for the middle-low group, since 4 times .25 equals 1. Each additional move, from middle to middle-high (from an odds of 1 to an odds of 4) and from middle-high to high (from 4 to 16), occasions another fourfold increase in the odds. So, as we proceed from lower to higher values of the independent variable, the odds of voting for any education group is four times the odds for the next-lower group. In the language of logistic regression, the relationship between the odds at one value of the independent variable compared with the odds at the next-lower value of the independent variable is called the **odds ratio**. Using this terminology to describe the "Odds of voting" column of Table 9-2, we would say that the odds ratio increases by 4 for each one-unit increase in education.

The pattern of odds shown in Table 9-2 may be described in another way. Instead of figuring out the odds ratio for each change in education, we could calculate the **percentage change in the odds** of voting for each unit change in education. This would be accomplished by seeing how much the odds increase and then converting this number to a percentage. Between low and middle-low, for example, the odds of voting go from .06 to .25, an increase of .19. The percentage change in the odds would be .19 divided by .06, which is equal to 3.17—a bit more than a 300-percent increase in the odds of voting. For the move from middle-low to middle we would have (1 − .25) / .25 = 3.00, another 300-percent increase in the odds of voting. In fact, the odds of voting increases by 300 percent for each additional unit increase in education: From middle to middle-high ([4 − 1] / 1 = 3.00) and from middle-high to high ([16 − 4] / 4 = 3.00). Using this method to describe the Table 9-2 data, we could conclude that the odds of voting increase by 300 percent for each one-unit increase in education.

Let's review what we have found so far. When we looked at the relationship between education and the probability of voting, we saw that an increase in the independent variable does not produce a consistent change in the dependent variable. In examining the relationship between education and the odds of voting, however, we saw that a unit change in education does produce a constant change in the odds ratio of voting—equal to 4 for each unit change in x. Alternatively, each change in the independent variable produces a consistent percentage increase in the odds of voting, a change equal to 300 percent for each unit change in x. What sort of model would summarize this consistent pattern?

The answer to this question lies at the heart of logistic regression. Logistic regression does not estimate the change in the probability of y for each unit change in x. Rather, it estimates the change in the *log of the odds of y* for each unit change in x. Consider the third column of numbers in Table 9-2. This column reports an additional conversion, labeled "Logged odds of voting." For low education, this number is equal to –2.8, for middle-low education it is equal to –1.4, for middle education 0, for the middle-high group +1.4, and for high education +2.8. Where did these numbers come from?

A logarithm, or log for short, expresses a number as an exponent of some constant or base. If we chose a base of 10, for example, the number 100 would be expressed as 2, since 100 equals the base of 10 raised to the power of 2 ($10^2 = 100$). We would say, "The base-10 log of 100 equals 2." Base-10 logs are called **common logarithms**, and they are used widely in electronics and the experimental sciences. Statisticians generally work with a different base, denoted as e.

The base e is approximately equal to 2.72. Base-e logs are called **natural logarithms** and are abbreviated *ln*. Using the base e, we would express the number 100 as 4.61, since 100 equals the base e raised to the power of 4.61 ($e^{4.61} \approx [100]$, or $\ln[100] \approx 4.61$). We would say, "The natural log of 100 equals 4.61." The five numbers in the Table 9-2 column "Logged odds of voting" are simply the natural logs of .06 ($e^{-2.8} = .06$), .25 ($e^{-1.4} = .25$), 1 ($e^0 = 1$), 4 ($e^{1.4} = 4$), and 16 ($e^{2.8} = 16$). Using conventional notation: $\ln(.06) = -2.8$, $\ln(.25) = -1.4$, $\ln(1) = 0$, $\ln(4) = 1.4$, and $\ln(16) = 2.8$.

These five numbers illustrate some general features of logarithmic transformations. Any number less than 1 has a negatively signed log. So to express .25 as a natural log, we would raise the base e to a negative power, –1.4. Any number greater than 1 has a positively signed log. To convert 4 to a natural log, we would raise e to the power of 1.4. And 1 has a log of 0, since e raised to the power of 0 equals 1. Natural log transformations of odds are often called logit transformations or **logits** for short. So the logit ("lowjit") of 4 is 1.4.[3]

You are probably unaccustomed to thinking in terms of odds instead of probabilities. And it is a safe bet that you are really unaccustomed to thinking in terms of the logarithmic transformations of odds. But stay focused on the "Logged odds of voting" column. Again apply the linear regression logic. Does a unit change in the independent variable, education, produce a consistent change in the log of the odds of voting? Well, going from low education to middle-low education, the logged odds increases from –2.8 to –1.4, an increase of 1.4. And going from middle-low to middle, the logged odds again increases by 1.4 (0 minus a negative 1.4 equals 1.4). From middle to middle-high and from middle-high to high—each one-unit increase in education produces an increase of 1.4 in the logged odds of voting.

Now there is the odd beauty of logistic regression. Although we may not use a linear model to estimate the effect of an independent variable on the probability of a binary dependent variable, we may use a linear model to estimate the effect of an independent variable on the logged odds of a binary dependent variable. Consider this plain-vanilla regression model:

$$\text{Logged odds } (y) = \hat{a} + \hat{b}\,(x).$$

As you know, the regression coefficient, $\hat{b}$, estimates the change in the dependent variable for each unit change in the independent variable. And the intercept, $\hat{a}$, estimates the value of the dependent variable when x is equal to 0. Using the numbers in the "Logged odds of voting" column to identify the values for $\hat{b}$ and $\hat{a}$, we would have:

$$\text{Logged odds (voting)} = -2.8 + 1.4 \text{ (education)}.$$

Review how this model fits the data. For the low-education group (coded 0 on education), the logged odds of voting would be $-2.8 + 1.4(0)$, which is equal to -2.8. For the middle-low education group (coded 1 on education): $-2.8 + 1.4(1)$, equal to -1.4. For the middle group (coded 2): $-2.8 + 1.4(2)$, equal to 0. And so on, for each additional one-unit increase in education. Clearly, this linear model nicely summarizes the relationship between education and the logged odds of voting.

Now if someone were to ask, "What, exactly, is the effect of education on the likelihood of voting?" we could reply, "For each unit increase in education there is an increase of 1.4 in the logged odds of voting." Although correct, this interpretation is not terribly intuitive—and is likely to occasion a quizzical look from our interlocutor. Therefore, we can use the regression coefficient, 1.4, to retrieve a more understandable number: the change in the odds ratio of voting for each unit change in education. How might this be accomplished? Remember that, as in any regression, all of the coefficients on the right-hand side are expressed in units of the dependent variable. Thus the intercept, a, is the logged odds of voting when x is 0. And the slope, b, estimates the change in the logged odds of voting for each unit change in education. Because logged odds are exponents of e, we can get from logged odds back to odds by raising e to the power of any coefficient in which we are interested. Accordingly, to convert the slope, 1.4, we would raise e to the power of 1.4. This exponentiation procedure, abbreviated *Exp(b)*, looks like this:

$$\text{Exp(b)} = \text{Exp}(1.4) = e^{1.4} = 4.$$

Now we have a somewhat more interpretable reply: "For each unit change in education, the odds ratio increases by four. Members of each education group are four times more likely to vote than are members of the next-lower education group."

Even more conveniently, we can translate the coefficient, 1.4, into a percentage change in the odds of voting. Here is the general formula:

$$\text{Percentage change in the odds of } y = 100 * (\text{Exp(b)} - 1).$$

For our example:

$$\text{Percentage change in the odds of voting} = 100 * [\text{Exp}(1.4) - 1] = 100 * (e^{1.4} - 1)$$
$$= 100 * (4 - 1) = 300.$$

Thus we now can say, "Each unit increase in education increases the odds of voting by 300 percent."

Note further that, armed with the logistic regression equation, "Logged odds (voting) = −2.8 + 1.4 (education)," we can estimate the odds of voting—and therefore the probability of voting—for each value of the independent variable. For the middle-low education group, for example, the logistic regression tells us that the logged odds of voting is −2.8 plus 1.4(1), equal to −1.4. Again, because −1.4 is the exponent of e, the odds of voting for this group would be Exp(−1.4), equal to .25. If the odds of voting is equal to .25, what is the probability of voting? We have already seen that:

$$Odds = Probability / (1 - Probability).$$

Following a little algebra:

$$Probability = Odds / (1 + Odds).$$

So for the middle-low group, the probability of voting would be:

$$.25 / (1 + .25) = .25 / 1.25 = .20.$$

By performing these reverse translations for each value of x—from logged odds to odds, and from odds to probabilities—we can retrieve the numbers in the "Probability of voting" column of Table 9-2. If we were to plot these retrieved probabilities of y for each value of x, we would end up with Figure 9-1.

Figure 9-1 Plotted Probabilities of Voting (y), by Education (x)

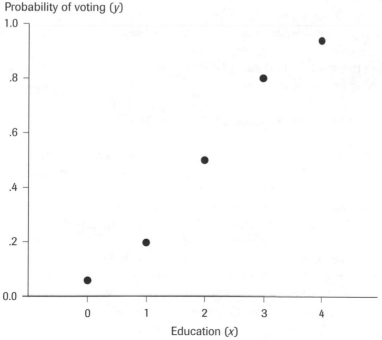

Note: Hypothetical data.

In the beginning we realized that the linear regression logic could not be accurately or appropriately applied to the nonlinear relationship between education and the probability of voting. But after transforming the dependent variable into logged odds, we could apply a linear model. So x bears a nonlinear relationship to the probability of y, but x bears a linear relationship to the logged odds of y. Furthermore, because the logged odds of y bears a nonlinear relationship to the probability of y, the logistic regression model permits us to estimate the probability of an occurrence for any value of x.

An S-shaped relationship, such as that depicted in Figure 9-1, is the visual signature of logistic regression. Just as OLS regression will tell us how well our data fit a linear relationship between an independent variable and the values of an interval-level dependent variable, logistic regression will tell us how well our data fit an S-shaped relationship between an independent variable and the probability of a binary dependent variable.[4] In the hypothetical education-voting example, the relationship worked out perfectly. The logistic regression returned the exact probabilities of voting for each value of education. By eyeballing the "Logged odds of voting" column of Table 9-2, we could easily identify the intercept and regression coefficient of the logistic regression equation. No prediction error. In the practical world of political research, of course, relationships are never this neat.

To apply what you have learned thus far—and to discuss some further properties of logistic regression—we enlist here a real-world dataset, the 1998 General Social Survey ($n = 2,605$), and reexamine the relationship between education and voting. We estimate the logistic regression equation as follows:

$$\text{Logged odds (voting)} = \hat{a} + \hat{b} \,(\text{education}).$$

The dependent variable is reported turnout in the 1996 presidential election. As in the hypothetical example, voters are coded 1 and nonvoters are coded 0. Unlike the hypothetical example, the independent variable, education, is measured in years of formal schooling. This is a more realistic interval-level variable, with values that range from 0 for no formal education to 20 for 20 years of education. Table 9-3 reports the results obtained from a logistic regression analysis using SPSS.

First consider the numbers in the row labeled "Coefficient estimates." Plugging these values into our equation, we would have:

$$\text{Logged odds (voting)} = -1.581 + .180 \,(\text{education}).$$

The coefficients tell us that, for individuals with no formal schooling, the estimated logged odds of voting is equal to -1.581, and each 1-year increment in education increases the estimated logged odds by .180. The value in the bottom-most row, "Exp(b)," translates logged odds back into odds and thus provides a more accessible interpretation. Every unit increase in education increases the odds ratio by 1.197. Individuals at any given level of education are about 1.2 times more likely to vote than are individuals at the next-lower level of education. That is, as we move from one value of education to the next-higher value, we would multiply the odds of voting by about 1.2. Perhaps the most intuitively appealing way to characterize the relationship is to estimate the percentage change in the odds of voting for each 1-year increase in education. As we have seen, this is accomplished by subtracting 1 from the exponent, 1.197, and multiplying by 100. Performing this calculation: $100 * (1.197 - 1) = 19.7$. Thus the odds of voting increase by about 20 percent for each 1-year increase in the independent variable.

Table 9-3   Education and Voting: Logistic Regression Coefficients
and Related Statistics

Logged odds (voting)	=	Intercept $\hat{a}$	+	Education $\hat{b}$
Coefficient estimate		−1.581		.180
Standard error				.016
Wald				128.85
Significance				.000
Exp(b)				1.197

Source: James A. Davis, Tom W. Smith, and Peter V. Marsden. *General Social Surveys, 1972–2002.* (Chicago: National Opinion Research Center [producer], 2003; Storrs, Conn.: Roper Center for Public Opinion Research, University of Connecticut/Ann Arbor: Inter-university Consortium for Political and Social Research [distributors], 2003).

Note: Displayed data are from the 1998 GSS. N = 2,605. Dependent variable is reported turnout in the 1996 presidential election. Independent variable is number of years of formal schooling.

What would the null hypothesis have to say about these results? As always, the null hypothesis claims that, in the population from which the sample was drawn, no relationship exists between the independent and dependent variables, that individuals' levels of education play no role in determining whether they vote. Framed in terms of the logistic regression coefficient, the null hypothesis says that, in the population, the true value of the coefficient is equal to 0, that a one-unit increase in the independent variable produces no change in the logged odds of voting. The null hypothesis also can be framed in terms of the odds ratio, Exp(b). As we have seen, the odds ratio tells us by how much to multiply the odds of the dependent variable for each one-unit increase in the independent variable. An odds ratio of less than 1 means that the odds decline as the independent variable goes up (a negative relationship). (For a discussion of negative relationships in logistic regression, see Box 9-1.) An odds ratio of greater than 1 says that the odds increase as the independent variable goes up (a positive relationship). But an odds ratio of 1 means that the odds do not change as the independent variable increases. Thus an odds ratio equal to 1 would be good news for the null hypothesis, because it would mean that individuals at any level of education are no more likely to vote than are individuals at the next-lower level of education. So if the logistic regression coefficient were equal to 0 or Exp(b) were equal to 1, then we would have to say that the independent variable has no effect on the dependent variable.[5] As it is, however, we obtained an estimated coefficient equal to .180, which is greater than 0, and an odds ratio equal to 1.197, which is greater than 1. But how can we tell if these numbers are statistically significant?

Notice that, just as in OLS regression, logistic regression has provided a standard error for the estimated slope, *b*. And, again like OLS, the standard error tells us how much prediction error is contained in the estimated coefficient.[6] Thus, according to Table 9-3, each 1-year increase in education produces a .180 increase in the logged odds of voting, give or take .016 or so. Whereas least squares regression computes a test statistic based on Student's *t*-distribution, logistic regression computes the Wald statistic, which follows a chi-square distribution.[7] Computer programs, such as SPSS, quite helpfully provide a *P*-value for Wald. Like any *P*-value, this number tells you the probability of obtaining the observed results, under the assumption that the null hypothesis is correct. A *P*-value equal to .000 says that, if

Box 9-1 How to Interpret a Negative Relationship in Logistic Regression

As we expected, Table 9-3 reveals a positive relationship between education and the likelihood of voting. Each 1-year increase in schooling increases the logged odds of voting by .180. Alternatively, each increment in education boosts the odds ratio by 1.2—a 20 percent increase for each increment in educational attainment. However, in your own or in others' research you will often encounter negative relationships, situations in which a unit increase in the independent variable is associated with a decrease in the logged odds of the dependent variable. (One of the exercises at the end of this chapter requires that you interpret a negative relationship.) Negative relationships can be a bit trickier in logistic regression than in OLS regression, so let's consider an example. Suppose we were to investigate the relationship between the likelihood of voting and the number of hours respondents spend watching television per day. In this situation, we might expect to find a negative relationship: The more television that people watch, the less likely they are to vote. In fact, we would obtain these estimates:

$$\text{Logged odds (voting)} = 1.013 - .091(\text{TV hours}).$$

Thus, each 1-hour increase in daily television watching occasions a decrease of .091 in the logged odds of voting. Obtaining the odds ratio, we would have: $\text{Exp}(-.091) = e^{-.091} = .913$. Positive relationships produce odds ratios of greater than 1, and negative relationships produce odds ratio of less than 1. How would you interpret an odds ratio of .913? Like this: Individuals watching any given number of hours of television per day are only about .9 times as likely to vote as are individuals who watch the next-lower number of hours. For example, people who watch 4 hours per day are .9 times as likely to vote as are people who watch 3 hours per day. Following the formula for percentage change in the odds: $100 * (.913 - 1) = -8.7$. Each additional hour spent in front of the television depresses the odds of voting by about 9 percent.[1]

---

1. Data for this analysis are from the 1998 General Social Survey. The independent variable, number of hours spent watching television, is based on the question "On the average day, about how many hours do you personally watch television?" The logistic regression analysis presented here ($n = 2,163$) produced a coefficient of $-.091$, with a standard error of .020 and a Wald statistic of 20.693 (significance = .000).

the null hypothesis is correct, then the probability of obtaining a regression coefficient of .180 is highly remote—clearly beyond the .05 standard. Therefore, we can safely reject the null hypothesis and conclude that education has a statistically significant effect on the likelihood of voting.

As you can see, in some ways logistic regression bears a kinship to OLS regression. In running OLS, we obtain an estimate for the linear regression coefficient that minimizes prediction errors. That is, OLS provides the best fit between the predicted values of the dependent variable and the actual, observed values of the dependent variable. OLS also reports a standard error for the regression coefficient, which tells us how much prediction error is contained in the regression coefficient. This information permits us to determine whether x has a significant effect on y. Similarly, logistic regression minimizes prediction errors by finding an estimate for the logistic regression coefficient that yields the maximum fit between the predicted probabilities of y and the observed probabilities of y. Plus it reports a standard error for this estimated effect.

However, a valuable statistic is missing from the analogy between OLS and logistic regression: R-square. As you know, R-square tells the researcher how completely the independent variable (or, in multiple regression, all of the independent variables) explains the dependent variable. In our current example, it certainly would be nice to know how completely the independent variable, education, accounts for the likelihood of voting. Does logistic regression provide an analogous statistic to R-square? Strictly speaking, the answer is no.[8] Even so, methodologists have proposed R-square-like measures that give an overall reading of the strength of association between the independent variables and the dependent variable. To understand these measures, we need to take a closer look at maximum likelihood estimation, the technique logistic regression uses to arrive at the best fit between the predicted probabilities of y and the observed probabilities of y.

FINDING THE BEST FIT: MAXIMUM LIKELIHOOD ESTIMATION

By way of introducing maximum likelihood estimation, it is helpful to recall the logic behind proportional reduction in error (PRE) measures of association, such as lambda or R-square. You will remember that a PRE measure first determines how well we can predict the values of the dependent variable *without* knowledge of the independent variable. It then compares this result with how well we can predict the dependent variable *with* knowledge of the independent variable. PRE uses the overall mean of the dependent variable to "guess" the dependent variable for each value of the independent variable. This guessing strategy produces a certain number of errors. PRE then figures out how many errors occur when the independent variable is taken into account. By comparing these two numbers—the number of errors without knowledge of the independent variable and the number of errors with knowledge of the independent variable—PRE determines how much predictive leverage the independent variable provides.

Maximum likelihood estimation (MLE) employs the same approach. MLE takes the sample-wide probability of observing a specific value of a binary dependent variable and sees how well this probability predicts that outcome for each individual case in the sample. At least initially, MLE ignores the independent variable. As in PRE, this initial strategy produces a number of prediction errors. MLE then takes the independent variable into account and determines if, by knowing the independent variable, these prediction errors can be reduced.

Consider a highly simplified illustration, which again uses education (x) to predict whether an individual voted (coded 1 on the dependent variable, y) or did not vote (coded 0 on y). MLE first would ask, "How well can we predict whether or not an individual voted *without* using education as a predictor?" For the sake of simplicity, suppose our sample consists of four individuals, as shown in Table 9-4. As you can see, two individuals voted (coded 1) and two did not (coded 0). Based only on the distribution of the dependent variable, what is the predicted probability of voting for each individual? MLE would answer this question by figuring out the sample-wide probability of voting and applying this prediction to each case. Since half the sample voted and half did not, MLE's initial predicted probability (labeled P) would be equal to .5 for each individual. Why .5? Because there is a .5 chance that any individual in the sample voted and a .5 chance that he or she did not vote. Let's label the

Table 9-4 Model 1's Predictions and Likelihoods: Not Using Education to Predict Voting

Individual	y^a	Predicted probability of voting (P)	Likelihood
A	1	.5	$P = .5$
B	1	.5	$P = .5$
C	0	.5	$1 - P = .5$
D	0	.5	$1 - P = .5$

Note: Hypothetical data.
[a] 1 = voted; 0 = did not vote.

model that gave rise to the initial predictions Model 1. Table 9-4 shows the predicted probabilities, plus some additional information, for Model 1.

How well, overall, does Model 1 predict the real values of y? MLE answers this question by computing a **likelihood function**, a number that summarizes how well a model's predictions fit the observed data. In computing this function, MLE first determines a likelihood for each individual case. An individual likelihood tells us how closely the model comes to predicting the observed outcome for that case. MLE then computes the likelihood function by calculating the product of the individual likelihoods, that is, by multiplying them together. The likelihood function can take on any value between 0—meaning the model's predictions do not fit the observed data at all—to 1, meaning the model's predictions perfectly fit the observed data.

Formally stated, the likelihood function is not beautiful to behold.[9] Practically applied to a small set of data, however, the function is not difficult to compute. If a case has an observed value of y equal to 1 (the individual voted), then the likelihood for that case is equal to P. So individuals A and B, with predicted probabilities equal to .5, have likelihoods equal to P, which is .5. If a case has an observed value of y equal to 0 (the individual did not vote), then the likelihood for that case is equal to $1 - P$. Thus individuals C and D, who have predicted probabilities of .5, have likelihoods equal to $1 - P$, or $1 - .5$, also equal to .5. The likelihoods for each individual are displayed in the right-most column of Table 9-4. The likelihood for Model 1 is determined by multiplying all the individual likelihoods together:

$$\text{Model 1 likelihood} = .5 * .5 * .5 * .5 = .0625.$$

MLE would use this number, .0625, as a baseline summary of how well we can predict voting without knowledge of the independent variable, education.[10]

In its next step, MLE would bring the independent variable into its calculations by specifying a logistic regression coefficient for education, recomputing the probabilities and likelihoods, and seeing how closely the new estimates conform to the observed data. Again, for the sake of illustration, suppose that these new estimates, which we'll call Model 2, yield the predicted probabilities displayed in Table 9-5.

Model 2, which takes into account the independent variable, does a better job than Model 1 in predicting the observed values of y. By using education to predict voting, Model 2 estimates probabilities equal to .9 and .8 for individuals A and B (who, in fact, voted), but probabilities of only .3 and .1 for individuals C and D (who, in fact, did not vote). Just as in the Model 1 procedure, the individual likelihoods for each case are equal to P for each of the

Table 9-5 Model 2's Predictions and Likelihoods: Using Education to Predict Voting

Individual	y[a]	Predicted probability of voting (P)	Likelihood
A	1	.9	$P = .9$
B	1	.8	$P = .8$
C	0	.3	$1 - P = .7$
D	0	.1	$1 - P = .9$

Note: Hypothetical data.

[a] 1 = voted; 0 = did not vote.

Table 9-6 Model 1 and Model 2 Compared

Model statistic	Model 1	Model 2
Likelihood	.0625	.4536
Logged likelihood (LL)	ln(.0625) = −2.78	ln(.4536) = −0.79
Model comparison:		
Model 1 LL minus Model 2 LL	−2.78 − (−0.79) = −1.99	
−2 * (Model 1 LL minus Model 2 LL)	−2(−1.99) = 3.98	

Note: Hypothetical data.

voters (for whom $y = 1$) and equal to $1 - P$ for each of the nonvoters (for whom $y = 0$). The individual likelihoods appear in the right-most column of Table 9-5. As before, the likelihood function for Model 2 is computed by multiplying the individual likelihoods together:

$$\text{Model 2 Likelihood} = .9 * .8 * .7 * .9 = .4536.$$

How much better is Model 2 than Model 1? Does using education as a predictor provide significantly improved estimates of the probability of voting? Now, MLE does not work directly with differences in model likelihoods. Rather it deals with the natural log of the likelihood, or **logged likelihood** (LL) of each model. Thus MLE would calculate the natural log of the Model 1 likelihood, calculate the natural log of the Model 2 likelihood, and then determine the difference between the two numbers. Table 9-6 shows these conversions, plus some additional calculations, for Model 1 and Model 2.

Examine the Table 9-6 calculations. As we found earlier, Model 2's likelihood (.4536) is greater than Model 1's likelihood (.0625). This increase is also reflected in the LLs of both models: The LL increases from −2.78 for Model 1 to −0.79 for Model 2. MLE makes the comparison between models by starting with Model 1's LL and subtracting Model 2's LL: −2.78 − (−0.79) = −1.99. Notice that if Model 2 did about as well as Model 1 in predicting y, then the two LLs would be similar, and the calculated difference would be close to 0.[11] As it is, MLE found a difference equal to −1.99.

So far, so good. But does the number −1.99 help us decide whether Model 2 is *significantly* better than Model 1? Yes, it does. It turns out that, with one additional calculation, the difference between two LLs follows a chi-square distribution. The additional calculation is achieved by multiplying the difference in LLs by −2. Doing so, of course, doubles the difference and reverses the sign: −2 (−1.99) = 3.98. This calculation, usually labeled in computer output as "Change in −2 Log Likelihood" or "Change in −2LL," is a chi-square test statistic, and MLE uses it to test the null hypothesis that the true difference between Model 1 and Model 2 is equal to 0. There is nothing mystical here. It is plain old hypothesis testing using chi-square. If the calculated value of the change in −2LL, equal to 3.98, could have occurred more frequently than 5 times out of 100, by chance, then we would not reject the null hypothesis. We would have to conclude that the education-voting relationship is not significant. However, if the chances of observing a chi-square value of 3.98 are less than or equal to .05, then we would reject the null hypothesis and infer that Model 2 is significantly better than Model 1. Using the appropriate degrees of freedom and applying a chi-square test, MLE would report a P-value of .046 for a test statistic of 3.98.[12] The P-value is less than .05, so we can reject the null hypothesis and conclude that education is a statistically significant predictor of the probability of voting.

MLE proceeds much in the way illustrated by this example. It first obtains a set of predictions and likelihoods based on a reduced model, that is, a model using only the sample-wide probability of y to predict the observed values of y for each case in the data. It then "tries out" a coefficient for the independent variable in the logistic regression model. MLE usually obtains the first "try out" coefficient by running a version of least squares regression using x to predict y. It enlists this coefficient to compute a likelihood, which it then compares with the likelihood of the reduced model. It then proceeds in an iterative fashion, using a complex mathematical algorithm to fine-tune the coefficient, computing another likelihood, and then another and another—until it achieves the best possible fit between the model's predictions and the observed values of the dependent variable.

MLE is the heart and soul of logistic regression. This estimation technique generates all of the coefficient estimates and other useful statistics that help the analyst draw inferences about the relationship between the independent and dependent variables. Let's return now to the GSS data and consider some of these additional statistics, as reported in Table 9-7. To enhance the comparison between the real-world data and the hypothetical example just discussed, the baseline model—the model estimated without taking into account education—

Table 9-7 Education and Voting: Model Comparison and Summary

Model statistic	Model 1: Education *not* included	Model 2: Education included
Logged likelihood (LL)	−1,627.98	−1,556.59
Model comparison:		
Model 1 LL minus Model 2 LL	−1,627.98 − (−1,556.59) = −71.39	
−2*(Model 1 LL minus Model 2 LL) or		
"Change in −2LL"	142.78	
Significance of change	.000	
Model 2 summary:		
Cox and Snell R-square	.053	
Nagelkerke R-square	.075	

Source: 1998 General Social Survey (GSS).

is called Model 1. Model 2 refers to the results obtained after the independent variable, education, is used to predict the likelihood of voting. Note the difference between the LLs of the models: When education is used to predict voting, the LL increases from −1627.98 to −1556.59. Is this a significant improvement? Yes, it is, at least according to the "Model comparison" numbers in the table. Subtracting Model 2's LL from Model 1's LL yields a difference of −71.39. Multiplying the difference by −2, labeled "Change in −2LL," gives us a chi-square test statistic of 142.78, well beyond the realm of the null hypothesis. Thus, compared with how well we can predict the dependent variable without knowledge of the independent variable, knowledge of respondents' education significantly improves our ability to predict the likelihood of voting.

Two additional points should be made about using "Change in −2LL" to evaluate logistic regression models. First, we may follow this procedure to assess the statistical significance of the relationship between the dependent variable and all of the independent variables included in the model. So if Model 2 had several predictors of voting—education, age, and race, for example—then the change in −2LL would provide a chi-square test for the null hypothesis that none of these variables is significantly related to the likelihood of voting. Second, the change in −2LL can be used as an alternative to Wald in evaluating the statistical significance of individual independent variables, provided that each variable is added to the estimation procedure in a stepwise fashion. Thus MLE would estimate the effect of education, report a chi-square statistic, and then add a second variable and tell us if this second variable made a significant improvement in the predictive power of the model. In fact, some methodologists recommend this procedure for testing the significance of individual logistic regression coefficients.[13]

Logistic regression enlists the change in likelihood function in yet another way—as the basis for *R*-square-type measures of association, two of which are reported in the Table 9-7 "Model 2 summary." These statistics are grounded on the intuitive PRE logic. Model 1's LL represents prediction error without knowing the independent variable. The difference between Model 1's LL and Model 2's LL represents the predictive leverage gained by knowing the independent variable. In conceptual terms, then, we could express the difference between the two models as a proportion of Model 1's LL:

$$R\text{-square} = (\text{Model 1 LL} - \text{Model 2 LL}) / (\text{Model 1 LL}).$$

If Model 2 did about as well as Model 1 in predicting voting—if the two models' LLs were similar—then *R*-square would be close to 0. If, by contrast, Model 2's LL were a lot higher than Model 1's LL, then *R*-square would approach 1.[14] The various *R*-square measures build on this conceptual framework and seek to adjust for its statistical inadequacies. Cox and Snell's *R*-square makes an adjustment based on sample size. Cox and Snell's *R*-square is somewhat conservative, however, because it can have a maximum value of less than 1. Nagelkerke's statistic adjusts Cox and Snell's *R*-square, yielding a measure that is usually higher. By and large, though, these two measures, and several others that you may encounter, give readings of strength that are pretty close to each other.[15]

So what are we to make of an *R*-square in the .05 to .07 range? Again, unlike least squares regression, MLE is not in the business of explaining variance in the dependent variable. So we cannot say something like, "Education explains about 5 percent of the variation in voter turnout." However, we know that *R*-square can assume values between 0 and 1, with 0 denoting a very weak relationship and 1 denoting a strong relationship. Thus we can say that education, while significantly related to the likelihood of voting, is not by itself a particularly strong predictive tool. From a substantive standpoint, this is not too surprising. You can probably think of several additional variables that might improve the predictive power of the logistic regression model. Age, race, political efficacy, strength of partisanship—all of these variables come to mind as other possible causes of voting. If we were running OLS, we could specify a multiple regression model and estimate the effect of each of these variables on the dependent variable. Happily, logistic regression also accommodates multiple predictors. We turn now to a discussion of logistic regression using more than one independent variable.

LOGISTIC REGRESSION WITH MULTIPLE INDEPENDENT VARIABLES

Thus far we have covered a fair amount of ground. You now understand the meaning of a logistic regression coefficient. You know how to interpret coefficients in terms of changes in the odds ratio, as well as the percentage change in the odds. You know how to evaluate the statistical significance of a logistic regression coefficient. Plus you have a basic understanding of MLE, and you can appreciate its central role in providing useful statistics, such as the change in $-2LL$, as well as R-square-type measures of association. So far, however, our substantive examples have been of a simple variety, with one independent variable. Yet political researchers are often interested in assessing the effects of several independent variables on a dependent variable. We often want to know whether an independent variable affects a dependent variable, controlling for other possible causal influences. In this section we show that the logistic regression model, much like the linear regression model, can be extended to accommodate multiple independent variables. We also illustrate how logistic regression models can be used to obtain and analyze the predicted probabilities of a binary variable. To keep things consistent with the previous examples—but to add an interesting wrinkle—we introduce a dummy independent variable into the education-voting model:

$$\text{Logged odds (voting)} = \hat{a} + \hat{b}_1 \text{ (education)} + \hat{b}_2 \text{ (partisan)}.$$

Education, as before, is measured in years of schooling, from 0 to 20. "Partisan" is a dummy variable that gauges strength of party identification: Strong Democrats and strong Republicans are coded 1 on this dummy, and all others (weak identifiers, independents, and independent leaners) are coded 0. From an empirical standpoint, we know that strongly partisan people, regardless of their party affiliation, are more likely to vote than are people whose partisan attachments are weaker. So we would expect a positive relationship between strength of partisanship and the likelihood of voting.

The coefficients in this model—$\hat{a}$, $\hat{b}_1$, and $\hat{b}_2$—are directly analogous to coefficients in multiple linear regression. The coefficient $\hat{b}_1$ will estimate the change in the logged odds of voting for each 1-year change in education, controlling for the effect of partisan strength. Similarly, $\hat{b}_2$ will tell us by how much to adjust the estimated logged odds for strong partisans, controlling for the effect of education. To the extent that education and partisan strength are themselves related, the logistic regression procedure will control for this, and it will estimate the partial effect of each variable on the logged odds of voting. And the intercept, $\hat{a}$, will report the logged odds of voting when both independent variables are equal to 0, for respondents with no schooling (for whom education = 0) and who are not strong party identifiers (partisan = 0). This point bears emphasizing: The logistic regression model specified above is a linear-additive model, and it is just like a garden-variety multiple regression model. The

Table 9-8 Education, Partisan Strength, and Voting: Logistic Regression Coefficients and Model Summary

Logged odds (voting)	=	Intercept $\hat{a}$	+	Education[a] $\hat{b}_1$	+	Partisan[b] $\hat{b}_2$
Coefficient estimate		−2.022		.194		1.539
Standard error				.017		.139
Wald				134.739		123.012
Significance				.000		.000
Exp(b)				1.214		4.659

Model summary:	
Change in −2LL	292.655
Significance of change[c]	.000
Cox and Snell R-square	.109
Nagelkerke R-square	.153

Source: 1998 General Social Survey (GSS).
Note: Dependent variable is reported turnout in the 1996 presidential election. $N = 2,539$.
[a] Education is number of years of formal schooling.
[b] Partisan coded 1 for strong partisans and 0 otherwise.
[c] Degrees of freedom = 2. Both independent variables included in complete model.

partial effect of education on the logged odds of voting is assumed to be the same for strong partisans and nonstrong partisans alike. And the partial effect of partisan strength on the logged odds of voting is assumed to be the same at all values of education. (This point becomes important in a moment, when we return to a discussion of probabilities.)

Table 9-8 reports the results of the analysis, using the GSS data. Plugging the coefficient values into the logistic regression model, we find:

Logged odds (voting) = −2.022 + .194 (education) + 1.539 (partisan).

Interpretation of these coefficients is by now a familiar task. When we control for partisan strength, each 1-year increase in education increases the logged odds of voting by .194. And, after we take into account the effect of education, being a strong partisan increases the logged odds of voting by 1.539. Turning to the odds ratios, reported in the "Exp(b)" row of Table 9-8, we can see that a unit increase in education multiplies the odds by about 1.2. And, when "partisan" is switched from 0 to 1, the odds ratio jumps by nearly 4.7. In other words, when we control for education, strong partisans are nearly five times more likely to vote than are weak partisans or independents. Framing the relationships in terms of percentage change in the odds: The odds of voting increase by about 21 percent for each incremental change in education and by 366 percent for the comparison between nonstrong partisans and partisans. Finally, according to the Wald statistics (and accompanying P-values), each independent variable is significantly related to the logged odds of voting.

Overall, how well does the model perform? Not too badly. The "Change in −2LL" chi-square statistic (292.655, P-value = .000) says that including both independent variables in the estimation procedure provides significant predictive improvement over the baseline know-nothing model.[16] And Cox and Snell (.109) and Nagelkerke (.153), while not spellbinding, suggest that education and partisanship together do a decent job of predicting voting, especially when compared with our earlier analysis (see Table 9-7), in which education was used as the sole predictor.

These results add up to a reasonably complete analysis of the relationships. Certainly it is good to know size and significance of the partial effects of education and partisan strength on the logged odds of voting, and it is convenient to express these effects in the language of odds ratios and the percentage change in odds. Often, however, the researcher wishes to understand his or her findings in the most intuitively meaningful terms: probabilities. We might ask, "What are the effects of the independent variables on the probability of voting? Although education and partisan attachments clearly enhance the odds of voting, by how much do these variables affect the probability that people will turn out?" These questions are perfectly reasonable, but they pose two challenges. First, in any logistic regression model—including the simple model with one independent variable—a linear relationship exists between x and the logged odds of y, but a nonlinear relationship exists between x and the probability of y. As we discussed at the beginning of this chapter, the marginal effect of x on the probability of y will not be the same for all values of x. Thus the effect of, say, a 1-year increase in education on the probability of voting will depend on where you "start" along the education variable. Second, in a logistic regression model with more than one independent variable, such as the model we just discussed, the independent variables bear a linear-additive relationship to the logged odds of y, but they bear an interactive relationship to the probability of y. This means, for example, that logistic regression will permit the relationship between partisan strength and the probability of voting to vary, depending on respondents' levels of education. So logistic regression might find a big marginal effect of partisan strength on the probability of voting among people with less education (who are less likely to vote) but a much weaker effect among the better educated (who are already quite likely to vote). Odd as it may sound, these challenges define some rather attractive features of logistic regression. Properly applied, the technique allows the researcher to work with probabilities instead of odds or logged odds and, in the bargain, to gain revealing substantive insights into the relationships being studied.

WORKING WITH PROBABILITIES

Let's return to the logistic regression model we just estimated and figure out how best to represent and interpret these relationships in terms of probabilities. The model will, of course, yield the predicted logged odds of voting for any combination of the independent variables. Just plug in values for education and the partisan dummy, do the math, and obtain an estimated logged odds of voting for that combination of values. As we saw earlier, logged odds can be converted back into odds and, in turn, odds can be translated into probabilities. These conversions—from logged odds to odds, and from odds to probabilities—form the basis of two commonly used methods for representing complex relationships in terms of probabilities. First, the researcher might calculate and compare the predicted probabilities of voting for a few illustrative values of education and partisan strength. With these calculations in hand, the researcher obtains a **probability profile** of the effects of the independent variables on the probability of voting. The probability profile approach works nicely for models having few independent variables. A second and perhaps more widely used method is to examine the effect of each independent variable on the predicted probability of voting, while holding the other independent variables constant at their sample averages. Using the

sample averages approach, we could show the effect of partisan strength for people with average education or the effect of education among people with average partisan strength. Let's consider both methods using the GSS data, beginning with the probability profile approach.

The Probability Profile Method

Suppose that we wanted to see what happens to the probability of voting as education increases from 8 years, to 12 years, to 16 years, to 20 years. Suppose further that we would like to find out how partisan strength affects the probability of voting at each of these values of education. To accomplish this, we would use the model's estimates to figure out the logged odds of voting for the first combination of independent variables—respondents with 8 years of education (education = 8) who are not strong partisans (partisan = 0)—and convert this estimate into a probability. To find the probability of voting for strongly partisan people who have 8 years of education, we would recompute the logged odds for partisans (partisan = 1) and convert this value into a probability. We would then move to the next illustrative value of the education variable, 12 years, and repeat the procedure. A complete probability profile, with estimated probabilities for each combination of the four values for "education" and two values for "partisan," would require eight sets of conversions. This sounds like a lot of mathematical drudgery—and it is. Fortunately, any data analysis software worth its salt will do the work for you.[17] But to see how it is done, and to make an interesting substantive comparison, let's work through two conversions. We will translate the logged odds of voting into probabilities of voting for people who share the same level of education, 8 years, but who differ in partisan attachment.

For convenience, here is the logistic regression equation we obtained earlier:

$$\text{Logged odds (voting)} = -2.022 + .194 \text{ (education)} + 1.539 \text{ (partisan)}.$$

Our first illustrative group, people with 8 years of schooling who are not strong partisans, have a value of 8 on the education variable and a value of 0 on the partisan variable. Using the logistic regression model to estimate the logged odds of voting for this group:

$$\text{Logged odds (voting)} = -2.022 + .194 \text{ (8)} + 1.539 \text{ (0)},$$
$$= -2.022 + 1.552 + 0 = -.47.$$

So the estimated logged odds of voting are equal to −.47. What are the odds of voting for this group? Well, odds can be retrieved by taking the exponent of logged odds. Doing so: Exp(−.47), which is equal to .625. Thus the odds are .625. Now let's get back to a probability. Recall the formula: Probability = Odds / (1 + Odds). Converting .625 to a probability of voting, we have .625 / 1.625 = .385. So the estimated probability of voting for weak partisans or independents with 8 years of education is equal to .385. Clearly there is a weak probability—a bit more than one chance in three—that these individuals voted. How about their strongly partisan counterparts who have the same level of education? These respondents will have a value of 8 on the education variable and a value of 1 on the partisan variable. For this combination of independent variables, the logged odds are as follows:

$$\text{Logged odds (voting)} = -2.022 + .194 \text{ (8)} + 1.539 \text{ (1)},$$
$$= -2.022 + 1.552 + 1.539 = 1.069.$$

Table 9-9 Predicted Probabilities of Voting for Four Values of Education and Two Values of Partisan Strength

Education	Strong partisan?	Predicted probability
8 years	No	.385
	Yes	.744
	Difference[a]	*.359*
12 years	No	.576
	Yes	.863
	Difference	*.287*
16 years	No	.747
	Yes	.932
	Difference	*.185*
20 years	No	.865
	Yes	.968
	Difference	*.103*

Source: 1998 General Social Survey (GSS).

Note: Number of cases (*N*) for each value of education and partisan strength are as follows: 8 years (63 No/18 Yes/ 81 total cases), 12 years (678/157/835), 16 years (321/84/405), and 20 years (48/19/67).

[a] Predicted probability for strong partisans minus predicted probability for nonstrong partisans.

If the logged odds are 1.069, then the odds would be Exp(1.069), equal to 2.912. Finally, the estimated probability is: 2.912 / 3.912 = .744. There are almost three chances in four that these people voted. Pretty remarkable. For people at this low level of educational attainment, partisan attachment has a very large effect on the probability of voting. How large? Subtracting .385 from .744 gives us .359. So for people with 8 years of education, partisan strength increases the probability of voting by a robust .359. What happens to this effect as education increases?

Table 9-9 presents the estimated probabilities of voting for all eight education and partisan groups. First consider the effect of partisan strength on the estimated probability of voting, controlling for education. At the lowest education level, as we have seen, this effect is quite large, .359. At the next chosen illustrative value, 12 years, the effect of the "partisan" variable is fair-sized, too—there is a difference of .287 between the predicted probability for strong partisans (.863) and the probability for people with weak partisan ties (.576). But notice that, as we ascend into higher values of education, most respondents, despite partisan strength, are quite likely to have voted. Thus the marginal effect of the "partisan" variable gets weaker and weaker. At the highest level of education, 20 years, the effect of partisan strength shrinks to .103. We can, of course, turn the analysis around and examine the effect of education, controlling for partisan strength. Viewed in this way, we can see that education has a more potent effect among those lacking a strong partisan commitment. Between the lowest education reference group (.385) and the highest (.865), the probability of voting increases by .48. Among strong partisans, however, this effect is more modest (about .22).

The Sample Averages Method

Constructing a probability profile, as we have just done, provides richness and detail, and it is a good method to use if you have two or three independent variables. However, in doing your own analyses, or in reading the research findings of other researchers, you are quite likely to encounter logistic regression models having many independent variables. Under these circumstances, the probability profile approach becomes a bit cluttered and confusing. Therefore, many researchers use the sample averages method. The sample averages method is centered on this question: If we were to hold all other independent variables constant at their sample-wide means, what is the effect of this particular independent variable on the estimated probability of the dependent variable? Applying this question to our voting example, we might first ask, "If we hold education constant at its sample mean, what is the effect of partisan strength on the estimated probability of voting?" We also could ask, "If we hold partisan strength constant at its sample mean, what is the effect of education on the estimated probability of voting?" The answers to these questions will help to illustrate the effects of partisan strength and education for the average respondent. Typically, researchers will present the change in the estimated probability of the dependent variable across the full range of a particular independent variable of interest, at the sample means of the other variables in the model. For example, we would want to present the full effect of going from the lowest education code (0 years) to the highest education code (20 years) at the sample average of partisan strength.[18]

How do we proceed? First, we need two numbers, the sample mean of education and the sample mean of partisan strength. For an interval-level variable, such as education, we find the arithmetic average. According to the GSS data, the mean number of years of formal schooling is 13.25 years. For a dummy variable, such as partisan strength, the mean is equal to the proportion of the sample who are strong partisans, that is, the proportion coded 1 on the dummy. In the GSS data, .221 of the sample falls into this category.[19] To figure out the effect of partisan strength at mean levels of education, we would enlist the estimates of the logistic regression model, enter 13.25 as the value of the education variable, and then estimate the probability for nonstrong partisans, respondents coded 0 on partisan strength:

$$\text{Logged odds (voting)} = -2.022 + .194\,(13.25) + 1.539\,(0),$$
$$= -2.022 + 2.571 + 0 = .549.$$

A logged odds equal to .549 translates into a probability of .634. So, for nonstrong partisans with average education, there are a bit more than 6 chances in 10 that they voted. We then would do the calculations again, this time switching "partisan" from 0 to 1:

$$\text{Logged odds (voting)} = -2.022 + .194\,(13.25) + 1.539\,(1),$$
$$= -2.022 + 2.571 + 1.539 = 2.088.$$

A logged odds of 2.088 converts to an odds of 8.069, which returns an estimated probability of .890, or about 9 chances in 10 that these respondents voted. Thus, by subtracting the nonstrong probability, .634, from the strong partisan probability, .890, we can figure out the full effect of partisan strength while holding education constant at its sample mean: .890 − .634 = .256. Table 9-10, which summarizes these calculations, also shows what happens to the estimated probability of voting across the full range of education, at the sample-wide

Table 9-10 Full Effects of Partisan Strength and Education on the Estimated Probability of Voting

	Independent variable (low value, high value)	
Estimated probability at	*Partisan strength* (0,1)	*Education* (0,20)
Low value	.634	.157
High value	.890	.900
Full effect	.256	.743

Source: 1998 General Social Survey (GSS).

Note: For partisan strength, probabilities calculated at education sample mean (13.25). For education, probabilities calculated at partisan strength sample mean (.221).

mean of partisan strength. If you opt for the sample averages method, you could use Table 9-10 as a template for presenting your own logistic regression results.

SUMMARY

A political researcher wants to explain why some people approve of same-sex marriage whereas others disapprove. Thinking that age plays a causal role, she hypothesizes that as age increases, the likelihood of disapproval will go up, that older people will be more likely than younger people to disapprove of same-sex marriage. A plausible and interesting idea. Consulting her survey dataset, the researcher finds a binary variable that will serve as the dependent variable (respondents who approve of same-sex marriage are coded 0 on this variable and those who disapprove are coded 1). She also finds an interval-level independent variable, age measured in years, from 18 to 99. So she has the hypothesis, the data, and the variables. Now what? Which analytic technique is best suited to this research problem? If this researcher is someone other than you, she may need to test her idea by collapsing age into three or four categories, retrieving the tool labeled cross-tabulation from her methods toolbox, and comparing the percentages of disapprovers across the collapsed categories of the independent variable. That might work okay. But what if she decides to control for the effects of several other variables that may shape individuals' approval or disapproval of same-sex marriage—such as education, sex, and partisanship? Cross-tabulation would become cumbersome to work with, and she may need to settle for an incomplete analysis of the relationships. The larger point, of course, is that this researcher's ability to answer an interesting substantive question is severely limited by the tools at her disposal.

If this researcher is you, however, you now know a far better approach to the problem. Reach into your toolbox of techniques, select the tool labeled logistic regression, and estimate this model: Logged odds (disapproval) = $a + b$ (age). The logistic regression coefficient, b, will tell you how much the logged odds of disapproval increase for each 1-year change in age. Of course, logged odds are not easily grasped. But by entering the value of b into your hand-held calculator and tapping the e^x key—or, better still, by examining the Exp(b) values in the computer output—you can find the odds ratio, the change in the odds of disapproving as age increases by 1 year. You can convert Exp(b) into a percentage change in the odds of disapproval. You can test the null hypothesis that b is equal to 0 by consulting the P-value of the Wald statistic. You could see how well the model performs by examining changes in the magnitude of $-2LL$ and reviewing the accompanying chi-square test. Several R-square-like measures, such as Cox-Snell and Nagelkerke, will give you a general idea of the strength of the relationship between age and the likelihood of disapproving of same-sex marriage. You can calculate and examine the predicted probabilities of disapproval for a few illustrative age groups and thus achieve a closer understanding of your results. If you are challenged by a skeptic who thinks you should have controlled for education and sex, you can reanalyze your model, controlling for these variables—and any other independent variables that might affect your results. By adding logistic regression to your arsenal of research techniques, you are now well prepared to handle any research question that interests you.

KEY TERMS

binary variable (p. 199)
common logarithms (p. 203)
likelihood function (p. 210)
logged likelihood (p. 211)
logits (p. 203)
maximum likelihood estimation (p. 209)

natural logarithms (p. 203)
odds (p. 201)
odds ratio (p. 202)
percentage change in the odds (p. 202)
probability profile (p. 216)
sample averages (p. 217)

EXERCISES[20]

1. Students of comparative politics are quite interested in questions of democratic development. Under what conditions are countries more likely (or less likely) to develop democratic forms of government? One idea has to do with the equitable—or inequitable—distribution of wealth. According to this reasoning, when economic resources are concentrated in the hands of a few, those controlling the wealth would much prefer a political system not open to popular rule. Thus, as economic inequality goes up, the likelihood of democracy will go down. Does this idea have empirical support? Below are the results of a logistic regression analysis of the relationship between type of government and economic inequality. The binary dependent variable is coded 1 for democratic countries and 0 for nondemocratic countries. The independent variable, economic inequality, is a 10-point scale, with high values denoting greater inequality:

Logged odds (democracy)	=	Intercept a	+	Economic inequality b
Coefficient estimate		3.538		−.944
Standard error				.249
Wald				14.325
Significance				.000
Exp(b)				.389

A. The logistic regression coefficient tells us that, for each one-unit increase in inequality, the logged odds of democracy declines by .944. Turn your attention to the odds ratio, Exp(b). This coefficient says that a country at one value of economic inequality is only about _____ times as likely to be a democracy as a country at the next-lower level of economic inequality.

B. Use the value of Exp(b) to compute a percentage change in the odds. According to your calculations, each unit increase in economic inequality decreases the odds of democracy by about how much?

C. State the null hypothesis for this relationship. What is your inferential decision—reject the null hypothesis or do not reject the null hypothesis?

2. Here is an extension of the idea you examined in Exercise 1. According to this broader idea, the inequitable distribution of any resource within a country will depress the likelihood of democracy. This could be a material resource, such as economic wealth, but it might be a symbolic resource, such as the value attached to language, religion, or ethnicity. So, for example, a country having several linguistic or ethnic groups, all of whom are vying for political power, may have a hard time establishing a democracy, because each disparate group may seek a form of governance that establishes its language or religion as dominant. By contrast, countries in which nearly all citizens share the same language or ethnicity might avoid these bitter disputes, thus making democracy easier to achieve. This is just an idea. Does it have merit? Here are the results of a logistic regression analysis that add a dummy variable, "homogeneous," to the model you interpreted in Exercise 1. The variable homogeneous takes on a value of 1 for countries having low levels of ethnic and linguistic diversity and a value of 0 for nonhomogeneous countries.

Logged odds (democracy) =	Intercept a	+	Economic inequality b_1	+	Homogeneous b_2
Coefficient estimate	3.033		−.906		1.033
Standard error			.289		.525
Wald			9.833		3.881
Significance			.002		.049
Exp(b)			.404		2.811

Model summary:

Change in −2LL	22.307
Significance of change	.000
Cox and Snell R-square	.241
Nagelkerke R-square	.322

Parts A–D present interpretations based on these results. For each part, (i) state whether the interpretation is correct or incorrect, and (ii) explain why the interpretation is correct or incorrect. For incorrect interpretations, be sure that your response in (ii) includes the correct interpretation.

A. Interpretation One: If we control for countries' homogeneity, each one-unit increase in economic inequality decreases the likelihood of democracy by about 40 percent.

B. Interpretation Two: If we control for countries' levels of economic inequality, homogeneous countries are about 2.8 times more likely to be democracies than are nonhomogeneous countries.

C. Interpretation Three: Compared with how well the model performs without including measures of economic inequality and homogeneity, inclusion of both of these independent variables provides a statistically significant improvement.

D. Interpretation Four: Economic inequality and homogeneity together explain about 32 percent of the variance in the likelihood of democracy.

10

Thinking Empirically,
Thinking Probabilistically

This book has covered only the basics—the essential skills you need to understand political research and to perform your own analysis. Even so, we have discussed a wide range of topics and methodological issues. The first five chapters dealt with the foundations of political analysis: defining and measuring concepts, describing variables, framing hypotheses and making comparisons, designing research, and controlling for rival explanations. In the last four chapters we considered the role of statistics: making inferences, gauging the strength of relationships, performing linear regression analysis, and interpreting logistic regression. As you read research articles in political science, discuss and debate political topics, or evaluate the finer points of someone's research procedure, the basic knowledge imparted in this book will serve you well.

This book also has tried to convey a larger vision of the enterprise of political analysis. Political scientists seek to establish new facts about the world, to provide rich descriptions and accurate measurements of political phenomena. Political scientists also wish to explain political events and relationships. In pursuit of these goals, researchers learn to adopt a scientific mindset toward their work, a scientific approach to the twin challenges of describing and explaining political variables. As you perform your own political analysis, you too are encouraged to adopt this way of thinking. Here are two recommendations. First, in describing new facts, try to think empirically. Try to visualize how you would measure the phenomenon you are discussing and describing. Be open to new ideas, but insist on empirical rigor. Political science, like all science, is based on empirical evidence. This evidence must be described and measured in such a way that others can do what you did and obtain the same results. Second, in proposing and testing explanations, try to think probabilistically. You are well aware of one reason that political researchers must rely on probabilities: Random samples are a fact of life for much political science. Another reason is political science deals with human behavior and human events, and so it is an inexact science. Let's briefly illustrate why it is important to think empirically. Let's also look at the reasons political scientists must think probabilistically.

THINKING EMPIRICALLY

The main projects of political science are to describe concepts and to analyze the relationships between them. But potentially interesting relationships are often obscured by vague, conceptual language. During a class meeting of a voting and elections course, for example, students were discussing the electoral dynamics of ballot initiatives, law-making vehicles used frequently in several states. Controversial proposals, such as denying state benefits to illegal immigrants or banning same-sex marriage, may appear on the ballot to be decided directly by voters. Near the end of the discussion, one student observed: "It appears to me that most ballot initiatives target specific groups. Most ballot initiatives, if they pass, decrease equality. Very few seem designed with egalitarian principles in mind." Now, this is an interesting, imaginative statement. Is it true? Without conceptual clarification, there is no way to tell.

Upon hearing statements such as the one made by this student, you have learned to insist that conceptual terms, like *equality* or *egalitarian principles*, be described in concrete language. How would one distinguish an egalitarian ballot initiative, an initiative that increases equality, from one that is not egalitarian, an initiative that decreases equality? Pressed to clarify her conceptual terms, the student settled on one defining characteristic of the degree of egalitarianism in a ballot initiative. The concept of egalitarianism, she said, is defined as the extent to which a ballot initiative would confer new legal rights on a specific group. So some initiatives, such as those that would confer protections on commercial farm animals, could be classified as more egalitarian, whereas others, like those making English the state's official language, could be classified as less egalitarian. With a workable measurement strategy in hand, this student could then turn to an empirical assessment of her in-class claim. Using clearly defined terms and reproducible findings, this student's research would enhance our understanding of these vehicles of direct democracy.[1]

An openness that is tempered by skepticism nurtures knowledge of the political world. This means that political scientists must sometimes revisit relationships and rethink established explanations. The search for truth is an ongoing process. Consider another example. For years, scholars of U.S. electoral behavior have measured voter turnout in presidential elections by dividing the number of people who voted by the size of the voting-age population. So if one wanted to describe trends in turnout, one would calculate the percentage of the voting-age population who voted in each year, and then track the numbers over time. Indeed, measured in this way, turnouts declined steadily after 1960, a high watermark of the twentieth-century. The relatively low turnout in presidential elections is one of the most heavily researched phenomena in U.S. politics, and it has fostered a search for explanatory factors. Some explanations have linked declining turnout to attitudinal variables, such as a weakened attachment to the political parties or an erosion of trust in government.

However, research by Michael P. McDonald and Samuel Popkin points to a potentially serious measurement problem with using the voting-age population as the basis for calculating turnout.[2] They show that, by using the voting-age population, researchers have been including large groups of ineligible people, such as felons or noncitizens. What is more, the number of ineligible persons has been increasing over time—for example, the percentage of noncitizens among the voting-age population went from 2 percent in 1966 to 8.6 percent in 2006. Once McDonald and Popkin adjust for these and other measurement errors, they show that, although turnout dropped after 1972 (when the eligible electorate was expanded to include eighteen-year-olds), there has been no downward trend.

How will this new measurement strategy affect empirical research on turnout? No doubt, in the coming years political scientists will debate this and other measurements. They will create new explanations and test new hypotheses. We can be sure of one thing: In the end, we will have expanded our knowledge of electoral participation in the United States. Rigorous, empirical debate will guarantee it.

THINKING PROBABILISTICALLY

Statistics occupies a position of central importance in the research process. A discussion of random error has come up in several topics in this book—measurement, sampling, hypothesis testing, and statistical significance. We have seen that the accuracy of measurements may be affected by haphazard factors that can be difficult to control. And, in drawing samples from large and unseen populations, political researchers consciously introduce random sampling error. You have learned how to estimate the size of this error. You know how to "give chance a chance" to account for your results. And you can recognize and apply accepted standards for rejecting the null hypothesis, the omnipresent spokesperson for random probability. These are key skills in probabilistic thinking.

But probabilistic thinking is important in political research in a larger sense, one that has less to do with measurement error and random samples. Rather, it involves an important difference between the way the scientific approach is applied in the social sciences, such as political science, and the way it is applied in the physical sciences, such as physics or astronomy. To appreciate this difference, consider political scientist Gabriel Almond's retelling of philosopher Karl Popper's famous "clouds and clocks" metaphor:

> Karl Popper . . . uses the metaphor of clouds and clocks to represent the commonsense notions of determinacy and indeterminacy in physical systems. He asks us to imagine a continuum stretching from the most irregular, disorderly, and unpredictable "clouds" on the left to the most regular, orderly, and predictable "clocks" on the right. As the best example of a deterministic system at the clock-extreme, Popper cites the solar system. . . . As an example of a system near the other, indeterminate, end of the continuum, he cites a cluster of gnats or small flies in which each insect moves randomly except that it turns back toward the center when it strays too far from the swarm. Near this extreme we would find gas clouds, the weather, schools of fish, human societies, and, perhaps a bit closer to the center, individual human beings and animals.[3]

Physical scientists seek to explain phenomena toward the clocks side of the continuum. They develop and test explanations by gathering empirical data and seeing if the data are consistent with their causal explanations. And they set a deterministic standard in evaluating their theories. Causal factor X must be found to determine outcome Y. Political scientists seek to explain political phenomena, activity more toward the clouds side of the continuum. Political scientists, too, propose causal explanations and marshal empirical, reproducible facts to test their theories. But political behavior, like all human activity, is different from the behavior of objects in the physical world. Human behavior is far more complex. It is difficult to know for certain if political behavior Y is being determined by causal factor X or by some other cause of which we are unaware and for which we have not accounted. Indeed, you have learned that the "How else?" question is the centerpiece of controlled comparisons in political research. Only by identifying plausible alternative causes, and controlling for their effects, can we gain confidence in the explanations that we propose. For this reason, political scientists set a probabilistic standard in evaluating their theories. Political scientists do not expect their theories to provide deterministic predictions of political behavior, but they do expect them to predict appreciably better than chance.[4] Thus, causal factor X must be found to increase the probability that people will choose political behavior Y instead of political behavior Z.

Probabilistic thinking should always accompany you when you observe the world and propose hypotheses to explain what you see. Consider a final illustration. In the 2004 presidential election, 49.4 percent of major-party voters reported voting for the Democratic candidate and 50.6 for the Republican. Based on these numbers, if you had to predict the vote choice of any randomly chosen major-party voter, you would do fairly well by flipping a coin—heads, they voted Democratic; tails, they voted Republican. But suppose you knew each individual's party identification, the direction and strength of their psychological attachment to one of the political parties. This knowledge would permit you to predict vote choice with far more certainty than chance alone. The percentage of Democratic voters declines systematically as you move across the values of the party identification variable, from a high of 96.7 percent among strong Democrats to a low of 1.8 percent among strong Republicans.[5] Thus, partisanship is a powerful predictor, perhaps the most powerful predictor we have in explaining this particular political behavior. But does party identification provide a deterministic explanation of vote choice? No, it does not. After all, some strong Democrats voted Republican, and some strong Republicans voted Democratic. What are we to make of these individuals, these voters who do not cooperate with our explanation? In political science, unexplained phenomena do not lead us to abandon the search for the underlying causes of human behavior. Quite the contrary, "facts that don't fit" animate the search for better probabilistic explanations. The truth is always out there, waiting for you to discover it.

Notes

Introduction

1. Maine (since 1972) and Nebraska (since 1996) award two electoral votes by statewide vote and the rest by congressional district.
2. See Neal R. Peirce and Lawrence D. Longley, *The People's President: The Electoral College in American History and the Direct Vote Alternative* (New Haven, Conn.: Yale University Press, 1981).

Chapter 1

1. Hanna Fenichel Pitkin, *The Concept of Representation* (Berkeley: University of California Press, 1972): 1–2 (emphasis in original).
2. Of course, you might want to use a concept to study different units of analysis. This is discussed below.
3. Many interesting and frequently discussed concepts have commonly accepted labels for these opposites. For example, we refer to political systems as "democratic" or "authoritarian," or individuals as "religious" or "secular." In this example, we will contrast "liberal" with "conservative."
4. Supreme Court Justice Potter Stewart, in *Jacobellis v. Ohio* (1964): "I have reached the conclusion . . . that under the First and Fourteenth Amendments criminal laws in this area are constitutionally limited to hard-core pornography. . . . I shall not today attempt further to define the kinds of material I understand to be embraced within that shorthand description; and perhaps I could never succeed in intelligibly doing so. But I know it when I see it, and the motion picture involved in this case is not that."
5. Liberalism may have additional dimensions. Racial issues, such as affirmative action, might form a separate dimension, and attitudes toward military force versus diplomacy in foreign policy may be separate as well. For a good introduction to this multidimensional concept, see William S. Maddox and Stuart A. Lilie, *Beyond Liberal and Conservative: Reassessing the Political Spectrum* (Washington, D.C.: Cato Institute, 1984).
6. Robert A. Dahl, *Polyarchy: Participation and Opposition* (New Haven, Conn.: Yale University Press, 1971).
7. For example, see Michael Coppedge, Angel Alvarez, and Claudia Maldonado, "Two Persistent Dimensions of Democracy: Contestation and Inclusiveness," *Journal of Politics* 70, no. 3 (July 2008): 632–647.
8. Among social scientists, cross-disciplinary debate exists concerning the measurement and dimensionality of social status. Political scientists generally prefer objective measures based on income, education, and (less frequently used) occupational status. See Sidney Verba and Norman Nie's classic, *Participation in America: Political Democracy and Social Equality* (New York: Harper and Row, 1972). Sociologists and social psychologists favor subjective measures, attributes gauged by asking individuals which social class

they belong to. Furthermore, social status may have a separate dimension based on status within one's community, as distinct from status in society as a whole. See research using the MacArthur Scale of Subjective Social Status, John D. and Catherine T. MacArthur Research Network on Socioeconomic Status and Health, www.macses.ucsf.edu/.

9. W. S. Robinson, "Ecological Correlations and the Behavior of Individuals," *American Sociological Review* 15, no. 3 (June 1950): 351–357. See also William Claggett and John Van Wingen, "An Application of Linear Programming to Ecological Influence: An Extension of an Old Procedure," *American Journal of Political Science* 37 (May 1993): 633–661.

10. Emile Durkheim, *Suicide* [1897], English translation (New York: Free Press, 1951). Durkheim finds that populations with higher proportions of Protestants have higher suicide rates than Catholic populations. However, see Frans van Poppel and Lincoln H. Day, "A Test of Durkheim's Theory of Suicide—Without Committing the 'Ecological Fallacy,' " *American Sociological Review* 61, no. 3 (June 1996): 500–507.

11. The term *operational definition*, universally used in social research, is something of a misnomer. An operational definition does not take the same form as a conceptual definition, in which a conceptual term is defined in empirical language. Rather, an operational definition describes a procedure for measuring the concept. *Measurement strategy* is probably a more descriptive term than *operational definition*.

12. The research on political tolerance is voluminous. This discussion is based mostly on the work of Samuel A. Stouffer, *Communism, Conformity and Civil Liberties* (New York: Wiley, 1966), and the conceptualization offered by John L. Sullivan, James Piereson, and George E. Marcus, "An Alternative Conceptualization of Tolerance: Illusory Increases, 1950s–1970s," *American Political Science Review* 73 (September 1979): 781–794. For further reading, see George E. Marcus, John L. Sullivan, Elizabeth Theiss-Morse, and Sandra L. Wood, *With Malice Toward Some* (New York: Cambridge University Press, 1995). For an excellent review of conceptual and measurement issues, see James L. Gibson, "Enigmas of Intolerance: Fifty Years after Stouffer's *Communism, Conformity, and Civil Liberties*," *Perspectives on Politics* 4, no. 1 (March 2006): 21–34.

13. The least-liked approach was pioneered by Sullivan, Piereson, and Marcus, "An Alternative Conceptualization of Tolerance." This measurement technology is more faithful to the concept of tolerance because it satisfies what Gibson terms "the objection precondition," the idea that "one cannot tolerate (i.e., the word does not apply) ideas of which one approves. Political tolerance is forbearance; it is the restraint of the urge to repress one's political enemies. Democrats cannot tolerate Democrats, but they may or may not tolerate Communists. Political tolerance, then, refers to allowing political activity . . . by one's political enemies." Gibson, "Enigmas of Intolerance," 22.

14. The term *Hawthorne effect* gets its name from a series of studies of worker productivity conducted in the late 1920s at the Western Electric Hawthorne Works in Chicago. Sometimes called reactive measurement effects, Hawthorne effects can be fairly durable, changing little over time. Test anxiety is an example of a durable reactive measurement effect. Other measurement effects are less durable. Some human subjects may initially respond to the novelty of being studied, and this effect may decrease if the subjects are tested again. The original Hawthorne effect was such a response to novelty. See Neil M. Agnew and Sandra W. Pyke, *The Science Game* (Englewood Cliffs, N.J.: Prentice-Hall, 1994), 159–160.

15. W. Phillips Shively argues that reliability is a necessary (but not sufficient) condition of validity. Using the metaphor of an archer aiming at a target, Shively describes four possible patterns: (A) a random scatter of arrows centered on an area away from the bull's-eye (high systematic error and high random error), (B) arrows tightly grouped but not on the bull's-eye (high systematic error and low random error), (C) a random scatter of arrows centered on the bull's-eye (low systematic error and high random error), and (D) arrows tightly grouped inside the bull's-eye (low systematic error and low random error). According to Shively, only the last pattern represents a valid measurement. Earl Babbie, however, argues that reliability and validity are separate criteria of measurement. Using a metaphor identical to Shively's, Babbie characterizes pattern C as "valid but not reliable" and pattern D as "valid and reliable." See W. Phillips Shively, *The Craft of Political Research*, 6th ed. (Upper Saddle River, N.J.: Pearson Prentice Hall, 2005), 48–49; Earl Babbie, *The Practice of Social Research*, 10th ed. (Belmont, Calif.: Thomson Wadsworth, 2004), 143–146.

16. On this and related points, see Edward G. Carmines and Richard A. Zeller, *Reliability and Validity Assessment* (Thousand Oaks, Calif.: SAGE Publications, 1979).

17. Other adjectives can be used to describe data designs. *Longitudinal study* is synonymous with panel study, except that longitudinal studies generally have many more measurement points. *Time series* describes a chronological series of cross-sections. Respondents a, b, and c are asked questions 1, 2, and 3. At a later time, respondents x, y, and z are asked questions 1, 2, and 3.

18. Lee J. Cronbach, "Coefficient Alpha and the Internal Structure of Tests," *Psychometrika* 16, no. 3 (September 1951): 297–334.

19. Most methodologists recommend minimum alpha coefficients of between .7 and .8. See Jum C. Nunnally and Ira H. Bernstein, *Psychometric Theory*, 3rd ed. (New York: McGraw-Hill, 1994); Carmines and Zeller, *Reliability and Validity Assessment*, 51.

20. This example is from Herbert Asher, *Polling and the Public: What Every Citizen Should Know*, 5th ed. (Washington, D.C.: CQ Press, 2001), 88–89. Asher notes that this question has been dropped from the American National Election Study.

21. Howard Schuman, Stanley Presser, and Jacob Ludwig, "Context Effects on Survey Responses to Questions about Abortion," *Public Opinion Quarterly* 45, no. 2 (Summer 1981): 216–223. Schuman, Presser, and Ludwig find the question-order effect on the "does not want any more children" item to be "both large and highly reliable," although "[t]he exact interpretation of the effect is less clear than its reliability" (p. 219). Responses to the question citing a "serious defect in the baby" were the same, regardless of where it was placed. For an excellent review and analysis of the abortion question-wording problem, see Carolyn S. Carlson, "Giving Conflicting Answers to Abortion Questions: What Respondents Say," paper presented at the annual meeting of the Southern Political Science Association, New Orleans, January 6–8, 2005.

22. An individual's susceptibility to question-order effects can be thought of as a durable unintended characteristic. Some people are more susceptible, others less so. If the questions are left in the same order for all respondents, then the answers of the susceptible respondents will be consistently measured, introducing bias into an overall measure of support for abortion rights. By randomizing the question order, question-order susceptibility will be inconsistently measured—some respondents will see the "serious defect" question first, others will see the "does not want any more children" question first—introducing random noise into the measure of abortion rights.

23. For a discussion of how the construct validity approach has been applied to the Graduate Record Examination, see Janet Buttolph Johnson and H. T. Reynolds, *Political Science Research Methods*, 6th ed. (Washington, D.C.: CQ Press, 2008), 99.

24. The interviewer asks, "Generally speaking, do you think of yourself as a Republican, a Democrat, an Independent, or what?" Respondents are given six choices: Democrat, Republican, Independent, Other Party, No Preference, and Don't Know. Those who choose Democrat or Republican are asked, "Would you call yourself a strong Democrat [Republican] or a not very strong Democrat [Republican]?" Those who choose Independent, Other Party, No Preference, or Don't Know are asked, "Do you think of yourself as closer to the Republican Party or to the Democratic Party?" Interviewers record these responses: Closer to Republican Party, Neither, or Closer to Democratic Party. Of the 1,807 people who were asked these questions in the 2000 American National Election Study, 1,776 were classified along the 7-point scale, 9 identified with another party, 17 were apolitical, and 5 did not give complete responses. See Nancy Burns, Donald R. Kinder, Steven J. Rosenstone, Virginia Sapiro, and the American National Election Studies, American National Election Study, 2000: Pre- and Post-Election Survey [computer file], 2nd version (Ann Arbor: University of Michigan, Center for Political Studies [producer], 2001; Inter-university Consortium for Political and Social Research [distributor], 2002).

25. Bruce E. Keith, David B. Magleby, Candice J. Nelson, Elizabeth Orr, Mark C. Westlye, and Raymond E. Wolfinger, *The Myth of the Independent Voter* (Berkeley: University of California Press, 1992).

26. Herbert F. Weisberg, "A Multidimensional Conceptualization of Party Identification," *Political Behavior* 2 (1980): 33–60.

27. The measurement problem illustrated by Table 1-1 is known as the *intransitivity problem*. For a concise review of the scholarly debate about intransitivity and other measurement issues, see Richard G. Niemi and Herbert F. Weisberg, *Controversies in Voting Behavior*, 4th ed. (Washington, D.C.: CQ Press, 2001), ch. 17.

Chapter 2

1. This helpful hint comes from Robert A. Bernstein and James A. Dyer, *An Introduction to Political Science Methods,* 3rd ed. (Englewood Cliffs, N.J.: Prentice-Hall, 1992), 3.

2. A higher level of measurement is sometimes discussed: ratio-level variables. A ratio variable has all the properties of an interval-level measure, plus it has a meaningful zero-point—the complete absence of the attribute being measured. As a practical matter, you will probably never encounter a research situation in which the distinction between interval and ratio measurements makes a difference.

3. Rensis Likert, "The Method of Constructing an Attitude Scale," in *Scaling: A Sourcebook for Behavioral Scientists,* ed. Gary M. Maranell, 233–243 (Chicago: Aldine Publishing, 1974).

4. Edward G. Carmines and Richard A. Zeller, *Reliability and Validity Assessment* (Thousand Oaks, Calif.: SAGE Publications, 1979): 46.

5. Likert scales require only that each constituent item bear an ordinal relationship with the measured characteristic. For the items displayed in Table 2-1, for example, Likert scaling assumes that as a respondent's level of egalitarianism increases, so will the probability of an egalitarian response to each individual item. Because Likert scales aggregate codes on a strictly additive basis, respondents who are assigned identical scores on a Likert scale may or may not have given identical responses to all the questions. Another technique, Guttman scaling, requires that each item bear an ordinal relationship with the measured characteristic *and* an ordinal relationship with all other items in the scale. Thus, respondents who are assigned identical scores on a Guttman scale must have given identical responses to all the questions. In the Guttman logic, if we cannot use a respondent's overall scale score to predict each individual answer, then we cannot be sure that the scale is tapping a single dimension of measurement. See Louis L. Guttman, "The Basis for Scalogram Analysis," in *Scaling: A Sourcebook for Behavioral Scientists,* 142–171. For an excellent discussion of Likert, Guttman, and other scaling techniques, see John P. McIver and Edward G. Carmines, *Unidimensional Scaling* (Thousand Oaks, Calif.: SAGE Publications, 1981).

6. For a review of this debate—and for an argument that ordinal scales can be modeled as interval variables—see Bruno D. Zumbo and Donald W. Zimmerman, "Is the Selection of Statistical Methods Governed by Level of Measurement?" *Canadian Psychology* 34, no. 4 (October 1993): 390–400. See also Jae-On Kim, "Multivariate Analysis of Ordinal Variables," *American Journal of Sociology* 81, no. 2 (September 1975): 261–298.

7. Bear F. Braumoeller, "Explaining Variance; Or, Stuck in a Moment We Can't Get Out Of," *Political Analysis* 14, no. 3 (Summer 2006): 268–290. In this thought-provoking article, Braumoeller argues that the dominant focus on "mean-centric" description and explanation has left us theory-poor in appreciating or explaining variation. For an exposition of how mean-centric thinking can lead to erroneous conclusions, see Stephen Jay Gould, *Full House: The Spread of Excellence from Plato to Darwin* (New York: Three Rivers Press, 1996).

8. To avoid confusion in terminology, we should note that a *proportion* is the raw frequency divided by the total frequency. A *percentage* is a proportion multiplied by 100. Barring rounding error, proportions total to 1.00 and percentages total to 100.

9. In describing a variable, whether one chooses to display the percentage of cases or the number of cases along the vertical axis of a bar chart is largely a matter of individual preference. In either case, the relative heights of the bars will be the same. When comparing the distributions of a variable for two or more subsets of cases—for example, comparing the distribution of religious attendance for whites with that of blacks—percentages must be used because the subsets have unequal numbers of cases.

10. Statistically speaking, a distribution that is perfectly symmetrical, one that has no skew, has a skewness equal to zero. Data analysis packages, such as SPSS or Stata, report a measure of skewness that takes on a positive or a negative value, indicating the direction of skew. SPSS provides a companion statistic, the standard error of skewness, that allows the researcher to determine if the skewness of the distribution departs too much from zero. One of Stata's commands, sktest, tests the hypothesis that a distribution is not significantly skewed.

11. Another graphic display, the histogram, is often used to depict interval-level variables. Histograms are similar to bar charts. Whereas a bar chart shows the percentages (or frequency) of cases in each value of a

variable, a histogram shows the percentage or frequency of cases falling into intervals of the variable. These intervals, called bins, compress the display, removing choppiness and gaps between the bars. Both bar charts and histograms work about equally well in helping the researcher to describe an interval-level variable.

12. Data for this exercise are from the 2006 General Social Survey. The job autonomy variable was constructed from responses to two questions: "I am given a lot of freedom to decide how to do my own work" (GSS variable, WKFREEDM), and "I have a lot of say about what happens on my job" (LOTOFSAY). The car manufacturer requirement question: "How much do you favor or oppose requiring car makers to make cars and trucks that use less gasoline? Do you strongly favor, favor, neither favor nor oppose, oppose, or strongly oppose such a requirement?" (GASREGS).

Chapter 3

1. The education–turnout explanation also suggests that people with less education are less politically aware and less efficacious than people with more education, and that awareness and efficacy are linked to turnout. An explanation can, and usually does, suggest more than one empirical relationship. In this chapter we discuss this aspect of explanations.

2. The 2004 American National Election Study asked respondents, "Do you think the federal government should make it more difficult for people to buy a gun than it is now, make it easier for people to buy a gun, or keep these rules about the same as they are now?" Of the 1,202 respondents, 59 percent said "more difficult" and 41 percent said "easier" or "keep about the same."

3. This is the origin of the term *dependent variable*. In a causal explanation, the values of the dependent variable are said to "depend on" the values of the independent variable.

4. Robert D. Putnam, *Bowling Alone: The Collapse and Revival of American Community* (New York: Simon & Schuster, 2000).

5. Leslie Lenkowsky, "Still 'Bowling Alone'?" *Commentary* (October 2000): 57.

6. Like many explanations for social or political variables, Putnam's explanation does not rely on a single independent variable. The changing age composition is very important to Putnam's explanation, but he also suggests that another long-term social trend—the rise in television as the primary source of entertainment—is causally linked to the decline in organized social and community interaction. Putnam regards the decline in civic organizations as a serious matter. He argues that such face-to-face interaction fosters social trust and participatory skills, essential elements for a healthy democratic society. These assertions are, of course, open to empirical test.

7. Malcolm Gladwell, *The Tipping Point: How Little Things Can Make a Big Difference* (Boston: Little, Brown, 2000), 5, 14.

8. As an example of an archetypical connector, Gladwell cites Lois Weisberg, commissioner of cultural affairs for the city of Chicago. By Gladwell's count, Weisberg has hundreds of contacts spanning ten different professional and social contexts: actors, writers, doctors, lawyers, park-lovers, politicians, railroad buffs, flea market aficionados, architects, and people in the hospitality industry. The point about connectors like Weisberg "is that by having a foot in so many different worlds, they have the effect of bringing them all together." Gladwell, *The Tipping Point*, 51.

9. Gladwell also discusses the role of other causal factors in the spread of contagions. A disease needs to change form, becoming less virulent and "sticking around" so that it can infect more people. Also, the behavior that supports the contagion must be a widely shared social norm within the demographic context in which it takes its highest toll.

10. Gladwell, *The Tipping Point*, 56–57.

11. Ibid., 33.

12. These typical errors in hypothesis writing are pointed out by Robert A. Bernstein and James A. Dyer, *An Introduction to Political Science Methods*, 3rd ed. (Englewood Cliffs, N.J.: Prentice-Hall, 1992), 11–12.

13. The catchphrase, "The economy, stupid," was coined by Clinton adviser James Carville during Clinton's successful 1992 campaign. The idea was to shift voters' attention toward the performance of the economy, where Republican incumbent George H.W. Bush was considered vulnerable, and away from foreign policy, where Bush enjoyed an advantage.

14. This proposed explanation is based on ideas from Harry Eckstein's classic study, *Division and Cohesion in Democracy: A Study of Norway* (Princeton, N.J.: Princeton University Press, 1966). An extensive literature has established that education is associated with turnout. However, as Steven Tenn puts it, the "literature has not determined . . . why this is the case." Tenn tests the civic education theory and finds it lacking. See Steven Tenn, "The Effect of Education on Voter Turnout," *Political Analysis* 15, no. 4 (Autumn 2007): 446–464. On the education-turnout relationship, see Raymond E. Wolfinger and Steven J. Rosenstone, *Who Votes?* (New Haven: Yale University Press, 1980).

15. Analysis and interpretation of interval-level dependent and independent variables are covered in Chapter 8. Chapter 9 covers research situations in which the dependent variable is a two-category nominal or ordinal variable and the independent variable is interval level.

16. The American National Election Study's 7-point party identification scale was discussed in Chapter 1. The scale classifies respondents as Strong Democrat, Weak Democrat, Independent-leaning Democrat, pure Independent, Independent-leaning Republican, Weak Republican, and Strong Republican. For the purpose of illustrating cross-tabulation analysis, the Weak and Strong Democrats are grouped together, as are the Weak and Strong Republicans. Independents are composed of the leaners and the pure Independents.

17. For this example we did not offer an explanation for the causal link between income and smoking. It is curious, though, that smoking is most prevalent among people least able to afford it. For a test of competing explanations for this puzzling connection, see Fred C. Pampel, "Inequality, Diffusion, and the Status Gradient in Smoking," *Social Problems* 49, no. 1 (February 2002): 35–57. Pampel's analysis of cross-national data suggests that patterns of smoking fit a diffusion process, in which smoking "emerges first among high status persons, spreads to the rest of the population, and then declines first among high status persons" (p. 52).

18. Freedom House, a nonprofit organization, ranks countries on two separate 7-point scales: political rights and civil liberties. The variable being discussed here was created by summing the average political rights score and the average civil liberties score for each country for the period 1980–2004, resulting in a scale ranging from 2 to 14. The final variable was rescaled to range from 0 (fewest political rights and freedoms) to 12 (most political rights and freedoms).

19. Two of the most popular data analysis packages, SPSS and Stata, use the format shown in Table 3-3.

20. Properly constructed hypotheses require that we describe the *tendency* of relationships, including relationships having nominal-level independent variables. If one were to say that women are more likely than men to favor gun control, one would be describing the tendency of the relationship—women are *more likely* to favor than are men. This logical usage makes perfect sense. But to say that the gender–gun opinions relationship has a "positive direction" or "negative direction" does not make sense, because gender is not an ordinal or interval variable.

21. In the real world of political research, relationships are rarely symmetrical. Figure 3-5 is no exception. You can see that respondents on the Republican side of the scale's midpoint are more likely to vote than are their counterparts on the Democratic side. Still, the relationship roughly approximates a V- or U-shaped pattern.

22. This explanation draws on an analysis by Daniele Checchi, Jelle Visser, and Herman G. van de Werfhorst, "Inequality and Union Membership: The Impact of Relative Earnings Position and Inequality Attitudes," Institute for the Study of Labor, University of Bonn, Bonn, Germany, Discussion Paper No. 2691, March 2007.

23. Kevin J. Coleman and Eric A. Fischer, "Elections Reform: Overview and Issues," *CRS [Congressional Research Service] Report for Congress,* January 2003, p. 5.

24. See Paul S. Herrnson and John C. Green, eds., *Multiparty Politics in America: Prospects and Performance,* 2nd ed. (Lanham, Md.: Rowman & Littlefield, 2002).

25. Robert Michels, *Political Parties: A Sociological Study of the Oligarchical Tendencies of Modern Democracy,* trans. Eden and Cedar Paul (New York: International Library Co., 1915), 241, reprinted and made available online by Batoche Books, Kirchener, Ontario, 2001.

26. Data are from the 2006 General Social Survey. The GSS gauges strength of approval by asking respondents how strongly they agreed or disagreed that "it is sometimes necessary to discipline a child with a

good, hard spanking." In this exercise, "strongly agree" means strong approval of spanking, and all other responses mean not strong approval.

27. The independent variable is based on the 2006 GSS variable NEWS: "How often do you read the newspaper—every day, a few times a week, once a week, less than once a week, or never?" The dependent variable, public policy knowledge, was constructed from two GSS questions measuring self-assessed knowledge of foreign policy (KNWFORGN) and economic policy (KNWECON). The question wording was: "We want to ask about how much information you have on various topics. For each of the following areas, please indicate whether you are very informed, somewhat informed, neither informed nor uninformed, somewhat uniformed, or very uninformed about the issues." Responses to both questions were coded in the "more informed" direction and summed. The final variable was rescaled to range from 0 (uninformed) to 8 (informed).

28. Economic liberalism was constructed from two 2006 GSS variables: whether the respondent favors cuts in government spending (CUTGOVT) and less government regulation of business (LESSREG). The question wording was: "Here are some things the government might do for the economy. Circle one number for each action to show whether you are in favor of it or against it." The GSS recorded five response categories, from strongly favor (code 1) to strongly against (code 5). The measure of economic liberalism was constructed by summing responses. This measure was then rescaled, yielding a scale that can range from 0 to 8.

Chapter 4

1. Edward R. Tufte, *Data Analysis for Politics and Policy* (Englewood Cliffs, N.J.: Prentice-Hall, 1974), 4. This story is a medical school recollection of E. E. Peacock Jr., quoted in *Medical World News* (September 1, 1972): 45.

2. See Donald P. Green and Alan S. Gerber, "Reclaiming the Experimental Tradition in Political Science," in *Political Science: The State of the Discipline,* 3rd ed., ed. Helen V. Milner and Ira Katznelson, 805–832 (New York: W. W. Norton & Co., 2002).

3. Donald R. Kinder and Thomas R. Palfrey, "On Behalf of an Experimental Political Science," in *Experimental Foundations of Political Science,* ed. Donald R. Kinder and Thomas R. Palfrey (Ann Arbor: University of Michigan Press, 1993), 5.

4. This requirement accords with the *counterfactual model of causation.* The main assumption of the counterfactual model is that individuals have two potential values of the dependent variable, one that corresponds to their observed value, given their value on the independent variable, and one that corresponds to the value that would be observed if they had a different value of the independent variable. "For example, for the causal effect of having a college degree rather than a high school degree on subsequent earnings, adults who have completed high school degrees have theoretical what-if earnings under the state 'have a college degree,' and adults who have completed college degrees have theoretical what-if earnings under the state 'have only high school degree.' These what-if potential outcomes are counterfactual." Stephen L. Morgan and Christopher Winship, *Counterfactuals and Causal Inference: Methods and Principles for Social Research* (New York: Cambridge University Press, 2007), 5.

5. James A. Davis, *The Logic of Causal Order* (Thousand Oaks, Calif.: SAGE Publications, 1985), 35 (emphasis in original).

6. Kinder and Palfrey, "On Behalf of an Experimental Political Science," 7. Random assignment does not completely guarantee that the two groups are identical, of course. By chance, subjects picked for the test group might still differ from control subjects. Methodologists have devised ways to minimize these random distortions. In one technique, called *blocking,* subjects are matched on premeasurements of the dependent variable and then randomly assigned to the treatment or control. See Donald T. Campbell and Julian C. Stanley's classic, *Experimental and Quasi-Experimental Designs for Research* (Chicago: Rand McNally, 1963), 15. See also R. Barker Bausell, *A Practical Guide to Conducting Empirical Research* (New York: Harper and Row, 1986), 95–96.

7. This description draws on Shanto Iyengar and Donald R. Kinder, *News That Matters: Television and American Opinion* (Chicago: University of Chicago Press, 1987); and Shanto Iyengar, Mark D. Peters, and

Donald R. Kinder, "Experimental Demonstrations of the 'Not-So-Minimal' Consequences of Television News Programs," in Kinder and Palfrey, *Experimental Foundations of Political Science*, 313–331.

8. Because premeasurements are almost always related to postmeasurements but are not related to the treatment, premeasurement also increases the precision of experimental findings. For example, suppose a participant ranks issue x as "most important" in the premeasurement phase. This propensity serves as a predictor of how the participant will rank issue x in the postmeasurement phase, giving the experimenter a baseline for evaluating the effect of the agenda-setting intervention.

9. Iyengar and Kinder, *News That Matters*, 18–19.

10. Iyengar, Peters, and Kinder, "Experimental Demonstrations," 321. These researchers are very much aware of questions about external validity. They suggest that the problem of external validity can be addressed by using a variety of methods, so-called methodological pluralism, to investigate the role of media.

11. This is sometimes called the *endogeneity problem*. When we make a simple comparison, such as comparing the turnout rates of the contacted and the uncontacted, we implicitly assume that the independent variable is exogenous—that it occurs independently and is not caused by the dependent variable. If the dependent variable is a cause of the independent variable, if endogeneity is occurring, then we run the risk of overestimating the effect of the independent variable on the dependent variable. For example, if most people who are contacted were already quite likely to vote (and most of the uncontacted were not going to vote anyway), then it would be erroneous to say that contact causes turnout.

12. Alan Gerber, Donald Green, and their collaborators have published extensively on these topics. The discussion here is based on Alan S. Gerber and Donald P. Green, "The Effects of Canvassing, Telephone Calls, and Direct Mail on Voter Turnout: A Field Experiment," *American Political Science Review* 94, no. 3 (September 2000): 653–663. For a review of much of their work and an extension of their model, see Alan S. Gerber, Donald P. Green, and Christopher W. Larimer, "Social Pressure and Voter Turnout: Evidence from a Large-Scale Field Experiment," *American Political Science Review* 102, no. 1 (February 2008): 33–48. For a collection of get-out-the-vote field experiments in a variety of settings, see *The Annals of the American Academy of Political and Social Science* 601 (September 2005).

13. Gerber, Green, and Larimer find that messages evoking social pressure are more likely to cause people to vote. See Gerber, Green, and Larimer, "Social Pressure and Voter Turnout."

14. Philip H. Pollock III and Bruce M. Wilson, "Evaluating the Impact of Internet Teaching: Preliminary Evidence from American National Government Classes," *PS: Political Science and Politics* 35, no. 3 (September 2002): 561–566.

15. To simplify the presentation, we will omit Independents from this part of the discussion.

16. This usage of causal language follows Davis, *The Logic of Causal Order*.

17. When a relationship is spurious, researchers sometimes refer to the relationship as an "artifact" of a rival cause, as an artificial creation of an uncontrolled variable.

18. This popular example is retold by Royce Singleton Jr., Bruce C. Straits, Margaret M. Straits, and Ronald McAllister, *Approaches to Social Research* (New York: Oxford University Press, 1988), 81.

19. In interpreting controlled comparisons using cross-tabulation analysis, particularly when one of the variables is nominal level, there is no universally accepted term for a set of relationships herein labeled *additive*. However, the term *additive* does have apposite usage in regression analysis, in which relationships are often modeled as "linear-additive." The "linear" part of linear-additive says that (controlling for Z) each unit increase in X has a consistent effect on Y, and that (controlling for X) each unit increase in Z has a consistent effect on Y. The "additive" part of linear-additive says that one determines the combined effects of X and Z on Y by summing or "adding up" their individual effects. Regression analysis is discussed in Chapter 8. In the current discussion, the term *additive* conveys the idea that each variable, the independent variable and the control variable, makes a distinct and consistent contribution to the explanation of the dependent variable.

20. Of the two interchangeable terms that describe this scenario—specification and interaction—specification perhaps has more intuitive appeal. The word *specification* tells us that the control variable Z specifies the relationship between X and Y, that the X→Y relationship depends on the specific value of Z. *Interaction*, which is frequently used in regression analysis (see Chapter 8), imparts the idea that the indepen-

dent effects of X on Y and Z on Y, when combined, interact to produce an effect on Y that is greater than the effect of each independent variable alone. Medication labels, for example, often warn that the effect of the drug on drowsiness, combined with alcohol, will be greater than simply adding the independent effect of taking the drug to the independent effect of consuming alcohol. To keep the presentation consistent, the term *interaction* is used throughout this book. However, interaction and specification are synonymous terms, and you will encounter both terms in political research.

Chapter 5

1. Data are from American National Election Study, American National Election Study, 2004: Pre- and Post-Election Survey [computer file], ICPSR04245-v1 (Ann Arbor: University of Michigan, Center for Political Studies [producer], 2004; Inter-university Consortium for Political and Social Research [distributor], 2004). To simplify the discussion, the analysis is confined to respondents claiming an affiliation (strong, weak, or leaning) with the Democratic Party or the Republican Party. The gun-control variable is based on ANES variables V043188 and V043190. V043188: "Do you think the federal government should make it more difficult for people to buy a gun than it is now, make it easier for people to buy a gun, or keep these rules about the same as they are now?" V043190 gauges the strength of the respondent's opinion: "How important is this issue to you personally? Extremely important, very important, somewhat important, not too important, or not at all important?" Respondents were first divided into two camps on V043188, "easier/same" and "more difficult." After recoding V043190 into four levels of importance (the "not too" and "not at all" responses were collapsed), V043188 was combined with V043190 to create an eight-category gun-control scale ranging from 1 (easier/same-extremely important) to 8 (more difficult-extremely important). The variable analyzed in Table 5-1 and Table 5-2 combines codes 1–6 ("Low" support) and codes 7–8 ("High" support).

2. The partisan differences we have been describing would be termed first-order partials because we have controlled for one other variable, gender. As more variables are taken into account, the designation increases in order. Comparing Democrats and Republicans, controlling for gender *and* region, for example, would produce a second-order partial.

3. It is tempting to simply average the two effects: $(30.5 + 34.7)/2 = 32.6$. This procedure produces an unweighted average, which works well when the control groups are of roughly equal size. Because sample percentages of women and men are always about the same, we could use an unweighted average. Often, however, the control groups are not of similar sizes. Therefore, a better approach is to weight each effect according to the size of each control group and then add the weighted effects together. In our example, there are 724 respondents, 413 women and 311 men. So women comprise 413/724, or .570, of the sample. Men comprise 311/724, or .430, of the sample. Weighting the partisanship effect for females: $(30.5)(.57) = 17.39$. Weighting the partisanship effect for males: $(34.7)(.43) = 14.92$. Summing the weights: $17.39 + 14.92 = 32.3$. For an intuitive and accessible discussion of causal effects in cross-tabulation analysis, see Ottar Hellevik, *Introduction to Causal Analysis: Exploring Survey Data by Crosstabulation*, 2nd ed. (Oslo: Norwegian University Press, 1988).

4. A more precise partial effect, obtained by weighting, is equal to 12.2 percentage points. Democrats are .522 of the sample $(378/724 = .522)$; Republicans are .478 of the sample $(346/724 = .478)$. Weighting the effects: for Democrats, $(10.2)(.522) = 5.32$; for Republicans, $(14.4)(.478) = 6.88$. Summing the weights: $5.32 + 6.88 = 12.2$.

5. There are 378 Democrats in Table 5-2, 236 females and 142 males. Thus, the female composition of Democrats is $236/378 = .624$, or 62.4 percent. There are 346 Republicans, 177 females and 169 males. The female composition of the Republicans is $177/346 = .512$, or 51.2 percent. Thus, there is an 11.2-point difference in the percentage of Democrats who are women and the percentage of Republicans who are women.

6. Data presented in Table 5-3 are from the 2004 American National Election Study. Abortion opinion is based on V045132. Respondents saying, "By law, a woman should always be able to obtain an abortion," are measured as "Always permit." All other valid responses are measured as "Not always permit." Salience is based on V045133: "How important is this issue to you personally?" Respondents saying that the

abortion issue is "extremely important" are measured as "High salience." All other valid responses are measured as "Low salience." Vote choice is based on V045026.

7. The interviewer frames the feeling thermometer questions with this preamble: "I'd like to get your feelings toward some of our political leaders and other people who are in the news these days. I'll read the name of a person and I'd like you to rate that person using something we call the feeling thermometer. Ratings between 50 degrees and 100 degrees mean that you feel favorable and warm toward the person. Ratings between 0 degrees and 50 degrees mean that you don't feel favorable toward the person and that you don't care too much for that person. You would rate the person at the 50-degree mark if you don't feel particularly warm or cold toward the person." After asking respondents about people, the interviewer introduces the group feeling thermometer questions by saying: "Still using the thermometer, how would you rate the following groups." The feeling thermometer for homosexuals is variable V045074 in the 2004 American National Election Study.

8. The egalitarianism scale was constructed from the following variables from the 2004 American National Election Study: V045212 (Our society should do whatever is necessary to make sure that everyone has an equal opportunity to succeed), V045213 (We have gone too far in pushing equal rights in this country), V045214 (One of the big problems in this country is that we don't give everyone an equal chance), V045215 (This country would be better off if we worried less about how equal people are), V045216 (It is not really that big a problem if some people have more of a chance in life than others), and V045217 (If people were treated more equally in this country we would have many fewer problems). All variables were coded in the more-egalitarian direction and summed, creating a scale ranging from 0 (low egalitarianism) to 24 (high egalitarianism). For the analysis presented in Table 5-4, respondents scoring 0–15 are classified as "Low" and those scoring 16–24 are classified as "High." The control variable, age, is based on variable V043250.

9. See Joe Bergeron, "Examining Determinants of American Support for Same-Sex Marriage," paper presented at the annual meeting of the American Political Science Association, Washington, D.C., September 1–4, 2005.

10. See Wilma Rule, "Women's Underrepresentation and Electoral Systems," *PS: Political Science and Politics* 27, no. 4 (December 1994): 689–692.

11. See Pippa Norris and Ronald Inglehart, "Cracking the Marble Ceiling: Cultural Barriers Facing Women Leaders," a Harvard University Report, John F. Kennedy School of Government (Cambridge, Mass.: Harvard University, January 13, 2008).

12. This idea is suggested by Andrew Reynolds, "Women in the Legislatures and Executives of the World: Knocking at the Highest Glass Ceiling," *World Politics* 51, no. 4 (July 1999): 547–572.

13. The variables on which Table 5-5 is based are from Democracy Crossnational Data (revised Spring 2008), Pippa Norris, John F. Kennedy School of Government, Harvard University. Data on the percentage of women in parliament are from 2005 and available from Inter-parliamentary Union, *Women in Parliament*, www.ipu.org. The variable that measures cultural acceptance of women in politics was compiled by Norris from the World Values Survey, 1995–2000 waves, www.worldvaluessurvey.org. Countries are coded from 1 (low acceptance) to 4 (high acceptance) based on the extent to which citizens disagreed with the statement, "On the whole, men make better political leaders than women do." Observed values range from 1.4 to 3.4. In Table 5-5, countries with codes between 1.4 and 2.7 have "Low" acceptance, and countries with codes between 2.8 and 3.4 have "High" acceptance.

14. Data for this exercise are from the 2006 General Social Survey. Authoritarianism is based on respondents' beliefs about childrearing practices: "If you had to choose, which thing on this list would you pick as the most important for a child to learn to prepare him or her for life?" The measure was calculated from the relative rankings of "to obey" and "to think for himself or herself." This approach to the measurement of authoritarianism is suggested by Karen Stenner, *The Authoritarian Dynamic* (Cambridge: Cambridge University Press, 2005).

15. Data for this exercise are from Shared Global Database (revised Fall 2004), Pippa Norris, John F. Kennedy School of Government, Harvard University, Cambridge, Mass., and from *Comparing Democracies 2: New Challenges in the Study of Elections and Voting*, ed. Lawrence LeDuc, Richard G. Niemi, and Pippa Norris (London: SAGE Publications, 2002).

16. The original study is considerably more sophisticated than described in this exercise. See James T. Wassell, Lytt I. Gardner, Douglas P. Landsittel, Janet J. Johnston, and Janet M. Johnston, "A Prospective Study of Back Belts for Prevention of Back Pain and Injury," *Journal of the American Medical Association* 284, no. 21 (December 6, 2000): 2727–2732.

Chapter 6

1. The terms *population characteristic* and *population parameter* are synonymous and will be used interchangeably in this chapter.
2. Much has been written and debated about the 1936 *Literary Digest* poll. See Maurice C. Bryson, "The Literary Digest Poll: Making of a Statistical Myth," *The American Statistician* 30, no. 4. (November 1976): 184–185; Peverill Squire, "Why the 1936 *Literary Digest* Poll Failed," *Public Opinion Quarterly* 52, no. 1 (Spring 1988): 125–133. See also Thomas B. Littlewood, *Calling Elections: The History of Horse-Race Journalism* (Notre Dame, Ind.: University of Notre Dame Press, 1998). Bryson and Squire argue that response bias was a bigger problem than selection bias.
3. Of course, purely random sampling is impractical for large populations. However, researchers have devised several random-based techniques for large sampling frames. An excellent treatment may be found in Herbert Asher, *Polling and the Public: What Every Citizen Should Know,* 6th ed. (Washington, D.C.: CQ Press, 2004). Also see S. K. Thomson, *Sampling* (New York: Wiley, 1992).
4. The simulations presented in Figure 6-2 and Figure 6-3 were created using the Stata program, bxmodel (version 1.2, January 31, 2006), written by Philip B. Ender, Statistical Computing and Consulting UCLA, Academic Technology Services.
5. Statisticians typically detest imprecision, but this rule of thumb is acceptable because it is conservative; that is, it slightly widens the bandwidth of the random error associated with the 95 percent confidence interval.
6. By dividing the summation of the squared deviations by $n - 1$, instead of n, we correct for the known tendency of the sample variance to underestimate the population variance. The correction is more pronounced for smaller samples (smaller values of n) than for larger samples (larger values of n).
7. Normal estimation may be used for samples of $n = 100$ or more. See David Knoke, George W. Bohrnstedt, and Alisa Potter Mee, *Statistics for Social Data Analysis,* 4th ed. (Belmont, Calif.: Wadsworth/Thomson, 2002), 128.

Chapter 7

1. The method described here assumes that the subsample variances (the variance of Democratic ratings among women and the variance of the Democratic ratings among men) are not equal. As a default, this is preferred to the alternative method, which assumes equal variances, because the unequal-variance approach provides a more conservative test.
2. David Knoke, George W. Bohrnstedt, and Alisa Potter Mee, *Statistics for Social Data Analysis,* 4th ed. (Belmont, Calif.: Wadsworth/Thomson, 2002), 128.
3. The term P-*value* is widely, though not universally, used. Computer output commonly reports P-values under the heading "significance" or its abbreviation, "sig." P-*value* and *significance* are synonymous terms. Both terms tell you the probability of obtaining the observed results if the null hypothesis is correct.
4. Statistically, one can think of a random sample of, say, 100 cases as equivalent to 100 random samples of size $n = 1$. Assigning each individual outcome the value 1 if it falls into a specific category of a variable, and coding it 0 if it does not fall into that category, we find the mean value by summing the 100 0s and 1s and dividing by 100. The result, of course, is a proportion.
5. In Chapter 6 we saw that the standard error of a sample proportion is equal to $\sqrt{pq} / \sqrt{n}$. Thus, the squared standard error is equal to pq/n.
6. Alan Agresti and Barbara Finlay, *Statistical Methods for the Social Sciences,* 3rd ed. (Upper Saddle River, N.J.: Prentice-Hall, 1997), 255. Chi-square is one member of a large family of nonparametric statistics, tests that make fewer assumptions about the population from which the sample is drawn.

7. This intuitive way of getting expected frequencies works okay, but it can introduce a fair amount of rounding error. A more precise method of obtaining expected frequencies is to multiply the column total by the row total and divide by the sample size. For example, the expected frequency (f_e) for the "Female-Force" cell would be the total number of cases in the "Female" column, 530, multiplied by the number of cases in the "Force" category of the dependent variable, 353, divided by the sample size, 1,041: (530 * 353) / 1,041 = 179.7.

8. George W. Bohrnstedt and David Knoke, *Statistics for Social Data Analysis,* 2nd ed. (Itasca, Ill.: F. E. Peacock, 1988), 124–126.

9. Bohrnstedt and Knoke analyze the birth-month distribution of 1,462 individuals from the 1984 General Social Survey. They obtain a calculated χ^2 of 11.39. For 11 degrees of freedom, the upper limit of random error, using the .05 threshold, is 19.68. Since a χ^2 of 11.39 could occur more frequently than five times out of 100 by chance, one cannot say that some months are more prevalent than are other months for being born.

10. *Yates' correction* is an adjustment often seen in computer calculations. It consists of either adding or subtracting .5 from an observed frequency to reduce the difference between f_o and f_e.

11. Leo A. Goodman and William H. Kruskal, "Measures of Associations for Cross Classifications," *Journal of the American Statistical Association* 49, no. 268 (December 1954): 732–764.

12. Robert H. Somers, "A New Asymmetric Measure of Association for Ordinal Variables," *American Sociological Review* 27, no. 6 (December 1962): 799–811.

13. For whites, $\chi^2 = 6.50$, and for blacks, $\chi^2 = 7.60$. Both have *P*-values of less than .05 (1 degree of freedom).

14. Cramer's V is χ^2, adjusted for sample size (n). It is the square root of the expression $\chi^2 / ((n)(\min(r-1, c-1))$, where $\min(r-1, c-1)$ tells us to plug in the number of rows minus 1 or the number of columns minus 1, whichever is smaller. There is another non-PRE measure based on chi-square, the *contingency coefficient* (C) that some researchers prefer. It returns values that are similar to Cramer's V. Since the interpretation of Cramer's V is somewhat more straightforward, it is generally recommended. See H.T. Reynolds, *Analysis of Nominal Data,* 2nd ed. (Thousand Oaks, Calif.: SAGE Publications, 1984): 46–49.

15. The three most widely used measures of ordinal association are gamma, Kendall's tau-*b*, and Somers' d_{yx}. All use the same numerator ($C - D$), and all agree that the summation of concordant and discordant pairs ($C + D$) belongs in the denominator. The measures part company on the question of how to treat observations that are tied. There are two ways that observations can be tied. Observations can have the same value on the independent variable (x) but differ on the dependent variable (tied on x, symbolized as T_x), or the same value on the dependent variable (y) but different values on the independent variable (tied on y, T_y). Gamma is a symmetric measure that ignores ties and reports the net number of concordant pairs as a proportion of all untied pairs: ($C - D$) / ($C + D$). Because it treats as irrelevant the information contained in ties, gamma tends to overestimate strength. Kendall's tau-b takes all ties into account: ($C - D$) / $[\sqrt{((C + D + T_x)(C + D + T_y))}]$. Kendall's tau-*b* is a symmetric measure and can only be used for square tables, that is, tables having the same number of columns and rows—attributes that limit its appeal. (Another symmetric measure, Kendall's tau-*c*, can be used for nonsquare tables.) Somers d_{yx} is an asymmetric measure that reports the surplus of concordant pairs as a proportion of all pairs that have different values on the independent variable.

16. Information on workday voting and nonworkday voting in 58 democracies was obtained from Lawrence LeDuc, Richard G. Niemi, and Pippa Norris, "Introduction: Comparing Democratic Elections," in *Comparing Democracies 2: New Challenges in the Study of Elections and Voting,* ed. LeDuc, Niemi, and Norris, 1–39 (London: SAGE Publications, 2002). See Table 1.3, pp. 13–15. Turnout data are from the Shared Global Database (revised Fall 2004), Pippa Norris, John F. Kennedy School of Government, Harvard University, Cambridge, Mass. Retrieved from http://ksghome.harvard.edu/~pnorris/Data/Data.htm. For each country, turnout is the average of all national legislative elections held during the 1990s.

17. Michael L. Ross, "Does Oil Hinder Democracy?" *World Politics* 53 (April 2001): 325–361. This quote page 325. This exercise compares the 15 most highly oil-dependent economies, as identified by Ross (Table 1, p. 326), with 20 other countries, which were randomly selected from the Shared Global Database (revised

Fall 2004), Pippa Norris, John F. Kennedy School of Government, Harvard University, Cambridge, Mass. The dependent variable, based on the Freedom House political rights and civil liberties scales (2003–2004), also was obtained from the Shared Global Database.

18. These data are a randomly selected subset of the 2004 American National Election Study. The death penalty variable is based on V043187, strength of respondent's support or opposition to the death penalty. Respondents who favor strongly are classified as "strong"; all other responses (favor not strongly, oppose not strongly, oppose strongly) are classified as "not strong." Level of education is based on V043252. To avoid confusion, marginal totals are not shown.

Chapter 8

1. In practice, regression is often used when the dependent variable is an ordinal-level measure that has many categories.

2. Michael S. Lewis-Beck, *Data Analysis: An Introduction* (Thousand Oaks, Calif.: SAGE Publications, 1995), 21.

3. The error term represents two sorts of error: left-out variables and random error. Left-out variables are independent variables, perhaps unknown to the investigator, that have a causal impact on the dependent variable but are not included in the regression model. In the studying–exam score example, it could be that additional differences between students—major, experience (or lack thereof) in the instructor's other courses, and so on—account for the regression model's inability to exactly predict students' exam scores. Random error involves chaotic or haphazard events that depress the scores of some students and elevate the scores of others.

4. The regression coefficient summarizes change in the dependent variable for each unit change in the independent variable. Computationally, $\hat{b}$ is equal to $\Sigma\,(x_i - \text{mean of } x)(y_i - \text{mean of } y)$ divided by $\Sigma\,(x_i - \text{mean of } x)^2$. The numerator summarizes the covariation of x and y. The denominator summarizes variation in x. The result is the average amount of change in y for each unit change in x.

5. The standard error of the estimated regression slope is based on how closely the predicted values of y approximate the actual values of y. Logically enough, if the regression line produces predicted values that are close to the actual values, the standard error of $\hat{b}$ is small. As the size of the prediction errors increases, so does the slope's standard error. See Edward R. Tufte, *Data Analysis for Politics and Policy* (Englewood Cliffs, N.J.: Prentice-Hall, 1974), 65–73.

6. OLS regression determines $\hat{a}$ by the formula (mean of y) $- \hat{b}$ (mean of x).

7. These labels are based on those used by Michael S. Lewis-Beck, *Applied Regression: An Introduction* (Thousand Oaks, Calif.: SAGE Publications, 1980).

8. Another measure of goodness of fit for the regression line, which has the unfortunate label *standard error of the estimate,* is less commonly used. Its appearance in computer output is a potential source of confusion, however, because its name sounds similar to a statistic that we use all the time, the *standard error of the regression coefficient*. This is a confusion that should be avoided.

9. See Barbara G. Tabachnick and Linda S. Fidell, *Using Multivariate Statistics,* 3rd ed. (New York: Harper-Collins, 1996), 164–165.

10. Regression analysis also produces a standard error and t-statistic for the y-intercept. The standard error of the y-intercept can be used to test the null hypothesis that the true value of the y-intercept in the population, α, is equal to 0. The usefulness of these statistics depends on the research problem at hand. In the current example, we are not interested in testing the hypothesis that α, the turnout rate of nonsouthern states, is 0. So the standard error of $\hat{a}$, and its accompanying t-ratio, have been omitted from Table 8-3.

11. You may have noticed that the t-ratio for the West dummy, -1.69, surpasses the 1.645 test of significance and, therefore, would permit us to reject the null hypothesis that the difference between midwestern states and western states is equal to 0. If we were testing the hypothesis (based on a more fully reasoned explanation) that states in the West will have lower turnouts than will states in the Midwest, then the regression results would indeed justify rejection of the null hypothesis at the .05 level of significance. As it is, we are pursuing a casual investigation of regional differences in turnout. You may also have noticed that, despite a t-ratio with a magnitude exceeding 1.645, the P-value for the West coefficient, .10, is

greater than .05, the region of the curve above the absolute value of $Z = 1.645$. As far as the author is aware, the regression results produced by all computer analysis programs, including the program that generated the Table 8-3 results, return two-tailed P-values, not one-tailed P-values. As discussed in Chapter 7, two-tailed values provide a stringent .025 one-tailed test of significance, making it harder to reject the null hypothesis. Because a .025 test is more conservative than a .05 test, most research articles report two-tailed P values and, in so doing, apply a .025 test of significance. Even so, if you wish to explicitly adopt the .05 standard in your regression analyses, you would derive one-tailed values by dividing two-tailed values in half.

12. John R. Zaller, *The Nature and Origins of Mass Opinion* (New York: Cambridge University Press, 1992).

13. Morris P. Fiorina, Samuel J. Abrams, and Jeremy C. Pope, *Culture War? The Myth of Polarized America,* 2nd ed. (New York: Pearson, 2006).

14. The data displayed in Table 8-6 are from the American National Election Study, 2004: Pre- and Post-Election Survey [computer file], ICPSR04245-v1 (Ann Arbor: University of Michigan, Center for Political Studies [producer], 2004; Inter-university Consortium for Political and Social Research [distributor], 2004). The abortion rights scale was constructed from V043179 (Does respondent favor/oppose government funds to pay for abortion?), V043181 (Does respondent favor/oppose ban on partial-birth abortions?), and V045132 (Abortion position: self-placement). V043179 and V045132 were recoded $(1 = 4, 2 = 3, 4 = 2, 5 = 1)$. V043181 was recoded $(1 = 1, 2 = 2, 4 = 3, 5 = 4)$. The variables were summed and rescaled to range from 0 (low prochoice) to 9 (high prochoice). The political knowledge dummy is based on correct answers to V045089 (identifies which party controls the House), V045090 (identifies which party controls the Senate), V045162 (identifies Dennis Hastert), V045163 (identifies William Rehnquist), V045164 (identifies Tony Blair), and V045165 (identifies Dick Cheney). In Table 8-6, respondents giving 0–3 correct answers are measured as "low" and those giving 4–6 correct answers are measured as "high." Party identification is based on V043116.

15. This brief exposition shows the main ideas behind the modeling of interaction. For an excellent guide to more advanced applications, see James Jaccard and Robert Turrisi, *Interaction Effects in Multiple Regression,* 2nd ed. (Thousand Oaks, Calif.: SAGE Publications, 2003).

16. One technique, called centering, is to recalculate the values of the independent variables as deviations from their means. Values below the mean will have negative deviations, those above the mean will have positive deviations, and those on the mean will have deviations equal to zero. Interaction variables can then be computed by calculating the products of the newly centered variables. Centered independent variables, and interaction variables computed using centered variables, have lower inter-correlations than the same variables in uncentered form. General treatments of centering may be found in Lawrence H. Boyd Jr. and Gudmund R. Iversen, *Contextual Analysis: Concepts and Statistical Techniques* (Belmont, Calif.: Wadsworth, 1979): 65–70. See also Leona S. Aiken and Stephen G. West, *Multiple Regression: Testing and Interpreting Interactions* (Thousand Oaks, Calif.: SAGE Publications, 1991): 28–36. For a discussion of dummy variable centering, see Anthony S. Bryk and Stephen W. Raudenbush, *Hierarchical Linear Models* (Thousand Oaks, Calif.: SAGE Publications, 1992): 25–28.

17. The regression results in Exercises 4 and 5 were obtained from an analysis of the 1998 General Social Survey. The GSS asks respondents whether or not a woman should be able to obtain a legal abortion under each of seven circumstances. The abortion scale was created by summing the number of "No" responses. The religious attendance dummy in Exercise 5 was coded 1 for people who attend "every week" or "more than once a week," and coded 0 otherwise. For the regression in Exercise 4, $n = 1,578$. For Exercise 5, $n = 1,557$. The abortion scale used in Exercises 4 and 5 differs from the prochoice scale analyzed in Tables 8-6 and 8-7. The prochoice variable is based on responses to the 2004 American National Election Study.

Chapter 9

1. Methodologists have developed several techniques that may be used to analyze binary dependent variables. One popular technique, probit analysis, is based on somewhat different assumptions than logistic regression, but it generally produces similar results. Logistic regression, also called logit analysis or logit

regression, is computationally more tractable than probit analysis and thus is the sole focus of this chapter. For a lucid discussion of the general family of techniques to which logistic regression and probit analysis belong, see Tim Futing Liao, *Interpreting Probability Models: Logit, Probit, and Other Generalized Linear Models* (Thousand Oaks, Calif.: SAGE Publications, 1994).

2. There are two statistically based problems with using OLS on a binary dependent variable, both of which arise from having only two possible values for the dependent variable. OLS regression assumes that its prediction errors, the differences between the predicted values of y and the actual values of y, follow a normal distribution. The prediction errors for a binary variable, however, follow a binomial distribution. More seriously, OLS also assumes *homoscedasticity* of these errors, that is, that the prediction errors are the same for all values of x. With a binary dependent variable, this assumption does not hold up. An accessible discussion of these problems may be found in Fred C. Pampel, *Logistic Regression: A Primer* (Thousand Oaks, Calif.: SAGE Publications, 2000), 3–10.

3. Because of this natural log transformation of the dependent variable, many researchers use the terms *logit regression* or *logit analysis* instead of logistic regression. Others make a distinction between logit analysis (used to describe a situation in which the independent variables are not continuous but categorical) and logistic regression (used to describe a situation in which the independent variables are continuous or a mix of continuous and categorical). To avoid confusion, we use *logistic regression* to describe any situation in which the dependent variable is the natural log of the odds of a binary variable.

4. Logistic regression will fit an S-shaped curve to the relationship between an interval-level independent variable and the probability of a dependent variable, but it need not be the same S-shaped pattern shown in Figure 9-1. For example, the technique may produce estimates that trace a "lazy S," with probabilities rising in a slow, nearly linear pattern across values of the independent variable. Or perhaps the relationship is closer to an "upright S," with probabilities changing little across the high and low ranges of the independent variable but increasing rapidly in the middle ranges.

5. If the logistic regression coefficient, $\hat{b}$, were equal to 0, then the odds ratio, $\text{Exp}(\hat{b})$, would be $\text{Exp}(0)$, or e^0, which is equal to 1.

6. Computer output also will report a standard error and test of significance for the intercept, $\hat{a}$. This would permit the researcher to test the hypothesis that the intercept is significantly different from 0. So if we wanted to test the null hypothesis that the logged odds of voting for individuals with no formal education (who have a value of 0 on the independent variable) was equal to 0—that is, that the odds of voting for this group was equal to 1—we would use the standard error of the intercept. Much of the time such a test has no practical meaning, and so these statistics have been omitted from Table 9-3.

7. The Wald statistic (named for statistician Abraham Wald) divides the regression coefficient by its standard error and then squares the result. The value of Wald follows a chi-square distribution with degrees of freedom equal to 1.

8. The estimation procedure used by logistic regression is not aimed at minimizing the sum of the squared deviations between the estimated values of y and the observed values of y. So the conventional interpretation of R-square, the percentage of the variation in the dependent variable that is explained by the independent variable(s), does not apply when the dependent variable is binary.

9. The likelihood function $= \Pi \{P_i^{yi} * (1 - P_i)^{1-yi}\}$. The expression inside the brackets says, for each individual case, to raise the model's predicted probability (P) to the power of y, and to multiply that number by the quantity 1 minus the predicted probability raised to power of $1 - y$. The symbol Π tells us to multiply all these individual results together. The formula is really not as intimidating as it looks. When y equals 1, the formula simplifies to ($P * 1$), since P raised to y equals P, and ($1 - P$) raised to $1 - y$ equals 1. Similarly when y equals 0, the formula simplifies to $1 - P$.

10. The baseline model is also called the reduced model because its predictions are generated without using the independent variable. Informally, we could also call it the "know-nothing model" because it does not take into account knowledge of the independent variable.

11. A likelihood model that uses the independent variable(s) to generate predicted probabilities is called the full model or complete model. In making statistical comparisons between models, some computer programs work with the log of the likelihood ratio, denoted $\ln(L1 / L2)$, in which L1 is the likelihood of

Model 1 (reduced model) and L2 is the likelihood of Model 2 (complete model). Taking the log of the likelihood ratio is equivalent to subtracting the logged likelihood of Model 2 from the logged likelihood of Model 1: $\ln(L1 / L2) = \ln(L1) - \ln(L2)$.

12. Degrees of freedom is equal to the difference between the number of independent variables included in the models being compared. Since Model 2 has one independent variable and Model 1 has no independent variables, degrees of freedom is equal to 1 for this example. We can, of course, test the null hypothesis the old-fashioned way, by consulting a chi-square table. The critical value of chi-square, at the .05 level with 1 degree of freedom, is equal to 3.84. Since the change in −2LL, which is equal to 3.98, exceeds the critical value, we can reject the null hypothesis.

13. The Wald statistic, commonly used to test the significance of logistic regression coefficients, is somewhat controversial. Some methodologists argue that, if the estimated logistic regression coefficients are large or if dummy independent variables are being used, Wald may underestimate the effect of an independent variable and thus bias inference too strongly toward Type II error. These researchers recommend a procedure based more directly on changes in the likelihood function. For a discussion of potential problems with the Wald statistic, see Scott Menard, *Applied Logistic Regression Analysis*, 2nd ed. (Thousand Oaks, Calif.: SAGE Publications, 2002), 43–48. See also J. Scott Long, *Regression Models for Categorical and Limited Dependent Variables* (Thousand Oaks, Calif.: SAGE Publications, 1997).

14. Logged likelihoods can be confusing. Remember that likelihoods vary between 0 (the model's predictions do not fit the data at all) and 1 (the model's predictions fit the data perfectly). This means that the logs of likelihoods can range from very large negative numbers (any likelihood of less than 1 has a negatively signed log) to 0 (any likelihood equal to 1 has a log equal to 0). So if Model 2 had a likelihood of 1—that is, it perfectly predicted voter turnout—then it would have a logged likelihood of 0. In this case, the conceptual formula for *R*-square would return a value of 1.0.

15. Cox and Snell's *R*-square and Nagelkerke's *R*-square are included in SPSS logistic regression output. Another measure, popular among political researchers, is Aldrich and Nelson's Pseudo *R*-square: (Change in −2LL) / (Change in −2LL + *N*), in which *N* is the sample size. Menard has proposed yet another measure, based on the correlation between the logistic regression's predicted probabilities of *y* and the actual values of *y*. Reassuringly, the Aldrich-Nelson statistic for the GSS data is equal to .052, and Menard's proposed measure returned a value of .051, both close to Cox-Snell (.053) and a bit lower than Nagelkerke (.075). See John H. Aldrich and Forrest D. Nelson, *Linear Probability, Logit, and Probit Models* (Thousand Oaks, Calif.: SAGE Publications, 1984), 54–58; and Menard, *Applied Logistic Regression Analysis*, 24–27.

16. The logged likelihoods for the reduced model and the complete model are not shown in Table 9-8. Rather, only the chi-square test statistic of interest, the change in −2LL, is reported. Note that, because no independent variables are included in the baseline model and two independent variables are included in the complete model, there are 2 degrees of freedom for the chi-square test.

17. If requested by the user, SPSS will enlist the logistic equation estimates to calculate and save predicted probabilities of the dependent variable for each case in the dataset. These predicted values become a new variable, which can be further examined and analyzed.

18. Researchers have proposed and debated a variety of ways to present and interpret estimated probabilities. For a discussion of several approaches, see Pampel, *Logistic Regression: A Primer*, 28–30.

19. It may seem odd to consider a 0–1 variable as having a mean or average in the conventional sense of the term. After all, respondents are coded either 0 or 1. There is no respondent who is coded ".221" on partisan strength. However, means can also be thought of as random probabilities, termed *expected values* by statisticians. If you were to pick any respondent at random from the GSS dataset, what is the probability that the case you chose would be a strong partisan? The answer is .221, the expected value, or mean value, of this variable.

20. Data used in these exercises are from the Shared Global Database (revised Fall 2004), Pippa Norris, John F. Kennedy School of Government, Harvard University, Cambridge, Mass. Retrieved from http://ksghome. harvard.edu/~pnorris/data/data.htm. Using Norris's definition, democratic countries are coded 1 and nondemocratic countries are coded 0. Economic inequality is measured by the gini coefficient, divided by

10. This measure, based on information provided by the World Bank, can vary between 0 (most equal) and 10 (least equal). The variable homogeneous is based on the cultural homogeneity index (1980), which Norris obtained from the State Failure Project Phase III. Countries having homogeneity scores at or above the median value are coded 1 on homogeneous. Those having homogeneity scores below the median are coded 0. Numbers of cases (n) is 102 for Exercise 1 and 81 for Exercise 2.

Chapter 10

1. This dialogue occurred during classroom discussion of Richard J. Ellis, "The States: Direct Democracy," in *The Elections of 2000,* ed. Michael Nelson (Washington, D.C.: CQ Press, 2001), 133–159.
2. Michael P. McDonald and Samuel Popkin, "The Myth of the Vanishing Voter," *American Political Science Review* 95, no. 4 (2001): 963–974. Recent data made available by McDonald, Department of Public and International Affairs, George Mason University, Fairfax, Virginia.
3. Gabriel A. Almond, *A Discipline Divided: Schools and Sects in Political Science* (Thousand Oaks, Calif.: SAGE Publications, 1990), 33.
4. Charles A. Lave and James G. March, *An Introduction to Models in the Social Sciences* (New York: Harper and Row, 1975), 34.
5. These data are from the 2004 American National Election Study.

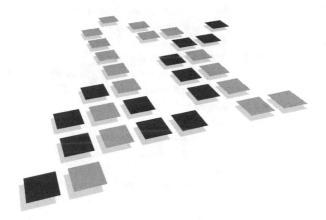

Index

Page references ending in *b* refer to boxes, *f* refer to figures, and *t* refer to tables.